Steam
Distribution
Systems Deskbook

by
James F. McCauley, P.E.

Published by
THE FAIRMONT PRESS, INC.
700 Indian Trail
Lilburn, GA 30047

Library of Congress Cataloging-in-Publication Data

McCauley, James F., 1935-
Steam distribution systems deskbook / by James F. McCauley.
   p. cm.
   Includes index.
   ISBN 0-88173-303-2
1. Steam engineering--Handbooks, manuals, etc. 2. Plant engineering
--Handbooks, manuals, etc. I. Title.
TJ277.M37   2000   511.3--dc21   99-052402
                                               CIP

Published by The Fairmont Press, Inc.
700 Indian Trail
Lilburn, GA 30047

Printed in the United States of America

10 9 8 7 6 5 4 3 2 1

ISBN 0-88173-303-2 FP

ISBN 0-13-026769-4 PH

While every effort is made to provide dependable information, the publisher, authors, and
editors cannot be held responsible for any errors or omissions.

Distributed by Prentice Hall PTR
Prentice-Hall, Inc.
A Simon & Schuster Company
Upper Saddle River, NJ 07458

Prentice-Hall International (UK) Limited, London
Prentice-Hall of Australia Pty. Limited, Sydney
Prentice-Hall Canada Inc., Toronto
Prentice-Hall Hispanoamericana, S.A., Mexico
Prentice-Hall of India Private Limited, New Delhi
Prentice-Hall of Japan, Inc., Tokyo
Simon & Schuster Asia Pte. Ltd., Singapore
Editora Prentice-Hall do Brasil, Ltda., Rio de Janeiro

Dedication

I offer my sincere appreciation to engineers and technicians for whom I have worked; for those who I have worked with, and for those who have worked for me. They have been talented bosses, team members, mentors, technical experts, competitors and friends.

Contents

Foreword

An energy conscious industrial firm automatically contributes to the preservation of our environment. Less energy consumption means less waste of irreplaceable fuels and fewer toxic emissions. It is well known that bringing good energy conservation practice to the work place not only reduces fuel costs but also reduces steam generator toxic emissions.

It follows that the less the amount of steam wasted, the less fuel is consumed by the steam plant. This book is dedicated to aid plant engineers and technicians in cost effective steam distribution and condensate systems management. This also produces less toxic waste by-products, as now required by government standards.

By practicing proven management and engineering techniques in steam and condensate transport, there will be less steam waste, less fuel consumption, and less toxic wastes. The return on investment in steam and condensate systems improvement can be considerable. This cost avoidance can often fund other needed projects.

Cost effective management of steam distribution and condensate systems saves valuable, non-replaceable fuel resources and reduces toxic wastes in our environment.

Chapter 1

Fundamentals of
Steam Engineering

FUNDAMENTALS OF STEAM ENGINEERING

Theory of Steam Generation

*I*f heat is added to ice, the effect will be to raise its temperature until the thermometer registers 32°F. When this point is reached, a further addition of heat does not produce an increase in temperature until all the ice has changed into water, or in other words, the ice melts. It has been found experimentally that 144 Btu are required to change 1 lb of ice into water at the constant temperature of 32°F. This quantity is called latent heat.

After the given quantity of ice, which for simplicity may be taken as 1 lb, has all been turned into water and more heat is added, then the temperature of the water will again increase, though not so rapidly as did that of the ice. While the addition of each Btu increases the temperature of ice 2°F, in the case of water an increase of only 1°F will take place for each Btu of heat added. This difference is due to the fact that the specific heat, or the resistance offered by ice to a change in temperature is one half that offered by water. That is, the specific heat of ice is 0.5 Btu/lb/°F.

If the water is heated in a vessel open to the atmosphere, its temperature will continue to rise until it reaches a temperature of about 212°F. This is the boiling point of water. More heating will not produce more temperature changes, but steam will issue from the vessel. It has been found that about 970 Btu are required to change 1 lb of water at atmospheric pressure and at 212°F into steam at the same temperature.

1

This quantity of heat which changes the physical state of water from that of a liquid to steam without changing the temperature is called the enthalpy of vaporization.

If the above operations are performed in a closed vessel, such as an ordinary steam boiler, water will boil at a higher temperature than 212°F, since the steam driven off cannot escape but is compressed, raising the pressure and, consequently, the temperature.

The fact that the boiling point of water depends on the pressure is well known. Thus, in a locality where the altitude is 6,000 ft. above sea level and the barometric pressure is 12.6 (psia) the boiling point of water is about 204°F as compared with 212°F at sea level where the barometric pressure is 14.7 psia.

When the pressure is increased to 60 psia, it will be found that the boiling point of water is 307.2°F. At 100 psia water will boil at 337.8°F and at 150 psia the temperature will be 365.8°F before steam will be formed.

To summarize, the saturation temperature of steam increases with increasing pressure.

Quality of Steam

Steam formed in contact with water is known as saturated steam, and contains some amount of water. It can be either negligible or the steam may be holding all the moisture it can for the given pressure.

In the first case, steam carries with it a certain amount of water which has not been evaporated. The percentage of this water determines the dryness or the quality of the steam (x); that is, if the steam contains 3 percent by weight, the steam is 97 percent dry, or x = 0.97. A stationary steam boiler, properly erected and operated should generate steam that is at least 98 percent dry. If there is more than 2 percent moisture, there is reason to believe that the boiler is improperly installed, inefficiently sized or improperly operated.

In the second condition, that is with dry steam, the vapor does not carry water that has not been evaporated; that is, it has 100% quality. Any loss of heat, however small, not accompanied by a corresponding reduction in pressure, will cause condensation, and wet steam will be the result. Saturated steam, whether wet or dry, has a definite temperature corresponding to its pressure.

An increase in temperature not accompanied by an increase in pressure will cause the steam to become what is known as superheated. One advantage of superheated steam is that its temperature may be reduced by the amount of superheat without causing condensation. This makes it possible to transmit the steam through mains and still have it dry at the time it reaches its point of use. Superheated steam may be obtained by passing saturated steam through coils of pipe in the path of the hot flue gases from the boiler to the chimney. An apparatus for superheating steam is called a superheater.

The pressure of steam will remain constant if it is used as fast as it is generated. If a turbine, for instance, uses steam too rapidly, the boiler pressure will drop. Similarly, if the fuel is burned at a constant rate and steam use decreases, the pressure of the steam in the boiler will increase.

Steam Tables

The relationship between pressure, volume, and temperature of saturated and superheated steam as well as the energy required to produce vaporization under various conditions have been determined experimentally. These experimental results have been expressed in the form of empirical equations from which steam tables have been computed.

Tables 1-1 and 1-2 give the most important properties of saturated steam which include:

1. Pressure (p) of saturated steam is measured in psia or inches of mercury for low pressures.

2. Temperature of saturated steam is measured in degrees Fahrenheit (F). This column shows the vaporization temperature or the boiling point, at each of the given pressures.

3. Specific volume. These include the specific volume of the saturated liquid (v_f) and the specific volume of saturated steam (v_g).

4. Enthalpy. The three columns for enthalpy are: the enthalpy of the liquid (h_f), enthalpy of evaporation (h_{fg}), and enthalpy of saturated steam (h_g). The enthalpy of the liquid (h_f) is the energy required to raise the temperature of 1 lb of water from the freezing point to the vaporization temperature, or boiling point, at the given pressure. The enthalpy of evaporation (h_{fg}) is the energy required to change 1 lb of water at its point of vaporization to dry saturated steam. The

enthalpy of saturated steam (h_g) is the sum of h_f and h_{fg} or the energy required to change 1 lb of water from the freezing point to dry saturated steam at the given pressure. Enthalpy is given in Btu/lb.

5.   Entropy. The three columns for entropy include entropy of the liquid (S_f), entropy of evaporation (S_{fg}), and entropy of saturated steam (s_g). Entropy is given in Btu/°F/lb.

6.   Internal energy is measured in Btu/lb. The internal energy of the liquid is termed (u_f), and (u_g) is the internal energy of the vapor.

SAMPLE CALCULATIONS

Example 1
Water at 200°F is pumped into a boiler in which the pressure is 155.3 lb psig. How much heat must be supplied by the fuel to evaporate each pound of water into dry steam?

Solution
A pressure of 155.3 lb psig = 155.3 + 14.7 = 170 lb psia. The enthalpy (h_g) required to evaporate 1 lb of water from the freezing point into dry steam at a pressure of 170 psia is 1196 (Table 1-2). Since the water pumped to the boiler has a temperature of 200°F, the total amount of heat to be supplied by the fuel to evaporate 1 lb of water into dry steam is

$$1196 - (200 - 32) = 1028 \text{ Btu}$$

Example 2
If the steam in Example 1, contained 3 percent moisture, calculate the heat which must be supplied by the fuel to evaporate each pound from feed water at 200°F

Solution
The enthalpy of the liquid (h_f), or the heat required to raise the temperature of a pound of water from 200°F to the boiling point corresponding to a pressure of 170 lb per square inch absolute is:

$$341.1 - (200 - 32) = 173.1 \text{ Btu.}$$

Table 1-1. Properties of Saturated Steam (temperature oriented).

Temp. Fahr., t	Abs. press., lb. per sq. in., p	Specific volume			Enthalpy			Entropy			Temp., Fahr., t
		Sat. liquid, v_f	Evap., v_{fg}	Sat. vapor, v_g	Sat. liquid, h_f	Evap., h_{fg}	Sat. vapor, h_g	Sat. liquid, S_f	Evap, S_{fg}	Sat. vapor, S_g	
32	0.08854	0.01602	3306	3306	0.00	1075.8	1075.8	0.0000	2.1877	2.1877	32
35	0.09995	0.01602	2947	2947	3.02	1074.1	1077.1	0.0061	2.1709	2.1770	35
40	0.12170	0.01602	2444	2444	8.05	1071.3	1079.3	0.0162	2.1435	2.1597	40
45	0.14752	0.01602	2036.4	2036.4	13.06	1068.4	1081.5	0.0262	2.1167	2.1429	45
50	0.17811	0.01603	1703.2	1703.2	18.07	1065.6	1083.7	0.0361	2.0903	2.1264	50
60	0.2563	0.01604	1206.6	1206.7	28.06	1059.9	1088.0	0.0555	2.0393	2.0948	60
70	0.3631	0.01606	867.8	867.9	38.04	1054.3	1092.3	0.0745	1.9902	2.0647	70
80	0.5069	0.01608	633.1	633.1	48.02	1048.6	1096.6	0.0932	1.9428	2.0360	80
90	0.6982	0.01610	468.0	468.0	57.99	1042.9	1100.9	0.1115	1.8972	2.0087	90
100	0.9492	0.01613	350.3	350.4	67.97	1037.2	1105.2	0.1295	1.8531	1.9826	100
110	1.2748	0.01617	265.3	265.4	77.94	1031.6	1109.5	0.1471	1.8106	1.9577	110
120	1.6924	0.01620	203.25	203.27	87.92	1025.8	1113.7	0.1645	1.7694	1.9339	120
130	2.2225	0.01625	157.32	157.34	97.90	1020.0	1117.9	0.1816	1.7296	1.9112	130
140	2.8886	0.01629	122.99	123.01	107.89	1014.1	1122.0	0.1984	1.6910	1.8894	140
150	3.718	0.01634	97.06	97.07	117.89	1008.2	1126.1	0.2149	1.6537	1.8685	150
160	4.741	0.01639	77.27	77.29	127.89	1002.3	1130.2	0.2311	1.6174	1.8485	160
170	5.992	0.01645	62.04	62.06	137.90	996.3	1134.2	0.2472	1.5822	1.8293	170
180	7.510	0.01651	50.21	50.23	147.92	990.2	1138.1	0.2630	1.5480	1.8109	180
190	9.339	0.01657	40.94	40.96	157.95	984.1	1142.0	0.2785	1.5147	1.7932	190
200	11.526	0.01663	33.62	33.64	167.99	977.9	1145.9	0.2938	1.4824	1.7762	200
210	14.123	0.01670	27.80	27.82	178.05	971.6	1149.7	0.3090	1.4508	1.7598	210
212	14.696	0.01672	26.78	26.80	180.07	970.3	1150.4	0.3120	1.4446	1.7566	212
220	17.186	0.01677	23.13	23.15	188.13	965.2	1153.4	0.3239	1.4201	1.7440	220
230	20.780	0.01684	19.365	19.382	198.23	958.8	1157.0	0.3387	1.3901	1.7288	230
240	24.969	0.01692	16.306	16.323	208.34	952.2	1160.5	0.3531	1.3609	1.7140	240
250	29.825	0.01700	13.804	13.821	218.48	945.5	1164.0	0.3675	1.3323	1.6998	250
260	35.429	0.01709	11.746	11.763	228.64	938.7	1167.3	0.3817	1.3043	1.6860	260
270	41.858	0.01717	10.044	10.061	238.84	931.8	1170.6	0.3958	1.2769	1.6727	270
280	49.203	0.01726	8.628	8.645	249.06	924.7	1173.8	0.4096	1.2501	1.6597	280
290	57.556	0.01735	7.444	7.461	259.31	917.5	1176.8	0.4234	1.2238	1.6472	290
300	67.013	0.01745	6.449	6.466	269.59	910.1	1179.7	0.4369	1.1980	1.6350	300
310	77.68	0.01755	5.609	5.626	279.92	902.6	1182.5	0.4504	1.1727	1.6231	310
320	89.66	0.01765	4.896	4.914	290.28	894.9	1185.2	0.4637	1.1478	1.6115	320
330	103.06	0.01776	4.289	4.307	300.68	887.0	1187.7	0.4769	1.1233	1.6002	330
340	118.01	0.01787	3.770	3.788	311.13	879.0	1190.1	0.4900	1.0992	1.5891	340
350	134.63	0.01799	3.324	3.342	321.63	870.7	1192.3	0.5029	1.0754	1.5783	350
360	153.04	0.01811	2.939	2.957	332.18	862.2	1194.4	0.5158	1.0519	1.5677	360
370	173.37	0.01823	2.606	2.625	342.79	853.5	1196.3	0.5286	1.0287	1.5573	370
380	195.77	0.01836	2.317	2.335	353.45	844.6	1198.1	0.5413	1.0059	1.5471	380
390	220.37	0.01850	2.0651	2.0836	364.17	835.4	1199.6	0.5539	0.9832	1.5371	390
400	247.31	0.01864	1.8447	1.8633	374.97	826.0	1201.0	0.5664	0.9608	1.5272	400
410	276.75	0.01878	1.6512	1.6700	385.83	816.3	1202.1	0.5788	0.9386	1.5174	410
420	308.83	0.01894	1.4811	1.5000	396.77	806.3	1203.1	0.5912	0.9166	1.5078	420
430	343.72	0.01910	1.3308	1.3499	407.79	796.0	1203.8	0.6035	0.8947	1.4982	430
440	381.59	0.01926	1.1979	1.2171	418.90	785.4	1204.3	0.6158	0.8730	1.4887	440
450	422.6	0.0194	1.0799	1.0993	430.1	774.5	1204.6	0.6280	0.8513	1.4793	450
460	466.9	0.0196	0.9748	0.9944	441.4	763.2	1204.6	0.6402	0.8298	1.4700	460
470	514.7	0.0198	0.8811	0.9009	452.8	751.5	1204.3	0.6523	0.8083	1.4606	470
480	566.1	0.0200	0.7972	0.8172	464.4	739.4	1203.7	0.6645	0.7868	1.4513	480
490	621.4	0.0202	0.7221	0.7423	476.0	726.8	1202.8	0.6766	0.7653	1.4419	490
500	680.8	0.0204	0.6545	0.6749	487.8	713.9	1201.7	0.6887	0.7438	1.4325	500
520	812.4	0.0209	0.5385	0.5594	511.9	686.4	1198.2	0.7130	0.7006	1.4136	520
540	962.5	0.0215	0.4434	0.4649	536.6	656.6	1193.2	0.7374	0.6568	1.3942	540
560	1133.1	0.0221	0.3647	0.3868	562.2	624.2	1186.4	0.7621	0.6121	1.3742	560
580	1325.8	0.0228	0.2989	0.3217	588.9	588.4	1177.3	0.7872	0.5659	1.3532	580
600	1542.9	0.0236	0.2432	0.2668	617.0	548.5	1165.5	0.8131	0.5176	1.3307	600
620	1786.6	0.0247	0.1955	0.2201	646.7	503.6	1150.3	0.8398	0.4664	1.3062	620
640	2059.7	0.0260	0.1538	0.1798	678.6	452.0	1130.5	0.8679	0.4110	1.2789	640
660	2365.4	0.0278	0.1165	0.1442	714.2	390.2	1104.4	0.8987	0.3485	1.2472	660
680	2708.1	0.0305	0.0810	0.1115	757.3	309.9	1067.2	0.9351	0.2719	1.2071	680
700	3093.7	0.0369	0.0392	0.0761	823.3	172.1	995.4	0.9905	0.1484	1.1389	700
705.4	3206.2	0.0503	0	0.0503	902.7	0	902.7	1.0580	0	1.0580	705.4

* Abridged from "Thermodynamic Properties of Steam," by Joseph H. Keenan and Frederick G. Keyes, John Wiley & Sons, Inc., New York.

Table 1-2. Properties of Saturated Steam (pressure oriented).

Abs. press., lb. per sq. in., p	Temp. Fuhr., t	Specific volume		Enthalpy			Entropy			Internal energy		Abs. press., lb. per sq. in., p
		Sat. liquid, v_f	Sat. vapor, v_g	Sat. liquid, h_f	Evap., h_{fg}	Sat. vapor, h_g	Sat. liquid, S_f	Evap., S_{fg}	Sat. vapor, S_g	Sat. liquid, u_f	Sat. vapor, u_g	
1.0	101.74	0.01614	333.6	69.70	1036.3	1106.0	0.1326	1.8456	1.9782	69.70	1044.3	1.0
2.0	126.08	0.01623	173.73	93.99	1022.2	1116.2	0.1749	1.7451	1.9200	93.98	1051.9	2.0
3.0	141.48	0.01630	118.71	109.37	1013.2	1122.6	0.2008	1.6855	1.8863	109.36	1056.7	3.0
4.0	152.97	0.01636	90.63	120.86	1006.4	1127.3	0.2198	1.6427	1.8625	120.85	1060.2	4.0
5.0	162.24	0.01640	73.52	130.13	1001.0	1131.1	0.2347	1.6094	1.8441	130.12	1063.1	5.0
6.0	170.06	0.01645	61.98	137.96	996.2	1134.2	0.2472	1.5820	1.8292	137.94	1065.4	6.0
7.0	176.85	0.01649	53.64	144.76	992.1	1136.9	0.2581	1.5586	1.8167	144.74	1067.4	7.0
8.0	182.86	0.01653	47.34	150.79	988.5	1139.3	0.2674	1.5383	1.8057	150.77	1069.2	8.0
9.0	188.28	0.01656	42.40	156.22	985.2	1141.4	0.2759	1.5203	1.7962	156.19	1070.8	9.0
10	193.21	0.01659	38.42	161.17	982.1	1143.3	0.2835	1.5041	1.7876	161.14	1072.2	10
14.696	212.00	0.01672	26.80	180.07	970.3	1150.4	0.3120	1.4446	1.7566	180.02	1077.5	14.696
15	213.03	0.01672	26.29	181.11	969.7	1150.8	0.3135	1.4415	1.7549	181.06	1077.8	15
20	227.96	0.01683	20.089	196.16	960.1	1156.3	0.3356	1.3962	1.7319	196.10	1081.9	20
25	240.07	0.01692	16.303	208.42	952.1	1160.6	0.3533	1.3606	1.7139	208.34	1085.1	25
30	250.33	0.01701	13.746	218.82	945.3	1164.1	0.3680	1.3313	1.6993	218.73	1087.8	30
35	259.28	0.01708	11.898	227.91	939.2	1167.1	0.3807	1.3063	1.6870	227.80	1090.1	35
40	267.25	0.01715	10.498	236.03	933.7	1169.7	0.3919	1.2844	1.6763	235.90	1092.0	40
45	274.44	0.01721	9.401	243.36	928.6	1172.0	0.4019	1.2650	1.6669	243.22	1093.7	45
50	281.01	0.01727	8.515	250.09	924.0	1174.1	0.4110	1.2474	1.6585	249.93	1095.3	50
55	287.07	0.01732	7.787	256.30	919.6	1175.9	0.4193	1.2316	1.6509	256.12	1096.7	55
60	292.71	0.01738	7.175	262.09	915.5	1177.6	0.4270	1.2168	1.6438	261.90	1097.9	60
65	297.97	0.01743	6.655	267.50	911.6	1179.1	0.4342	1.2032	1.6374	267.29	1099.1	65
70	302.92	0.01748	6.206	272.61	907.9	1180.6	0.4409	1.1906	1.6315	272.38	1100.2	70
75	307.60	0.01753	5.816	277.43	904.5	1181.9	0.4472	1.1787	1.6259	277.19	1101.2	75
80	312.03	0.01757	5.472	282.02	901.1	1183.1	0.4531	1.1676	1.6207	281.76	1102.1	80
85	316.25	0.01761	5.168	286.39	897.8	1184.2	0.4587	1.1571	1.6158	286.11	1102.9	85
90	320.27	0.01766	4.896	290.56	894.7	1185.3	0.4641	1.1471	1.6112	290.27	1103.7	90
95	324.12	0.01770	4.652	294.56	891.7	1186.2	0.4692	1.1376	1.6068	294.25	1104.5	95
100	327.81	0.01774	4.432	298.40	888.8	1187.2	0.4740	1.1286	1.6026	298.08	1105.2	100
110	334.77	0.01782	4.049	305.66	883.2	1188.9	0.4832	1.1117	1.5948	305.30	1106.5	110
120	341.25	0.01789	3.728	312.44	877.9	1190.4	0.4916	1.0962	1.5878	312.05	1107.6	120
130	347.32	0.01796	3.455	318.81	872.9	1191.7	0.4995	1.0817	1.5812	318.38	1108.6	130
140	353.02	0.01802	3.220	324.82	868.2	1193.0	0.5069	1.0682	1.5751	324.35	1109.6	140
150	358.42	0.01809	3.015	330.51	863.6	1194.1	0.5138	1.0556	1.5694	330.01	1110.5	150
160	363.53	0.01815	2.834	335.93	859.2	1195.1	0.5204	1.0436	1.5640	335.39	1111.2	160
170	368.41	0.01822	2.675	341.09	854.9	1196.0	0.5266	1.0324	1.5590	340.52	1111.9	170
180	373.06	0.01827	2.532	346.03	850.8	1196.9	0.5325	1.0217	1.5542	345.42	1112.5	180
190	377.51	0.01833	2.404	350.79	846.8	1197.6	0.5381	1.0116	1.5497	350.15	1113.1	190
200	381.79	0.01839	2.288	355.36	843.0	1198.4	0.5435	1.0018	1.5453	354.68	1113.7	200
250	400.95	0.01865	1.8438	376.00	825.1	1201.1	0.5675	0.9588	1.5263	375.14	1115.8	250
300	417.33	0.01890	1.5433	393.84	809.0	1202.8	0.5879	0.9225	1.5104	392.79	1117.1	300
350	431.72	0.01913	1.3260	409.69	794.2	1203.9	0.6056	0.8910	1.4966	408.45	1118.0	350
400	444.59	0.0193	1.1613	424.0	780.5	1204.5	0.6214	0.8630	1.4844	422.6	1118.5	400
450	456.28	0.0195	1.0320	437.2	767.4	1204.6	0.6356	0.8378	1.4734	435.5	1118.7	450
500	467.01	0.0197	0.9278	449.4	755.0	1204.4	0.6487	0.8147	1.4634	447.6	1118.6	500
550	476.94	0.0199	0.8424	460.8	743.1	1203.9	0.6608	0.7934	1.4542	458.8	1118.2	550
600	486.21	0.0201	0.7698	471.6	731.6	1203.2	0.6720	0.7734	1.4454	469.4	1117.7	600
650	494.90	0.0203	0.7083	481.8	720.5	1202.3	0.6826	0.7548	1.4374	479.4	1117.1	650
700	503.10	0.0205	0.6554	491.5	709.7	1201.2	0.6925	0.7371	1.4296	488.8	1116.3	700
750	510.86	0.0207	0.6092	500.8	699.2	1200.0	0.7019	0.7204	1.4223	498.0	1115.4	750
800	518.23	0.0209	0.5687	509.7	688.9	1198.6	0.7108	0.7045	1.4153	506.6	1114.4	800
850	525.26	0.0210	0.5327	518.3	678.8	1197.1	0.7194	0.6891	1.4085	515.0	1113.3	850
900	531.98	0.0212	0.5006	526.6	668.8	1195.4	0.7275	0.6744	1.4020	523.1	1112.1	900
950	538.43	0.0214	0.4717	534.6	659.1	1193.7	0.7355	0.6602	1.3957	530.9	1110.8	950
1000	544.61	0.0216	0.4456	542.4	649.4	1191.8	0.7430	0.6467	1.3897	538.4	1109.4	1000
1100	556.31	0.0220	0.4001	557.4	630.4	1187.8	0.7575	0.6205	1.3780	552.9	1106.4	1100
1200	567.22	0.0223	0.3619	571.7	611.7	1183.4	0.7711	0.5956	1.3667	566.7	1103.0	1200
1300	577.46	0.0227	0.3293	585.4	593.2	1178.6	0.7840	0.5719	1.3559	580.0	1099.4	1300
1400	587.10	0.0231	0.3012	598.7	574.7	1173.4	0.7963	0.5491	1.3454	592.7	1095.4	1400
1500	596.23	0.0235	0.2765	611.6	556.3	1167.9	0.8082	0.5269	1.3351	605.1	1091.2	1500
2000	635.82	0.0257	0.1878	671.7	463.4	1135.1	0.8619	0.4230	1.2849	662.2	1065.6	2000
2500	668.13	0.0287	0.1307	730.6	360.5	1091.1	0.9126	0.3197	1.2322	717.3	1030.6	2500
3000	695.36	0.0346	0.0858	802.5	217.8	1020.3	0.9731	0.1885	1.1615	783.4	972.7	3000
3206.2	705.40	0.0503	0.0503	902.7	0	902.7	1.0580	0	1.0580	872.9	872.9	3206.2

* Abridged from "Thermodynamic Properties of Steam," by Joseph H. Keenan and Frederick G. Keyes, John Wiley & Sons, Inc., New York.

The enthalpy required to vaporize a pound of water into dry steam at 170 psia, after the boiling point is reached, or the enthalpy of evaporation (h_{fg}), is 854.9 Btu.

Since the steam in this example contains 3 percent moisture, it is 97 percent dry, and the heat required to vaporize it is:

$$854.9 \times 0.97 = 829.2 \text{ Btu.}$$

The total heat required to change 1 lb of water at 200°F into steam, that is three percent wet, and at a pressure of 170 psia, is:

$$173.1 + 829.2 = 1002.3 \text{ Btu.}$$

Example 3

What is the volume of 1 lb of steam at 150 psia, if it is 20% wet?

Solution

Dry steam at a pressure of 150 psia has a volume of 3.015 cu. ft./ lb The volume of 1 lb of steam which is 20 percent wet, or 80 percent dry, at a pressure of 150 psia is:

$$3.015 \times .80 = 2.41 \text{ cu. ft.}$$

A more complete set of steam tables are presented in Appendix B, Steam Tables.

CALORIMETERS

The quality, or the percent of moisture in saturated steam x, is determined by a calorimeter. There are three types of steam calorimeters in general use—the throttling calorimeter, the separating calorimeter, and the electrical calorimeter.

The throttling calorimeter is an accurate instrument for measuring the amount of moisture in steam, if the moisture content of the steam is not too great. This instrument depends upon the fact that steam, nearly dry, becomes superheated when its pressure is reduced by throttling. This happens because slightly wet saturated steam at high pressure contains more heat than at low pressure. A simple type of throttling calorimeter is illustrated in Figure 1-1. Steam enters at the orifice, discharging into chamber C, in which a thermometer T is inserted. The throttling calorimeter is unsuitable for measuring the quality of steam that contains more than 3 or 4 percent moisture.

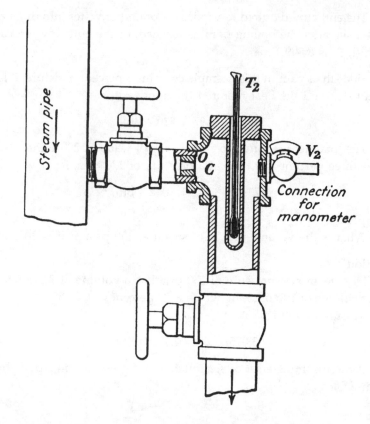

Figure 1-1. Throttling Steam Calorimeter.

The electrical or electronic calorimeter is used in today's plant operation, where a remote readout instrument can be installed in a control room for easy monitoring.

The separating calorimeter is used for performance testing of equipment, as during acceptance tests. Its operation is also dependent on the different heat values which define the liquid (water) and the gas (steam). Along with this calorimeter, the throttling calorimeter is also used extensively in performance testing.

Detailed discussion of calorimeters is beyond the scope of this text since many of the characteristics vary with different manufacturers. For more information on calorimeters, consult a heating instrumentation handbook or refer to manufacturers' literature.

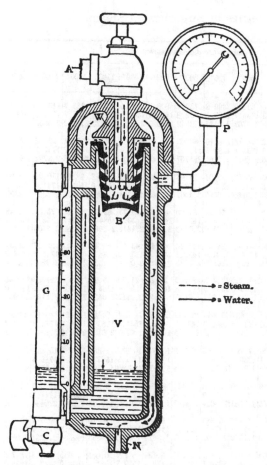

Figure 1-2. Separating Steam Calorimeter.

SUPERHEATED STEAM

Superheated steam has a temperature which is higher than the saturation temperature corresponding to its pressure as taken from the steam tables (Table 1-2). Thus, if steam at a pressure of 250 psia has a temperature of 600°F, the degree of superheat may be found as follows:

The saturation temperature corresponding to 250 lb, by Table 1-2, is equal to 400.95°F. Thus, the degree of superheat is

$$600 - 400.95 = 199°F$$

Table 1-3 gives the properties of superheated steam. At each pressure and at superheated steam temperatures up to 1600°F values are given for the specific volume (v), enthalpy (h), and entropy (s) of superheated steam. The saturation temperatures are given in parentheses below each pressure.

The following examples will illustrate the use of Table 1-3.

Example 1

Calculate the temperature and the specific volume of steam at a pressure of 220 psia and 200°F superheat.

Solution

Steam at a pressure of 220 psia and 200°F superheat has a temperature of 389.9 + 200 = 589.9°F, and a specific volume of 2.737 cu. ft.

Table 1-3. Properties of Superheated Steam.

Abs.press., lb. per sq. in. (sat. temp.)	200°	300°	400°	500°	600°	700°	800°	900°	1000°	1100°	1200°	1400°	1600°
1 (101.74) *v*	392.6	452.3	512.0	571.6	631.2	690.8	750.4	809.9	869.5	929.1	988.7	1107.8	1227.0
h	1150.4	1195.8	1241.7	1288.3	1335.7	1383.8	1432.8	1482.7	1533.5	1585.2	1637.7	1745.7	1857.5
s	2.0512	2.1153	2.1720	2.2233	2.2702	2.3137	2.3542	2.3923	2.4283	2.4625	2.4952	2.5566	2.6137
5 (162.24) *v*	78.16	90.25	102.26	114.22	126.16	138.10	150.03	161.95	173.87	185.79	197.71	221.6	245.4
h	1148.8	1195.0	1241.2	1288.0	1335.4	1383.6	1432.7	1482.6	1533.4	1585.1	1637.7	1745.7	1857.4
s	1.8718	1.9370	1.9942	2.0456	2.0927	2.1361	2.1767	2.2148	2.2509	2.2851	2.3178	2.3792	2.4363
10 (193.21) *v*	38.85	45.00	51.04	57.05	63.03	69.01	74.98	80.95	86.92	92.88	98.84	110.77	122.69
h	1146.6	1193.9	1240.6	1287.5	1335.1	1383.4	1432.5	1482.4	1533.2	1585.0	1637.6	1745.6	1857.3
s	1.7927	1.8595	1.9172	1.9689	2.0160	2.0596	2.1002	2.1383	2.1744	2.2086	2.2413	2.3028	2.3598
14.696 (212.00) *v*		30.53	34.68	38.78	42.86	46.94	51.00	55.07	59.13	63.19	67.25	75.37	83.48
h		1192.8	1239.9	1287.1	1334.8	1383.2	1432.3	1482.3	1533.1	1584.8	1637.5	1745.5	1857.3
s		1.8160	1.8743	1.9261	1.9734	2.0170	2.0576	2.0958	2.1319	2.1662	2.1989	2.2603	2.3174
20 (227.96) *v*		22.36	25.43	28.46	31.47	34.47	37.46	40.45	43.44	46.42	49.41	55.37	61.34
h		1191.6	1239.2	1286.6	1334.4	1382.9	1432.1	1482.1	1533.0	1584.7	1637.4	1745.4	1857.2
s		1.7808	1.8396	1.8918	1.9392	1.9829	2.0235	2.0618	2.0978	2.1321	2.1648	2.2263	2.2834
40 (267.25) *v*		11.040	12.628	14.168	15.688	17.198	18.702	20.20	21.70	23.20	24.69	27.68	30.66
h		1186.8	1236.5	1284.8	1333.1	1381.9	1431.3	1481.4	1532.4	1584.3	1637.0	1745.1	1857.0
s		1.6994	1.7608	1.8140	1.8619	1.9058	1.9467	1.9850	2.0212	2.0555	2.0883	2.1498	2.2069
60 (292.71) *v*		7.259	8.357	9.403	10.427	11.441	12.449	13.452	14.454	15.453	16.451	18.446	20.44
h		1181.6	1233.6	1283.0	1331.8	1380.9	1430.5	1480.8	1531.9	1583.8	1636.6	1744.8	1856.7
s		1.6492	1.7135	1.7678	1.8162	1.8605	1.9015	1.9400	1.9762	2.0106	2.0434	2.1049	2.1621
80 (312.03) *v*			6.220	7.020	7.797	8.562	9.322	10.077	10.830	11.582	12.332	13.830	15.325
h			1230.7	1281.1	1330.5	1379.9	1429.7	1480.1	1531.3	1583.4	1636.2	1744.5	1856.5
s			1.6791	1.7346	1.7836	1.8281	1.8694	1.9079	1.9442	1.9787	2.0115	2.0731	2.1303
100 (327.81) *v*			4.937	5.589	6.218	6.835	7.446	8.052	8.656	9.259	9.860	11.060	12.258
h			1227.6	1279.1	1329.1	1378.9	1428.9	1479.5	1530.8	1582.9	1635.7	1744.2	1856.2
s			1.6518	1.7085	1.7581	1.8029	1.8443	1.8829	1.9193	1.9538	1.9867	2.0484	2.1056
120 (341.25) *v*			4.081	4.636	5.165	5.683	6.195	6.702	7.207	7.710	8.212	9.214	10.213
h			1224.4	1277.2	1327.7	1377.8	1428.1	1478.8	1530.2	1582.4	1635.3	1743.9	1850.0
s			1.6287	1.6869	1.7370	1.7822	1.8237	1.8625	1.8990	1.9335	1.9664	2.0281	2.0854
140 (353.02) *v*			3.468	3.954	4.413	4.861	5.301	5.738	6.172	6.604	7.035	7.895	8.752
h			1221.1	1275.2	1326.4	1376.8	1427.3	1478.2	1529.7	1581.9	1634.9	1743.5	1855.7
s			1.6087	1.6683	1.7190	1.7645	1.8063	1.8451	1.8817	1.9163	1.9493	2.0110	2.0683
160 (363.53) *v*			3.008	3.443	3.849	4.244	4.631	5.015	5.396	5.775	6.152	6.906	7.656
h			1217.6	1273.1	1325.0	1375.7	1426.4	1477.5	1529.1	1581.4	1634.5	1743.2	1855.5
s			1.5908	1.6519	1.7033	1.7491	1.7911	1.8301	1.8667	1.9014	1.9344	1.9962	2.0535
180 (373.06) *v*			2.649	3.044	3.411	3.764	4.110	4.452	4.792	5.129	5.466	6.136	6.804
h			1214.0	1271.0	1323.5	1374.7	1425.6	1476.8	1528.6	1581.0	1634.1	1742.9	1855.2
s			1.5745	1.6373	1.6894	1.7355	1.7776	1.8167	1.8534	1.8882	1.9212	1.9831	2.0404
200 (381.79) *v*			2.361	2.726	3.060	3.380	3.693	4.002	4.309	4.613	4.917	5.521	6.123
h			1210.3	1268.9	1322.1	1373.6	1424.8	1476.2	1528.0	1580.5	1633.7	1742.6	1855.0
s			1.5594	1.6240	1.6767	1.7232	1.7655	1.8048	1.8415	1.8763	1.9094	1.9713	2.0287
220 (389.86) *v*			2.125	2.465	2.772	3.066	3.352	3.634	3.913	4.191	4.467	5.017	5.565
h			1206.5	1266.7	1320.7	1372.6	1424.0	1475.5	1527.5	1580.0	1633.3	1742.3	1854.7
s			1.5453	1.6117	1.6652	1.7120	1.7545	1.7939	1.8308	1.8656	1.8987	1.9607	2.0181
240 (397.37) *v*			1.9276	2.247	2.533	2.804	3.068	3.327	3.584	3.839	4.093	4.597	5.100
h			1202.5	1264.5	1319.2	1371.5	1423.2	1474.8	1526.9	1579.6	1632.9	1742.0	1854.5
s			1.5319	1.6003	1.6546	1.7017	1.7444	1.7839	1.8209	1.8558	1.8889	1.9510	2.0084
260 (404.42) *v*				2.003	2.330	2.582	2.827	3.067	3.305	3.541	3.776	4.242	4.707
h				1262.3	1317.7	1370.4	1422.3	1474.2	1526.3	1579.1	1632.5	1741.7	1854.2
s				1.5897	1.6447	1.6922	1.7352	1.7748	1.8118	1.8467	1.8799	1.9420	1.9995

Table 1-3 (Cont'd).

Abs.press., lb. per sq. in. (sat. temp.)		200°	300°	400°	500°	600°	700°	800°	900°	1000°	1100°	1200°	1400°	1600°
							Temperature-degrees Fahrenheit							
280 (411.05)	v				1.9047	2.156	2.392	2.621	2.845	3.066	3.286	3.504	3.938	4.370
	h				1260.0	1316.2	1369.4	1421.5	1473.5	1525.8	1578.6	1632.1	1741.4	1854.0
	s				1.5796	1.6354	1.6834	1.7265	1.7662	1.8033	1.8383	1.8716	1.9337	1.9912
300 (417.33)	v				1.7675	2.005	2.227	2.442	2.652	2.859	3.065	3.269	3.674	4.078
	h				1257.6	1314.7	1368.3	1420.6	1472.8	1525.2	1578.1	1631.7	1741.0	1853.7
	s				1.5701	1.6268	1.6751	1.7184	1.7582	1.7954	1.8305	1.8638	1.9260	1.9835
350 (431.72)	v				1.4923	1.7036	1.8980	2.084	2.266	2.445	2.622	2.798	3.147	3.493
	h				1251.5	1310.9	1365.5	1418.5	1471.1	1523.8	1577.0	1630.7	1740.3	1853.1
	s				1.5481	1.6070	1.6563	1.7002	1.7403	1.7777	1.8130	1.8463	1.9086	1.9663
400 (444.59)	v				1.2851	1.4770	1.6508	1.8161	1.9767	2.134	2.290	2.445	2.751	3.055
	h				1245.1	1306.9	1362.7	1416.4	1469.4	1522.4	1575.8	1629.6	1739.5	1852.5
	s				1.5281	1.5894	1.6398	1.6842	1.7247	1.7623	1.7977	1.8311	1.8936	1.9513

		500°	550°	600°	620°	640°	660°	680°	700°	800°	900°	1000°	1200°	1400°	1600°
450 (456.28)	v	1.1231	1.2155	1.3005	1.3332	1.3652	1.3967	1.4278	1.4584	1.6074	1.7516	1.8928	2.170	2.443	2.714
	h	1238.4	1272.0	1302.8	1314.6	1326.2	1337.5	1348.8	1359.9	1414.3	1467.7	1521.0	1628.6	1738.7	1851.9
	s	1.5095	1.5437	1.5735	1.5845	1.5951	1.6054	1.6153	1.6250	1.6699	1.7108	1.7486	1.8177	1.8803	1.9381
500 (467.01)	v	0.9927	1.0800	1.1591	1.1893	1.2188	1.2478	1.2763	1.3044	1.4405	1.5715	1.6996	1.9504	2.197	2.442
	h	1231.3	1266.8	1298.6	1310.7	1322.6	1334.2	1345.7	1357.0	1412.1	1466.0	1519.6	1627.6	1737.9	1851.3
	s	1.4919	1.5280	1.5588	1.5701	1.5810	1.5915	1.6016	1.6115	1.6571	1.6982	1.7363	1.8056	1.8683	1.9262
550 (476.94)	v	0.8852	0.9686	1.0431	1.0714	1.0989	1.1259	1.1523	1.1783	1.3038	1.4241	1.5414	1.7706	1.9957	2.219
	h	1223.7	1261.2	1294.3	1306.8	1318.9	1330.8	1342.5	1354.0	1409.9	1464.3	1518.2	1626.6	1737.1	1850.6
	s	1.4751	1.5131	1.5451	1.5568	1.5680	1.5787	1.5890	1.5991	1.6452	1.6868	1.7250	1.7946	1.8575	1.9155
600 (486.21)	v	0.7947	0.8753	0.9463	0.9729	0.9988	1.0241	1.0489	1.0732	1.1899	1.3013	1.4096	1.6208	1.8279	2.033
	h	1215.7	1255.5	1289.9	1302.7	1315.2	1327.4	1339.3	1351.1	1407.7	1462.5	1516.7	1625.5	1736.3	1850.0
	s	1.4586	1.4990	1.5323	1.5443	1.5558	1.5667	1.5773	1.5875	1.6343	1.6762	1.7147	1.7846	1.8476	1.9056
700 (503.10)	v		0.7277	0.7934	0.8177	0.8411	0.8639	0.8860	0.9077	1.0108	1.1082	1.2048	1.3853	1.5641	1.7405
	h		1243.2	1280.6	1294.3	1307.5	1320.3	1332.8	1345.0	1403.2	1459.0	1513.9	1623.5	1734.8	1848.8
	s		1.4722	1.5084	1.5212	1.5333	1.5449	1.5550	1.5665	1.6147	1.6573	1.6963	1.7666	1.8299	1.8881
800 (518.23)	v		0.6154	0.6779	0.7006	0.7223	0.7433	0.7635	0.7833	0.8763	0.9633	1.0470	1.2088	1.3662	1.5214
	h		1229.8	1270.7	1285.4	1299.4	1312.9	1325.9	1338.6	1398.6	1455.4	1511.0	1621.4	1733.2	1847.5
	s		1.4467	1.4863	1.5000	1.5129	1.5250	1.5366	1.5476	1.5972	1.6407	1.6801	1.7510	1.8146	1.8729
900 (531.98)	v		0.5264	0.5873	0.6089	0.6294	0.6491	0.6680	0.6863	0.7716	0.8506	0.9262	1.0714	1.2124	1.3509
	h		1215.0	1260.1	1275.9	1290.9	1305.1	1318.8	1332.1	1393.9	1451.8	1508.1	1619.3	1731.6	1846.3
	s		1.4216	1.4653	1.4800	1.4938	1.5066	1.5187	1.5303	1.5814	1.6257	1.6656	1.7371	1.8009	1.8595
1000 (544.61)	v		0.4533	0.5140	0.5350	0.5546	0.5733	0.5912	0.6084	0.6878	0.7604	0.8294	0.9615	1.0893	1.2146
	h		1198.3	1248.8	1265.9	1281.9	1297.0	1311.4	1325.3	1389.2	1448.2	1505.1	1617.3	1730.0	1845.0
	s		1.3961	1.4450	1.4610	1.4757	1.4893	1.5021	1.5141	1.5670	1.6121	1.6525	1.7245	1.7886	1.8474
1100 (556.31)	v			0.4532	0.4738	0.4929	0.5110	0.5281	0.5445	0.6191	0.6866	0.7503	0.8716	0.9885	1.1031
	h			1236.7	1255.3	1272.4	1288.5	1303.7	1318.3	1384.3	1444.5	1502.2	1615.2	1728.4	1843.8
	s			1.4251	1.4425	1.4583	1.4728	1.4862	1.4989	1.5535	1.5995	1.6405	1.7130	1.7775	1.8363
1200 (567.22)	v			0.4016	0.4222	0.4410	0.4586	0.4752	0.4909	0.5617	0.6250	0.6843	0.7967	0.9046	1.0101
	h			1223.5	1243.9	1262.4	1279.6	1295.7	1311.0	1379.3	1440.7	1499.2	1613.1	1726.9	1842.5
	s			1.4052	1.4243	1.4413	1.4568	1.4710	1.4843	1.5409	1.5879	1.6293	1.7025	1.7672	1.8263
1400 (587.10)	v			0.3174	0.3390	0.3580	0.3753	0.3912	0.4062	0.4714	0.5281	0.5805	0.6789	0.7727	0.8640
	h			1193.0	1218.4	1240.4	1260.3	1278.5	1295.5	1369.1	1433.1	1493.2	1608.9	1723.7	1840.0
	s			1.3639	1.3877	1.4079	1.4258	1.4419	1.4567	1.5177	1.5666	1.6093	1.6836	1.7489	1.8083

Table 1-3 (*Cont'd*).

Abs. press., lb. per sq. in. (sat. temp.)		500°	550°	600°	620°	640°	660°	680°	700°	800°	900°	1000°	1200°	1400°	1600°
1600 (604.90)	*v*				0.2733	0.2936	0.3112	0.3271	0.3417	0.4034	0.4553	0.5027	0.5906	0.6738	0.7545
	h				1187.8	1215.2	1238.7	1259.6	1278.7	1358.4	1425.3	1487.0	1604.6	1720.5	1837.5
	s				1.3489	1.3741	1.3952	1.4137	1.4303	1.4964	1.5476	1.5914	1.6669	1.7328	1.7926
1800 (621.03)	*v*					0.2407	0.2597	0.2760	0.2907	0.3502	0.3986	0.4421	0.5218	0.5968	0.6693
	h					1185.1	1214.0	1238.5	1260.3	1347.2	1417.4	1480.8	1600.4	1717 3	1835.0
	s					1.3377	1.3638	1.3855	1.4044	1.4765	1.5301	1.5752	1.6520	1.7185	1.7786
2000 (635.82)	*v*					0.1936	0.2161	0.2337	0.2489	0.3074	0.3532	0.3935	0.4668	0.5352	0.6011
	h					1145.6	1184.9	1214.8	1240.0	1335.5	1409.2	1474.5	1596.1	1714.1	1832.5
	s					1.2945	1.3300	1.3564	1.3783	1.4576	1.5139	1.5603	1.6384	1.7055	1.7660
2500 (668.13)	*v*							0.1484	0.1686	0.2294	0.2710	0.3061	0.3678	0.4244	0.4784
	h							1132.3	1176.8	1303.6	1387.8	1458.4	1585.3	1706.1	1826.2
	s							1.2687	1.3073	1.4127	1.4772	1.5273	1.6088	1.6775	1.7389
3000 (695.36)	*v*								0.0984	0.1760	0.2159	0.2476	0.3018	0.3505	0.3966
	h								1060.7	1267.2	1365.0	1441.8	1574.3	1698.0	1819.9
	s								1.1966	1.3690	1.4439	1.4984	1.5837	1.6540	1.7163
3206.2 (705.40)	*v*									0.1583	0.1981	0.2288	0.2806	0.3267	0.3703
	h									1250.5	1355.2	1434.7	1569.8	1694.6	1817.2
	s									1.3508	1.4309	1.4874	1.5742	1.6452	1.7080
3500	*v*								0.0306	0.1364	0.1762	0.2058	0.2546	0.2977	0.3381
	h								780.5	1224.9	1340.7	1424.5	1563.3	1689.8	1813.6
	s								0.9515	1.3241	1.4127	1.4723	1.5615	1.6336	1.6968
4000	*v*								0.0287	0.1052	0.1462	0.1743	0.2192	0.2581	0.2943
	h								763.8	1174.8	1314.4	1406.8	1552.1	1681.7	1807.2
	s								0.9347	1.2757	1.3827	1.4482	1.5417	1.6154	1.6795
4500	*v*								0.0276	0.0798	0.1226	0.1500	0.1917	0.2273	0.2602
	h								753.5	1113.9	1286.5	1388.4	1540.8	1673.5	1800.9
	s								0.9235	1.2204	1.3529	1.4253	1.5235	1.5990	1.6640
5000	*v*								0.0268	0.0593	0.1036	0.1303	0.1696	0.2027	0.2329
	h								746.4	1047.1	1256.5	1369.5	1529.5	1665.3	1794.5
	s								0.9152	1.1622	1.3231	1.4034	1.5066	1.5839	1.6499
5500	*v*								0.0262	0.0463	0.0880	0.1143	0.1516	0.1825	0.2106
	h								741.3	985.0	1224.1	1349.3	1518.2	1657.0	1788.1
	s								0.9090	1.1093	1.2930	1.3821	1.4908	1.5699	1.6369

* Abridged from "Thermodynamic Properties of Steam," by Joseph H. Keenan and Frederick G. Keyes, John Wiley & Sons, Inc., New York.

Example 2

Calculate the enthalpy and entropy at the conditions in Example 1.

Solution

Enthalpy (h) = 1314.0 Btu

Entropy (S) = 1. 6590 Btu/°F/hr

For more complete superheat information see Appendices B and C.

Thermodynamic Processes for Vapors

Vapors undergo pressure, volume, and temperature changes. When any of these changes occur extrinsically on a quantity of steam, the values of the intrinsic enthalpy, entropy and internal energy also change. Another way of stating this is to say: if any change occurs in either the temperature, pressure or volume, then all the intrinsic characteristic values of entropy, enthalpy, and internal energy will change. These changes may be studied graphically by means of: pressure-volume P-V, entropy-temperature T-S, and entropy-enthalpy S-H diagrams.

An entropy-temperature diagram for steam is illustrated in Figure 1-3. The saturation line separates the fluid into either liquid or steam. Above this saturation line, steam is considered superheated. As previously stated, the superheated steam is 100 percent dry and can hold additional heat energy if it is removed from the vessel containing saturated steam and is heated more in a separate container. A point defined as the critical point exists on the top of the saturation curve. The critical temperature is a condition where water is changed to steam without the addition of enthalpy (heat) for evaporation. Fluid will pass from a liquid to a gas without the sudden change of volume encountered below the

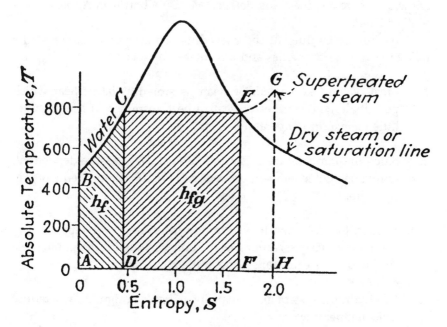

Figure 1-3. Temperature-entropy Diagram for Steam.

critical point.

The entropy-enthalpy diagram, called the Mollier chart, is given in Figure 1-4. In this chart the ordinate represents the enthalpy of steam above 32°F and the abscissa the entropy. The saturation curve marks the boundary between the superheated and saturated regions. In the region of the wet steam, lines of constant quality and pressure are drawn. In the superheated region, lines of equal superheats appear. In both the saturated and superheated regions, lines of constant pressure are drawn and the absolute pressure, in pounds per square inch of the various curves, is labeled.

To illustrate the use of the Mollier diagram, calculate the enthalpy of 1 lb of dry steam at a pressure of 150 psia.

This is found to be 1,195 as compared with 1,194.1 as given in Table 1-2.

If the steam in the above problem has a temperature of 500°F, the enthalpy by the chart is 1,275 as compared with 1,274.1 as given in Table 1-1.

The Mollier chart (Figure 1-4) is particularly useful for obtaining the approximate enthalpy of a vapor if its pressure and quality are known. A more complete and detailed Mollier Chart is in Appendix C - Mollier Chart.

An understanding of the preceding helps explain some of the following unique properties and advantages of steam:

1.   Most of the heat content of steam is stored as latent heat which permits large quantities of heat to be transmitted efficiently with little change in temperature.

2.   Since the temperature of saturated steam is pressure-dependent, only a negligible temperature reduction occurs from a reduction in pressure.

3.   Friction loss as steam flows through the system is independent of insulation thickness, as long as the boiler maintains the initial pressure and the steam traps remove the condensate.

4.   In a hydropic system inadequate insulation can significantly reduce fluid temperature.

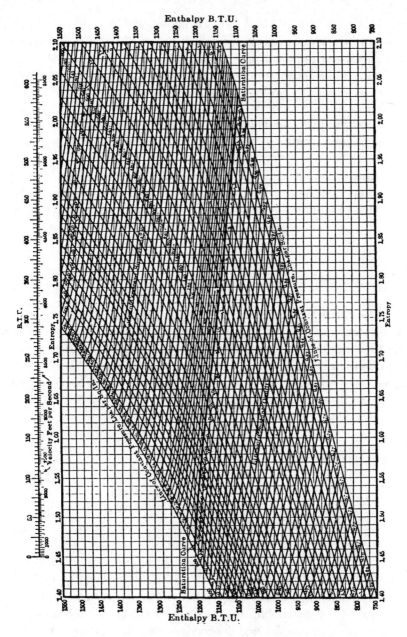

Figure 1-4. Enthalpy-Entropy Diagram for Steam.

5.   Steam, as with all fluids, flows from areas of high pressure to areas
      of low pressure. Heat dissipation causes the vapor to condense,
      creating a reduction in pressure caused by a change in specific
      volume of steam and water. Steam volume is 1600 times higher
      than water at a given temperature and pressure.

6.   As steam gives up its latent heat at the point of use, it can be at the
      same temperature as the utility being heated.

7.   After its initial use, some high pressure steam and large quantities
      of condensate are usually discharged to lower pressure systems.
      This permits the condensate to flash to steam again and be used
      over at a lower pressure.

FLASH STEAM

Discharging condensate to a lower pressure (as when a steam trap
passes condensate to the return system), the condensate contains more
heat than necessary to maintain the liquid phase at the lower pressure.
This excess heat causes some of the liquid to vaporize or "flash" to steam
at the lower pressure. The amount of liquid that flashes to steam can be
calculated by the following equation:

$$\text{Percent Flash Steam} = \frac{100\left(h_{f1} - h_{f2}\right)}{h_{fg2}}$$

where
h_{f1} = Enthalpy of liquid at pressure P_1
h_{f2} = Enthalpy of liquid at pressure P_2
h_{fg2} = Latent heat of vaporization at pressure P_2

Flash steam contains significant and useful heat energy that can be
recovered and used. This re-evaporation of condensate can be controlled
(minimized) by subcooling the condensate within the terminal equipment before it discharges into the condensate return piping. The amount
of subcooling should not be so large that a significant increase in heat
transfer surface is required. Flash steam and flash tanks and their appurtenances are discussed in a later chapter.

THE RANKINE STEAM CYCLE

This cycle is the standard for comparing steam plant economies. It is illustrated on the pressure-volume P-V and temperature-entropy TS diagrams in Figures 1-5 and 1-6.

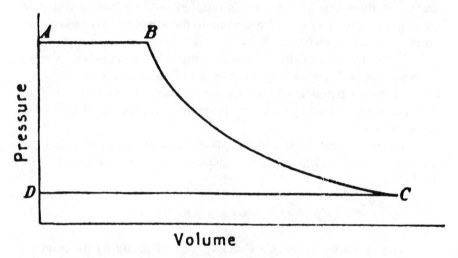

Figure 1-5. Pressure-volume Diagram for the Rankine Cycle.

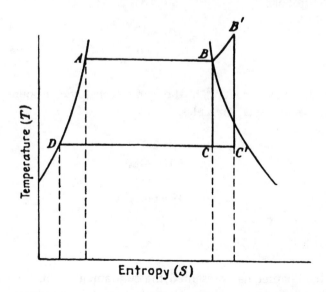

Figure 1-6. Temperature Entropy Diagram for the Rankine Cycle.

In this cycle, steam is admitted at constant pressure (AB) and is expanded adiabatically (BC) to the exhaust pressure. It is then exhausted at constant pressure (CD). In this cycle, it is assumed that the prime mover or other steam energy consuming process has no losses due to radiation, conduction, throttling, clearance, or friction. In the ideal Rankine Cycle the working substance is restored to the initial condition by raising the feed water from the exhaust to the temperature of the engine inlet, in a steam generator.

If the steam is initially superheated, the point B on the temperature-entropy-diagram (Figure 1-6) is changed to B' and C to C'.

Problems pertaining to the Rankine Cycle may best be solved by the use of steam tables (Appendix B) or by means of the Mollier chart (Appendix C).

As an illustration, if a steam plant operates, non-condensing, between a pressure of 200 psia and a temperature of 600°F, the work of the cycle is:

$$W = 1321 - 1094 = 227 \text{ Btu/lb/hr}$$

The efficiency of the Rankine Cycle is calculated by dividing the heat converted into work (W) by the heat supplied above the feed-water temperature, or

$$e = \frac{AW}{h_1 - h_f}$$

In the above equation, h_1 shows the enthalpy of a pound of steam, as found from the steam tables.

$$e = \frac{227}{1321 - 180.1}$$

$$e = 19.9 \text{ percent}$$

CLOSURE

This chapter has presented the fundamentals of practical steam engineering. The six primary characteristics of steam's intrinsic and extrinsic properties were discussed in what is hoped will refresh the engineer's knowledge.

It is shown how the Rankine Cycle defines the parameters within which steam engines and turbines comply.

Fluid characteristics of steam were separated with respect to liquid, saturated vapor, and superheated gas. Three terms have been given values: water (h_f), vapor (h_{fg}), and superheated steam (h_g). These terms are used in the Mollier Chart and steam tables. Other properties, volume, entropy, and internal energy are also available in the steam tables. It should be noted that the steam tables are the more accurate of the two but the Mollier Chart offers a diagram on which a steam cycle can be outlined graphically.

Chapter 2

Basic Steam Systems

STEAM

*O*f the 40 quadrillion Btu of energy consumed in industry every year over 50 percent is in the form of steam at one time or an other. Upon examination, it is understandable how this comes about. Steam has many advantages over its competitors for heating, driving machinery, and incorporating itself into processes that may not be possible without it.

Some advantages of steam are:

• Steam flows through the system unaided by external energy sources such as pumps.

• Terminal units can be added to or removed from the system without making basic changes to the system design.

• Steam components can be repaired or replaced by closing the steam supply without the difficulties associated with draining and refilling a water system.

• Steam is pressure-temperature dependent; therefore, the system temperature can be controlled by varying either system pressure or temperature.

• Steam can be distributed throughout a heating system with little change in temperature.

In view of the above advantages, steam is applicable to the following facilities:

- Where heat is required for processes and comfort heating, such as in industrial plants, hospitals, restaurants, dry-cleaning plants, laundries, and commercial buildings.

- Where the heating medium must travel great distances, such as in facilities with scattered building locations.

- Where intermittent changes in heat load occur.

STEAM SYSTEM DESIGN

Because of the various codes and regulations governing the design and operation of boilers and pressure vessels, steam systems are classified according to operating pressure. Low-pressure systems operate up to 15 psig, and high-pressure systems operate over 15 psig. There are many sub-classifications, especially for heating systems such as one- and two-pipe, gravity, vacuum, or variable vacuum return systems. However, these subclassifications relate to the distribution system or temperature-control method. Regardless of classification, all steam systems include a source of steam, a distribution system, and terminal equipment, where steam is used as the source of power or heat.

STEAM SOURCE

Steam can be generated directly by boilers using oil, gas, coal, wood, or solid waste, solar, nuclear, or electrical energy as a heat source, and heat can also be supplied indirectly by recovering heat from processes or equipment such as gas turbines and diesel or gas engines. The cogeneration of electricity and steam should always be considered for facilities or industries that have year-round steam requirements. Where steam is used as a power source (such as in turbine-driven equipment), the extraction exhaust or steam may be used in heat transfer equipment for process and space heating.

STEAM SYSTEMS

Steam can be provided by a facility's own boiler or cogeneration plant or can be purchased from a central steam plant (independently owned or otherwise) serving an industrial or commercial complex or other specific areas. This distinction can be very important. A facility with its own boiler plant usually has a closed loop system and requires the condensate to be as hot as possible when it returns to the boiler. But, condensate return pumps require a few degrees of subcooling to prevent cavitation or flashing of condensate to vapor in the pump impellers. The degree of subcooling will vary, depending on the hydraulic design or characteristics of the pump in use.

Central, independently owned steam plants often do not take back condensate, so it is discharged by the using facility and results in an open loop system. If a central plant does take back condensate, it rarely gives credit for its heat content. Except for losses to the atmosphere, which for the most part can be eliminated, every pound of steam eventually becomes a pound of condensate. If condensate is returned at 180°F and a heat recovery system reduces this temperature to 80°F, 100,000 Btu are recovered from every 1000 lbs of steam purchased. The heat remaining in the condensate represents 10 to 15 percent of the heat purchased from the central plant. Using this heat effectively can reduce steam and heating costs considerably.

BOILERS

Fired and waste heat boilers are usually constructed and labeled according to the ASME Boiler and Pressure Vessel Code since pressures normally exceed 15 psig. Boiler selection is based on the combined loads, including heating processes and equipment that use steam, hot water generation, and piping losses.

The American Society of Mechanical Engineers standards are used to test and rate most low-pressure heating boilers that have net and gross ratings. In smaller systems, selection is based on a net rating Larger system selection is made on a gross load basis. The occurrence and nature of the load components, with respect to the total load, determines the number of boilers used in an installation. Figures 2-1 and 2-2 show the configuration of a fire-tube boiler and a water-tube boiler respectively. Fire-tube boilers are used almost exclusively in low pressure service. The water-tube boiler is used when high steam loads and pressures are to be satisfied. Trends in using automatic control are to totally control the various processes needed to successfully operate a fire-tube boiler. Water-tube boilers have not attained as great a level of automation as fire-tube boilers. This situation results from state laws and municipal mandates that require a certain number of boiler personnel be on duty for every hour a high pressure boiler is in operation. Therefore, processes that are automatically controlled on a fire-tube boiler are turned over to respective personnel on high pressure boilers.

HEAT RECOVERY AND WASTE HEAT BOILERS

Steam can be generated by waste heat, such as exhaust from fuel-fired engines and turbines. Figure 2-3 schematically shows a typical recuperator. They are used to recover heat from devices such as gas turbines or other heat processes to supply steam for other heat processes. Where the quantity of steam generated by the waste heat boiler is not steady or ample enough to satisfy the facility's steam requirements, a conventional boiler is used to generate supplemental steam. Steam turbine extraction steam from predetermined turbine stages can be used for central steam distribution or exhaust steam from a back pressure turbine can be directed to a steam distribution network. In almost all recovery systems, a centrally operated steam boiler must be incorporated to provide

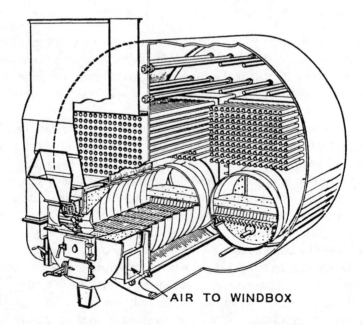

AIR TO WINDBOX

Figure 2-1. Coal-fired Fire Tube Boiler with Twin Furnaces.

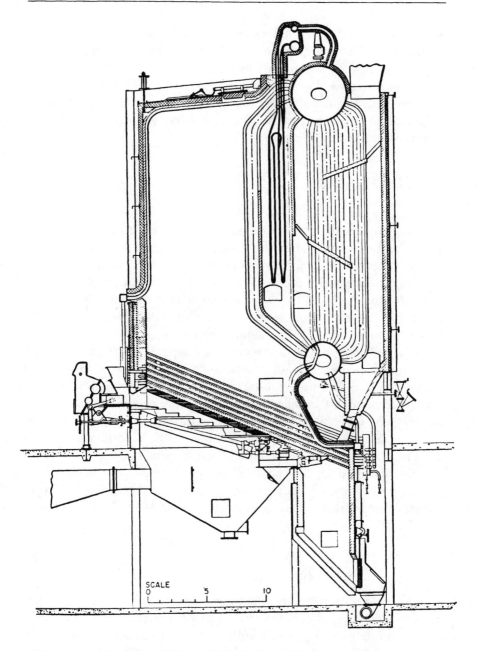

Figure 2-2. Coal-fired Water Tube Boiler with Predent Superheater.

for conditions when the prime movers are at low load and steam loads are high. A typical heat recovery unit is shown in Figure 2-3 for a gas turbine plant. Cogeneration methods are shown in Figures 2-4 and 2-5.

From a technical point all of the above systems constitute what can be called cogeneration of mechanical and heat energy. However the term cogeneration has come to mean the cooperative generation of electrical energy with a utility company.

HEAT EXCHANGERS

Heat exchangers are used in most steam systems. Steam-to-water heat exchangers (sometimes called converters or storage tanks with steam heating elements) are used to heat domestic hot water and to supply the terminal equipment in hot-water heating systems. These heat exchangers are usually the shell-and-tube type, where the steam is admitted to the shell and the water is heated as it circulates through the tubes. Condensate coolers (water-to-water) are sometimes used to subcool the condensate while reclaiming the heat energy.

Water-to-steam heat exchangers are a type of steam generator, are used in high-temperature water systems to provide process steam. Such heat exchangers generally consist of a tube bundle, through which the hot water circulates, installed in a tank or pressure vessel, where steam is generated. Often times industrial processes require that high temperature chemicals be cooled. To do this water can be used to absorb sufficient heat from the chemicals in the cooler to generate steam. The steam can then be used for low pressure service such as space heating, low temperature industrial processes, etc. All heat exchangers should be constructed and labeled according to the applicable ASME Boiler and Pressure Vessel Code.

BOILER CONNECTION

Many municipalities require double-wall construction in shell-and-tube heat exchangers between steam and potable water. Recommended boiler connections for pumped and gravity return systems are shown in Figures 2-6, 2-7, and 2-8. It is important to check local codes for specific legal requirements.

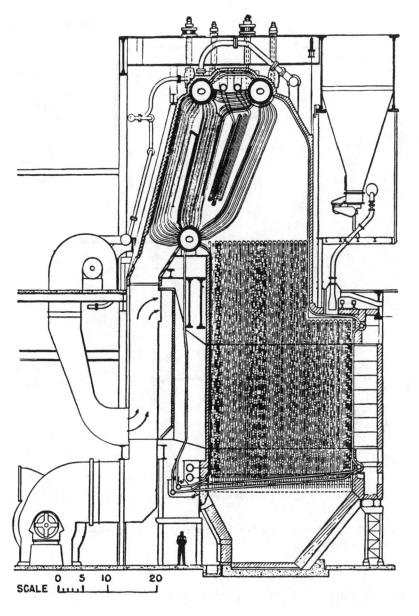

Figure 2-3. Water Tube Recuperator with Supplementary Coal Burner.

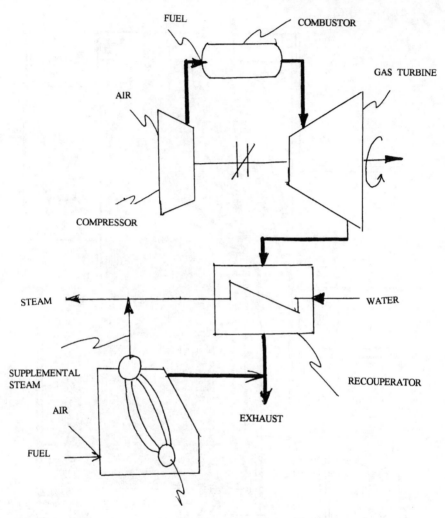

Figure 2-4. Cogenerating Gas Turbine with Recuperator and Auxiliary Boiler.

TRANSPORTING A SOURCE OF ENERGY

Steam is energy and like other commodities must be transported from its source, the boiler, to its point of use, a turbine, heater, steam engine, or industrial process. As with other forms of energy, steam requires energy to transport it. Gasoline is delivered to the station from the

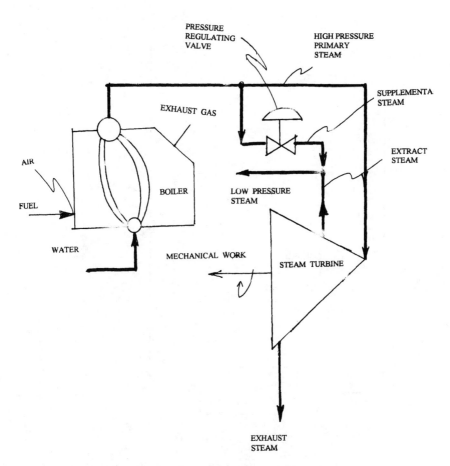

Figure 2-5. Single Extraction Steam Cogenerating Plant.

refinery by truck. The truck uses diesel fuel to propel it. This costs money. Electricity passes through wires from the power plant to the user. In traveling through the wires some of the electrical energy is dissipated. A voltage drop or loss of potential to do work is experienced in the electricity. As with the diesel fueled truck, this loss of potential to do work costs money.

Steam is transported through pipes. Like electricity and the truck energy is used to propel the respective commodities. The energy needed for the steam transportation comes from the steam's potential to do work and is experienced as a drop in pressure in the steam itself. This is synonymous with the electricity's voltage drop in its transportation.

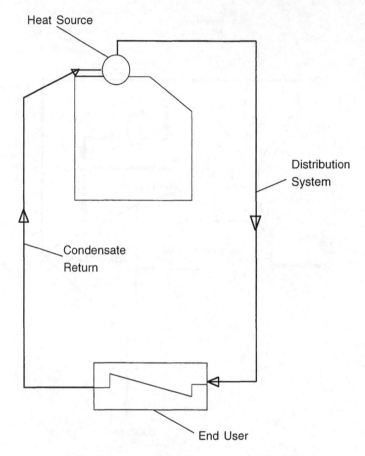

Figure 2-6. Closed Cycle Steam Heating Plant.

CENTRAL STEAM SYSTEMS

A central steam distribution system distributes thermal energy from a central source to commercial and/or industrial consumers for use in space heating, water heating, and/or process heating. The energy is transferred through steam lines. Thus, consumers obtain thermal energy from a distribution medium rather than generate it on-site at each facility.

A central steam system consists of three primary components: the central plant, the distribution network, and the user's system as shown in Figure 2-6. The central source of production may be any type of boiler,

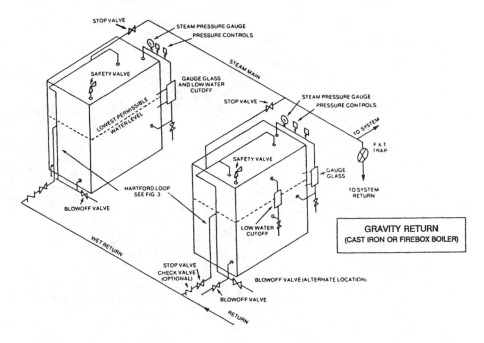

Figure 2-7. Multiple Cast Iron, Low Pressure Boiler Arrangement with Gravity Condensate Returns.

refuse incinerator, geothermal source, solar energy, or thermal energy developed as a by-product of electrical generation. The latter approach, called cogeneration, has a high energy utilization efficiency.

The second component is the piping network that conveys the energy. The piping for these systems usually consists of a combination of preinsulated and field-insulated pipe in concrete tunnel, trench, direct burial, or above ground applications. These networks are fairly simple in industry but may require substantial permitting and coordinating with nonusers of the system. Because, they are expensive, it is important to maximize their use.

The third component is the end use system, where the steam is dissipated to complete the process and do useful work or provide energy for a useful thermo-fluids process.

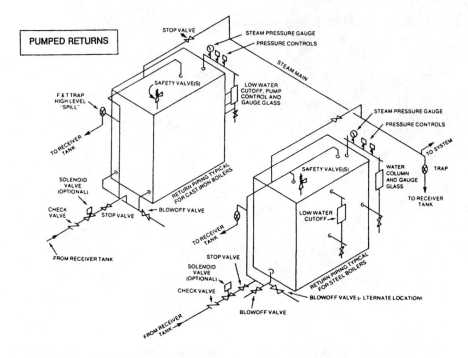

Figure 2-8. Multiple Cast Iron Boilers with Pumped Condensate Returns. Some low pressure steam boilers have check valves between boiler and step valve at each outlet to prevent backflow into the unfired boiler. Another safety feature is using two valves in series to prevent accidental cross flow between fired and idle boilers.

CENTRAL STEAM SYSTEMS APPLICATIONS

Central steam systems are best used where the thermal load density and the annual load factor is high. A high load density is important to cover the capital investment of a steam distribution network which can constitute as much as 50 percent of the investment for the entire central steam system. High load factor is important because of the large capital investment. These factors make central steam systems most attractive for large industrial and commercial facilities utilizing steam year round.

Groups of businesses aligned in close physical proximity can often times use a central steam plant for space heating and/or industrial processes. A high density cluster of facilities with high thermal loads is required to successfully utilize a central steam system whether the system is owned by the facility/s or by an independent steam producer.

In today's environmental scenario, central plants offer easier control of pollutants and toxic waste than individual units spread out through a complex. A central plant, for instance, that burns high sulfur coal can economically remove the noxious and sulfur emissions, where individual plants could not. Similarly, thermal energy from municipal solid waste can provide an economical source of energy to an environmentally sound community.

ECONOMIC BENEFITS

The following economic benefits and disadvantages should be considered before deciding on a central steam distribution system.

Operating Personnel

One of the primary advantages to a building owner is that operating personnel for the heating system can be reduced or eliminated. Most municipal codes require operating personnel to be on site when high-pressure boilers are in operation. Some older systems require trained operating personnel to be in the boiler/mechanical room at all times.

When thermal energy is brought into the building as a utility, the need for skilled engineers and technicians is reduced. Depending on the sophistication of the building heating control system, there may be an opportunity to reduce or minimize operating personnel considerably.

Insurance

Both property and liability insurance costs are significantly reduced with the elimination of a boiler in the mechanical room since risk of a fire or accident is reduced. Usable space: Usable space in the building increases when a boiler or related equipment are no longer necessary. The noise associated with boilers in buildings is also eliminated.

Equipment Maintenance

With less mechanical equipment, there is proportionally less equipment maintenance, resulting in less expense and a reduced maintenance staff. Higher thermal efficiency: A larger central plant can achieve higher thermal and emission efficiencies than can several smaller units. When environmental regulations mandate, additional pollution control equipment is usually more economical for larger plants. Partial load performance of central plants is more efficient than that of many isolated small systems, because the larger plants can operate one or more steam producing modules as the combined load requires and output can be regulated. Central plants generally have efficient base load units and less costly peaking equipment for use in extreme loads or emergencies.

Wider Range of Available Fuels

Smaller heating plants are usually designed for one fuel type. Generally, the fuel is limited to gas or oil. Central steam plants can operate on less expensive coal or refuse. Large facilities can often be designed for more than one fuel, i.e. coal and oil, or oil and gas, gas and coal.

Energy Source Economics

When an existing facility is the steam source, the available temperature and pressure is predetermined. If exhaust or extraction steam from an existing electrical generating turbine is used to provide thermal energy, the condition and quantity of the steam is determined by the turbine load. Steam from the main boiler/s can be used directly to supplement extraction steam when electrical loads are low.

In the above instance, a steam boiler/turbine plant is used. If a new central plant is being considered, a decision for cogeneration or thermal energy must be made, i.e. an electrical generator's prime mover must be selected for cogeneration. Another example of a cogeneration system is a diesel or gas turbine-driven generator with heat recovery equipment. The engine drives a generator to produce electricity, and heat is recovered from the exhaust, engine cooling, and lubrication cooling systems. Another commonly used configuration is a gas turbine used to drive an electric generator with the gas turbine exhaust releasing heat in a recuperative steam generator.

Steam Plant System Outputs

The selection of temperature and pressure is crucial because it can dramatically affect the economic feasibility of a central steam design. If the temperature and/or pressure level chosen is too low, a potential customer base can be eliminated from the system. One the other hand, if there is no demand for high-temperature industrial processes, then a low-temperature system usually provides the lowest delivered energy cost. Low pressure steam can be used for space heating or cooling, that is using steam directly for heating and steam absorption chillers for cooling.

The availability and location of fuel sources must also be considered in optimizing the economic design of a central steam system. For example, a coal burning gas boiler might not be feasible where abundant sources of coal are not available.

Initial Capital Investment

Initial capital investment is contingent upon: the availability of funds, the expected return on investment, and the amortization of debt to optimize the investment. Careful conceptual thought, planning and design must be exercised.

Concept

The concept should include existing facility/ies and equipment, the market for steam, the physical proximity of the steam consumers or Load Density, steam demand, also called Load Factor, capital investment and operating cost to produce steam, availability of primary energy sources or energy transportation costs and a realistic evaluation of the cost to charge the consumer for steam. Other considerations are the market for electricity as well as steam either for the steam using customer or for sale to an electric utility company.

Planning

Although procurement and construction costs usually account for most of the initial capital investment, neglect in any of the other areas could make the difference between success or failure of the project. In planning, three areas are generally reviewed. First, the technical feasibility of a central steam system must be considered. Experienced power plant and central heating system engineering know-how is required.

Secondly, financial feasibility such as interest rates and the bond market must be considered. Lastly, political feasibility must be considered, particularly if a municipality or government agency is involved in a central steam system installation. Historically, successful central steam systems have had to have the political backing and support of the community.

Design

The distribution system accounts for a significant portion of the initial investment. Its design depends on steam temperature and pressure and careful routing of the steam distribution network. Careful techno/economic analysis must be given to each step in the design phase, charge orders should be kept to a minimum once construction begins.

Procurement and Construction

As with any engineering project should be cost effective. To guarantee this, contract changes either for new or used equipment should be kept to a minimum.

Critical path construction procedures should be followed. This should ensure timely delivery of manufactured and pre-constructed components to facilitate on time construction completion. Field changes usually increase the final cost and delay start-up. Even a small delay in start-up can adversely affect the economics of the project.

In the critical path scheduling reasonable lead time must be allowed for equipment delivery if construction completion targets are to be met. One of the most unpredictable construction problems incurred with new steam distribution system installation is interference with existing utilities. This is particularly true with existing underground utilities which are sometimes so old that as-built plans and specifications are not available or if available, outdated. A distribution system running in a new industrial park is simpler and requires less time to install than a system being installed in an established industrial facility.

Consumer Interconnect

Interconnect costs are usually borne by the consumer. High interconnect costs may favor an in-building plant instead of a central steam system. This is usually not the case where the steam consumers have a high density. Reaching all the industrial and/or commercial facilities consumers with the central steam system piping network should be the intent of the planners and designers when economically feasible. The objective should be to have as many consumers as possible connected to the central steam network and this is done primarily through economies.

COMMON STEAM SYSTEM CONSIDERATIONS

Pressure and Pressure Drop

Consider Figure 2-10. Steam traveling through a pipe in the direc-

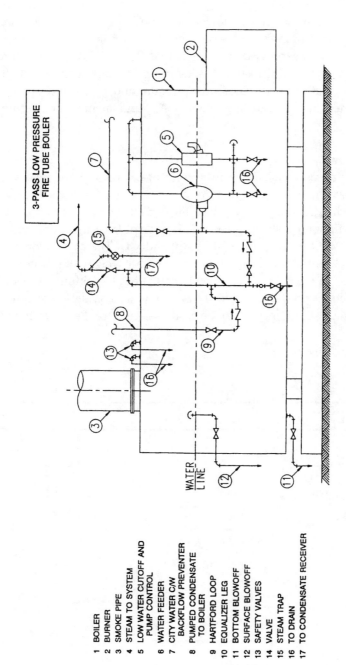

3-PASS LOW PRESSURE
FIRE TUBE BOILER

WATER
LINE

1 BOILER
2 BURNER
3 SMOKE PIPE
4 STEAM TO SYSTEM
5 LOW WATER CUTOFF AND
   PUMP CONTROL
6 WATER FEEDER
7 CITY WATER C/W
   BACKFLOW PREVENTER
8 PUMPED CONDENSATE
   TO BOILER
9 HARTFORD LOOP
10 EQUALIZER LEG
11 BOTTOM BLOWOFF
12 SURFACE BLOWOFF
13 SAFETY VALVES
14 VALVE
15 STEAM TRAP
16 TO DRAIN
17 TO CONDENSATE RECEIVER

Figure 2-9. Schematic of Boiler Connections and Fittings

tion of the arrow will have a lower pressure at point B then at point A. The difference in pressure between point A and B is known as the pressure drop and is expressed symbolically as P. The P is representative of the energy loss in the steam. It is used up in the transportation from point A to point B. A discussion of pressure and pressure drop and its measurement is included in Appendix A, Fundamentals of Thermodynamics.

Steam Flow

With an understanding of the concept of pressure, the relationship between steam flow and pressure, velocity and pipe size can now be considered. Attention is invited to Figure 2-11, which shows pressure drop -VS-flow for a steam pipe. This is excerpt from the more complete PAGE-KONZO charts presented in Appendix D. The chart, Figure 2-11, shows Pressure Drop (P) as the ordinate and weight flow rate as the abscissa. Note the ordinate and abscissa are both logarithmic scales. This permits plotting the steam velocity in ft/min as a straight line. And the nominal pipe size in inches, can be easily selected for the steam flow and

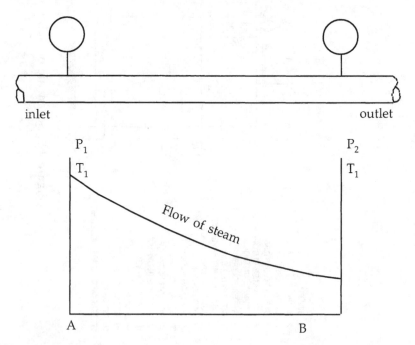

Figure 2-10. Illustration of Pressure Drop in Steam Piping.

velocity within an acceptable pressure drop per one hundred foot of pipe.

To explain the use of the chart, assume 6,000 lbs/hr of steam at 100 psig, dry saturated, must be transported from the source to a point 1,000 linear feet away as nearly dry at the point of delivery as possible. In practice, some pressure drop must be expected. A 2.5 psi, ΔP would produce practically no loss of the 100 percent steam quality. This is determined from the Mollier Chart (Appendix C). And it is acceptable for measuring quality for steam used in industrial or commercial processes requiring dry steam. So 2.5 psi, ΔP for 1,000 ft. Of pipe is acceptable and this equates to 0.25 psi/100 ft of pipe. Using Figure 2-11, it can be determined that a 5-inch pipe is required. It will produce a velocity of 3,000 ft/min in the pipe. This velocity is most likely acceptable in most piping installations.

Vibration

In many piping designs, vibration must be considered. The author recognizes this but the study of vibration is beyond the scope of this text. Attention is invited to the many handbooks and text book dedicated to the study of vibration especially in piping networks. The author does not warrant that the velocities in the charts are acceptable for all steam transmission purposes. This chapter seeks to illustrate the relationship between pressure drop, pipe size, and flow rate for use in operation and maintenance problems only.

Using 5-inch pipe to transport 6,000 lbs/hr of steam 1,000 ft, would be very expensive. By selecting a 3.5-inch pipe, the cost can be reduced considerably. Not only is the pipe cheaper, but valving, fittings, and supports will also be less costly.

However, by reducing the pipe diameter from 5-inch to 3.5-inch, a reduction of 30 percent, the pressure drop has quadrupled from 2.5 psi to 10 psi. However, from the Mollier Chart (Appendix C), the moisture content has increased by only 1.0 percent.

For the pressure ranges shown in the PAGE KONZO charts a good average is a 1.0 percent change in moisture content for each 10 psi pressure drop for steam traveling through a pipe.

Unfortunately, it is not an acceptable pressure drop possible to estimate an average for steam in piping since the acceptable pressure drop per 100 ft in a 300-ft length is different from say a 3,000-ft length at a given operating pressure. For most utility or industrial processes the

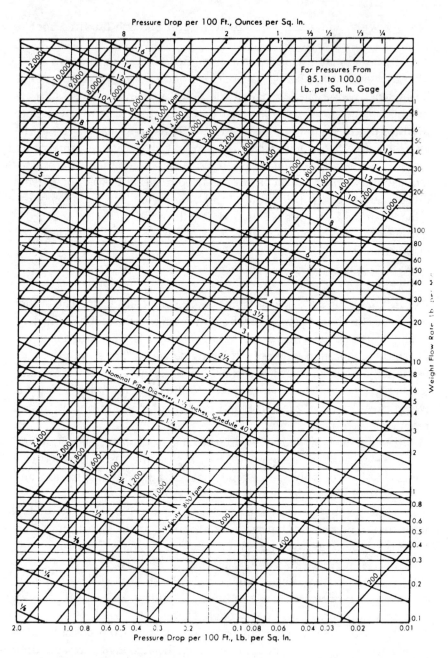

Figure 2-11. Excerpt from Page-Konzo Steam Flow Charts.

pressure drop discussed above is what can be expected from steam's internal friction, the effect of turbulent flow, and the friction of the steam with the pipe wall.

It is interesting to note that here steam temperature is not a function of pressure but is dependent upon the amount of insulation used. For this reason insulation is very important and is treated separately in Chapter 6. From the temperature-entropy chart, Figure 2-12, (Also included in Appendix C) it can be seen that if saturated steam at 100 psig cools at constant temperature (330°F) from an enthalpy of 1188 Btu/lb to an enthalpy of 1150 Btu/lb its quality will change from 100 percent to 96 percent.

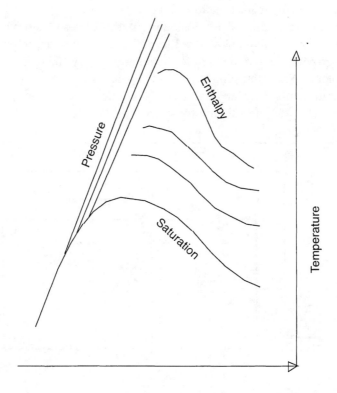

Figure 2-12. Temperature-entropy Chart for Steam.

STEAM TRAPS

Steam traps are used to remove condensate from steam lines, thereby maintaining some degree of dryness as the steam flows from the boiler to its point of use. They are devices which permit the passage of water or condensate but will not, when operating properly, pass steam. A typical steam trap installation is shown in Figure 2-13. When used in steam lines it is particularly important that they be properly selected and sized for start up loads. The current trend is to use a conventional trap and oversize it to handle not only the on-line condensate loads but also start-up loads. However, for some years now, the U.S. Navy and utility companies have been using an unconventional orifice steam traps with no moving parts in their ships and steam plants, respectively. The success they have experienced lends some encouragement to the industrial and commercial steam user to follow suit.

Steam traps are usually installed in steam lines at intervals of 100 to 300 Ft. On installations where the piping is run in a trench, tunnel or above ground, the requirement for installation every 100 ft. is usually not a problem. When the piping is direct buried steam traps can only be located in manholes. Many design considerations will influence the locations of manholes and therefore, steam trap locations.

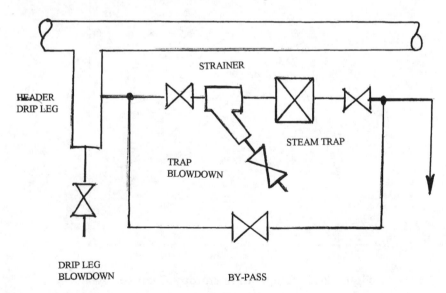

Figure 2-13. Steam Trap Installed on Steam Main Drip Leg.

Effective steam trapping techniques and equipment can maintain the moisture level in steam to one percent or lower. Steam traps for both steam piping and equipment are discussed at length in a later chapter.

STEAM PIPING LAYOUT CONSIDERATIONS

Ideally steam lines should be arranged to maintain as small a pressure drop as possible, contingent upon the size of pipe and pressure that can economically be used in the system. The larger the steam pipe size, the less the pressure drop will be per linear foot but construction costs go up dramatically from say a 4-inch to an 8-inch diameter pipe. Also, as the operating pressure increases so must the pipe wall thickness increase. Often times pipe material and structural configurations will be effected.

Chapter 4 addresses piping layout and aesthetics in detail. Chapter 3 discusses steam piping standards and design criteria.

BOILER PIPING

Boiler Outlet Piping

Small boilers usually have one steam outlet connection sized to reduce steam velocity to minimize carry-over of water into supply lines. Large boilers can have several outlets that minimize boiler water entrainment.

Figure 2-14 shows piping connections to the steam header. Although some engineers prefer to use an enlarged steam header for additional storage space. If there is not a sudden demand for steam, then an oversized header is a disadvantage. The boiler header can be the same size as the boiler connection or the pipe used on the steam main. This facilitates faster response to load changes The horizontal runouts from the boiler to the header should be sized by calculating the heaviest load that will be placed on them. The runouts should be sized on the same basis as the building mains. Any change in size after the vertical uptakes should be made by reducing elbows.

Boiler Return Piping

Cast-iron boilers have return lappings on both sides, while steel boilers may have one or two return lappings. Where two lappings are

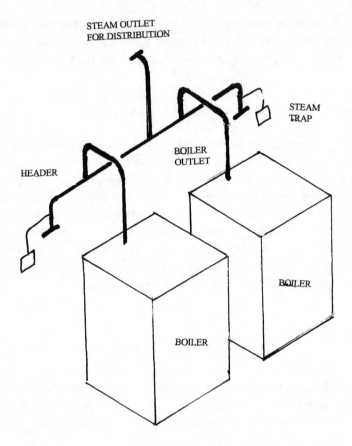

STEAM OUTLET
FOR DISTRIBUTION

STEAM
TRAP

BOILER
OUTLET

HEADER

BOILER

BOILER

Figure 2-14. Arrangement of Boiler Piping to a Header.

provided, both should be used to effect proper circulation through the boiler. Condensate to boilers can be returned by a pump or a gravity return system. Return connections, shown for a multiple-boiler gravity return installation, may not always maintain correct water levels in all boilers. Extra controls or accessories may be required.

Recommended return piping connections, for systems using gravity returns, are detailed in Figure 2-7. Dimension A must be a minimum of 28 in. for each 1 psig maintained at the boiler to provide the pressure required to return the condensate to the boiler. To provide a reasonable safety factor, make Dimension A a minimum of 14 in. for small systems with piping sized for pressure drop of 1/8 psi, and a minimum of 28 in. for larger systems with piping sized for a pressure drop of 0.5 psi. The

Hartford Loop protects against a low water condition, which can occur if a leak develops. Certain local building codes require extra safety elements that are specific to their location. Because of hydraulic pressure limitations, gravity return systems are only suitable for systems operating at a boiler pressure between 0.5 to 1 psig. However, since these systems have minimum mechanical equipment and low initial installed cost, they are widely used. There are many good assembly and operating manuals offered by product manufacturers that provide additional information on piping for gravity systems. They provide additional design information appropriate for many small systems. Typical, manufacturer recommended piping connections for steam boilers with pump-returned condensate are shown in Figure 2-8.

Common practice provides an individual condensate or boiler feedwater pump for each boiler. Pump operation is controlled by the boiler water level control on each boiler. However, one pump may be connected to supply the water to each boiler from a single manifold by using feedwater control valves regulated by the individual boiler water level controllers. When such systems are used, the condensate return pump runs continuously to pressurize the return header.

Return piping should be sized to accommodate total load. The line between the pump and boiler should be sized for a very small pressure drop at the maximum pump discharge flow rate.

Chapter 3

Steam and Condensate System Piping Structure

PIPING STANDARDS

*I*ndustrial piping has been standardized primarily by two societies. The American Society for Testing Materials (ASTM) and the American Standards Association (ASA) are the two primary classifiers of industrial steam and condensate piping. The ASTM has the aim of "Promoting Knowledge of Materials of Engineering and Standardization of Specifications and Methods of Testing." It is concerned primarily with materials used in piping. The ASA deals with all piping systems. It standardizes dimensions, sets permissible stress values as functions of temperature, has established working formulas for determination of wall thickness for pressures and material at a given temperature. The ASA has established a number of codes, one of which is the "Code for Pressure Pipe."

Piping is constructed by several methods some of these are electric-fusion welding, electric resistance welding, submerged arc electric welding, extruded piping, and others. A chart of allowable stresses for piping in power systems is given in Table 3-1 and Table 3-2. A table of physical properties of piping is included in Table 3-3.

Some materials used in steam piping are common steel, carbon-molybdenum steel, chromium-molybdenum steel, and stainless steel.

Table 3-1. Allowable Stresses for Pipe in Power Piping Systems.

Material*	ASTM specification	Grade	Minimum ultimate tensile strength	Values of S, psi, for temperatures in degrees F not to exceed †							
				-20 to 100	200	300	400§	450	500	600	650
Welded material:											
Furnace welded carbon steel											
Lap welded	A120			8,800	8,600	8,200	7,600	7,600			
Butt welded	A120			6,500	6,350	6,100	5,850	5,700			
Automatically welded austenitic stainless steel											
18% Cr-8% Ni-Ti	A312	TP321	75,000	15,950	15,950	14,450	13,450		12,900	12,650	12,600
18% Cr-8% Ni-Cb	A312	TP347									
Seamless material:											
Carbon steel	A120										
5% Cr-½% Mo	A335	P5	60,000	10,800	10,600	10,200	9,800	9,600			
	A335	P5b									
	A369	FP5									
18% Cr-8% Ni-Ti	A312			15,000	15,000	15,000	15,000		14,500	14,000	13,700
	A376										
18% Cr-8% Ni-Cb	A312	TP321	75,000	18,750	18,750	17,000	15,800		15,200	14,900	14,850
	A376	TP347									
Seamless:											
Red brass	B43			8,000	8,000	7,000	3,000				
Copper 2 in. and smaller	B42			6,000	5,500	4,750	3,000				
Copper, over 2 in.	B42			6,000	5,500	4,750	3,000				
Copper tubing	B75			6,000	5,500	4,750	3,000				
Annealed	B88		30,000	6,000	5,500	4,750	3,000				
Bright annealed	B68		30,000	6,000	5,500	4,750	3,000				
Copper brazed steel	A254	Class 1	42,000	6,000	5,500	4,750	3,000				
		Class 2	42,000	3,600	3,300	2,850	1,800				
Cast iron‡											
Centrifugally cast	F-SB WW-P-421	Types 1 & 2		6,000	6,000	6,000	6,000	6,000			
Metal molds	ASA A21.6			6,000	6,000	6,000	6,000	6,000			
Sand-lined molds	ASA B21.8			6,000	6,000	6,000	6,000	6,000			
Pit cast	ASA A21.2			4,000	4,000	4,000	4,000	4,000			

* Pipe in accordance with API specification 5L may be used.

† The several types and grades of pipe tabulated above shall not be used at temperatures in excess of the maximum temperatures for which the S values are indicated. (See also specific requirements for service conditions contemplated.) Allowable S values for intermediate temperatures may be obtained by interpolation.

‡ Cast-iron pipe shall not be used for lubricating oil lines for machinery and in any case not for oil having a temperature above 300 F.

§ For steam at 250 psi (406 F) the values given may be used.

Table 3-2. Allowable Stresses for Pipe in Power Piping Systems.

(ASA Code for Pressure Piping, 1963)

[Where welded construction is used, consideration should be given to the possibility of graphite formation in the following steels: carbon steel above 775 F; carbon-molybdenum steel above 800 F; chrome-molybdenum steel (with chromium under 0.60) above 975 F.]

Material[a]	ASTM specification	Grade	Identification symbol	Minimum ultimate tensile strength	Values of S psi for temperatures in deg F not to exceed[b]											
					-20 to 650	700	750	800	850	900	950	1000	1050	1100	1150	1200
Welded material																
Furnace welded:																
Lap welded:																
Carbon steel	A53	....	....	45,000	9,000											
Wrought iron	A72	....	....	40,000	8,000											
Butt welded																
Carbon steel	A53	....	....	45,000	6,750											
Wrought iron	A72	....	....	40,000	6,000											
Electric fusion welded:																
Carbon steel	A134	A245 A	....	48,000	8,000											
		A245 B	....	52,000	9,600											
		A245 C	....	55,000	10,100											
		A283 A	....	45,000	8,300											
		A283 B	....	50,000	9,200											
		A283 C	....	55,000	10,100											
		A283 D	....	60,000	10,100											
	A139	A[d]	....	48,000	9,600	9,250	8,300									
		B[d]	....	60,000	12,000	11,350	9,950									
	A155[c]	A285 A	C45	45,000	10,100	9,800	8,700	7,500	5,950							
		A285 B	C50	50,000	11,250	10,900	9,900	8,450	6,550							
		A285 C	C55	55,000	12,400	11,900	10,850	9,200	7,000							
Killed carbon steel		A201 A	KC55	55,000	12,400	11,900	10,850	9,200	7,000							
		A201 B	KC60	60,000	13,500	12,900	11,650	9,700	7,000							
		A212 A	KC65	65,000	14,600	13,950	12,450	10,250	7,000							
		A212 B	KC70	70,000	15,750	14,950	13,250	10,800	7,000							
Carbon-molybdenum steel		A204 A	CM65	65,000	14,600	14,600	14,600	14,100	12,950	11,250					
		A204 B	CM70	70,000	15,750	15,750	15,750	15,200	13,500	11,450					
		A204 C	CM75	75,000	16,850	16,850	16,850	16,200	14,300	11,700					
½% Cr-½% Mo steel		A301 A	½CR	65,000	14,600	14,600	14,600	14,100	12,950	11,250	9,000	5,600			
1% Cr-½% Mo steel		A301 B	1 CR	60,000	13,500	13,500	13,500	13,250	12,750	11,800	9,900	6,750	4,500	2,500	
1¼% Cr-½% Mo steel		A335 P11	1¼CR	60,000	13,500	13,500	13,500	13,500	12,950	11,800	9,900	7,000	4,950	3,600	
2¼% Cr-1% Mo steel		A335 P22	2¼CR	60,000	13,500	13,500	13,500	13,500	12,950	11,800	9,900	7,000	5,200	3,750		2,700

Table 3-2 (Cont'd).

(ASA Code for Pressure Piping, 1963)

[Where welded construction is used, consideration should be given to the possibility of graphite formation in the following steels: carbon steel above 775 F; carbon-molybdenum steel above 800 F; chrome-molybdenum steel (with chromium under 0.60) above 975 F.]

Material	ASTM specification	Grade	Identification symbol	Minimum ultimate tensile strength	Values of S psi for temperatures in deg F not to exceed											
					-20 to 650	700	750	800	850	900	950	1000	1050	1100	1150	1200
Electric resistance welded: Carbon steel	A53	A*	.....	48,000	10,200	9,900	9,100									
	A53	B*	.....	60,000	12,750	12,200	11,000									
	A135	A*	.....	48,000	10,200	9,900	9,100									
	A135	B*	.....	60,000	12,750	12,200	11,000									
Automatically welded stainless steel: 18% Cr-8% Ni-Ti 18% Cr-8% Ni-Cb	A312	TP321} TP347}	.....	75,000	Note f	12,550	12,500	12,350	12,150	12,000	11,750	11,500	11,150	8,750	6,450	4,250
Seamless material Carbon steel	A53	A	.....	48,000	12,000	11,650	10,700	9,000	7,100	5,000						
	A53	B	.....	60,000	15,000	14,350	12,950	10,800	7,800	5,000						
	A106	A	.....	48,000	12,000	11,650	10,700	9,000	7,100	5,000						
	A106	B	.....	60,000	15,000	14,350	12,950	10,000	7,800	5,000						
	A83 A179 A192	Type A Low carbon	.....	(47,000)	10,750	11,450	10,550	9,000	7,100	5,000						
	A210	....	.....	(47,000) 60,000	15,000	14,350	12,950	10,800	7,800	5,000						
Carbon molybdenum	A335 A369	P 1} FP 1}	.....	55,000	13,750	13,750	13,750	13,450	13,150	12,500	10,000					
Cr Mo ½% Cr-½% Mo	A335 A369	P 2} FP 2}	.....	55,000	13,750	13,750	13,750	13,450	13,150	12,500	10,000	6,250	5,000			
1% Cr ½% Mo	A335 A369	P12} FP12}	.....	60,000	15,000	15,000	15,000	14,750	14,200	13,100	11,000	7,500	5,000	2,800		
1¼% Cr-½% Mo	A335 A369	P11} FP11}	.....	60,000	15,000	15,000	15,000	15,000	14,400	13,100	11,000	7,800	5,500	4,000		

Table 3-2 (Cont'd).

Material	Spec	Type		Tensile												
2¼% Cr-1% Mo	A213 A335 A369	T22 P22 F*P22	.....	60,000	15,000	15,000	15,000	14,400	13,100	11,000	7,800	5,800	4,200	3,000		
3% Cr-1% Mo	A213 A335 A369	T21 P21 F*P21	.....	60,000	15,000	14,500	13,900	13,200	12,000	9,000	7,000	5,500	4,000	2,700		
5% Cr-½% Mo	A335 A369 A335	P 5 F*P 5 P 5b	.....	60,000	Note/	13,400	13,100	12,800	12,400	11,500	10,000	7,300	5,200	3,300	2,200	1,500
				60,000	Note/	13,400	13,100	12,800	12,400	10,900	9,000	5,500	3,500	2,500	1,800	1,200
Stainless steel: 18% Cr-8% Ni-Ti	A213 A312 A376	TP321	.....	75,000	Note/	14,800	14,700	14,550	14,300	14,100	13,850	13,500	13,100	10,300	7,600	5,000
18% Cr-8% Ni-Cb	A213 A312 A376	TP347	.....	75,000	Note/	14,800	14,700	14,550	14,300	14,100	13,850	13,500	13,100	10,300	7,600	5,000

a Pipe in accordance with API specification 5L may be used as specified.

b The several types and grades of pipe tabulated above shall not be used at temperatures in excess of the maximum temperatures for which the S values are indicated. (See also specific requirements for service conditions contemplated.) Allowable S values for intermediate temperatures may be obtained by interpolation.

c The values tabulated are for Class 2 pipe. For Class 1 pipe which is heat treated and radiographed, these stresses may be increased by the ratio of 0.95 divided by 0.90.

d If plate material having physical properties other than stated in Sec. 6 of the ASTM specification A139 is used in the manufacture of ordinary electric-fusion-welded steel pipe, the allowable stress shall be taken as 0.20 times the tensile strength for temperatures of 450 F and below.

e For electric-resistance-welded pipe for applications where the temperature is below 650 F, and where pipe furnished under this classification is subjected to supplemental tests and/or heat treatments as agreed to by the supplier and the purchaser, and whereby such supplemental tests and/or heat treatments demonstrate the strength characteristics of the weld to be equal to the minimum tensile strength specified for the pipe, the S values equal to the corresponding seamless grades may be used.

f See Table 1 for values from -20 to 650 F.

Table 3-3. Physical Properties of Pipe.

(Grinnell Co., Inc.)

Nominal pipe size, O.D., in.	Schedule number a	b	c	Wall thickness, in.	I.D., in.	Inside area, sq in.	Metal area, sq in.	Sq ft outside surface, per ft	Sq ft inside surface, per ft	Weight per ft, lb	Weight of water per ft, lb	Moment of inertia, in.⁴	Section modulus, in.³	Radius gyration, in.
⅛ 0.405	..	..	10S	0.049	0.307	0.0740	0.0548	0.106	0.0804	0.186	0.0321	0.00088	0.00437	0.1271
	40	Std	40S	0.068	0.269	0.0568	0.0720	0.106	0.0705	0.245	0.0246	0.00106	0.00525	0.1215
	80	XS	80S	0.095	0.215	0.0364	0.0925	0.106	0.0563	0.315	0.0157	0.00122	0.00600	0.1146
¼ 0.540	..	..	10S	0.065	0.410	0.1320	0.0970	0.141	0.1073	0.330	0.0572	0.00279	0.01032	0.1694
	40	Std	40S	0.088	0.364	0.1041	0.1250	0.141	0.0955	0.425	0.0451	0.00331	0.01230	0.1628
	80	XS	80S	0.119	0.302	0.0716	0.1574	0.141	0.0794	0.535	0.0310	0.00378	0.01395	0.1547
⅜ 0.675	..	..	10S	0.065	0.545	0.2333	0.1246	0.177	0.1427	0.423	0.1011	0.00586	0.01737	0.2169
	40	Std	40S	0.091	0.493	0.1910	0.1670	0.177	0.1295	0.568	0.0827	0.00730	0.02160	0.2090
	80	XS	80S	0.126	0.423	0.1405	0.2173	0.177	0.1106	0.739	0.0609	0.00862	0.02554	0.1991
½ 0.840	..	..	10S	0.083	0.674	0.357	0.1974	0.220	0.1765	0.671	0.1547	0.01431	0.0341	0.2692
	40	Std	40S	0.109	0.622	0.304	0.2503	0.220	0.1628	0.851	0.1316	0.01710	0.0407	0.2613
	80	XS	80S	0.147	0.546	0.2340	0.320	0.220	0.1433	1.088	0.1013	0.02010	0.0478	0.2505
	160	..	..	0.187	0.466	0.1706	0.383	0.220	0.1220	1.304	0.0740	0.02213	0.0527	0.2402
	..	XXS	..	0.294	0.252	0.0499	0.504	0.220	0.0660	1.714	0.0216	0.02425	0.0577	0.2192
¾ 1.050	..	..	5S	0.065	0.920	0.665	0.2011	0.275	0.2409	0.684	0.2882	0.02451	0.0467	0.349
	..	..	10S	0.083	0.884	0.614	0.2521	0.275	0.2314	0.857	0.2661	0.02970	0.0566	0.343
	40	Std	40S	0.113	0.824	0.533	0.333	0.275	0.2157	1.131	0.2301	0.0370	0.0706	0.334
	80	XS	80S	0.154	0.742	0.432	0.435	0.275	0.1943	1.474	0.1875	0.0448	0.0853	0.321
	160	..	..	0.218	0.614	0.2961	0.570	0.275	0.1607	1.937	0.1284	0.0527	0.1004	0.304
	..	XXS	..	0.308	0.434	0.1479	0.718	0.275	0.1137	2.441	0.0641	0.0579	0.1104	0.284
1 1.315	..	..	5S	0.065	1.185	1.103	0.2553	0.344	0.310	0.868	0.478	0.0500	0.0760	0.443
	..	..	10S	0.109	1.097	0.945	0.413	0.344	0.2872	1.404	0.409	0.0757	0.1151	0.428
	40	Std	40S	0.133	1.049	0.864	0.494	0.344	0.2746	1.679	0.374	0.0874	0.1329	0.421
	80	XS	80S	0.179	0.957	0.719	0.639	0.344	0.2520	2.172	0.311	0.1056	0.1606	0.407
	160	..	..	0.250	0.815	0.522	0.836	0.344	0.2134	2.844	0.2261	0.1252	0.1903	0.387
	..	XXS	..	0.358	0.599	0.2818	1.076	0.344	0.1570	3.659	0.1221	0.1405	0.2137	0.361
1¼ 1.660	..	..	5S	0.065	1.530	1.839	0.326	0.434	0.401	1.107	0.797	0.1038	0.1250	0.564
	..	..	10S	0.109	1.442	1.633	0.531	0.434	0.378	1.805	0.707	0.1605	0.1934	0.550
	40	Std	40S	0.140	1.380	1.496	0.669	0.434	0.361	2.273	0.648	0.1948	0.2346	0.540
	80	XS	80S	0.191	1.278	1.283	0.881	0.434	0.335	2.997	0.555	0.2418	0.2913	0.524

Table 3-3 (Cont'd).

Nom. Size / O.D.	Sched. No.	Desig.	SS												
1¼ 1.066	160		..	0.250	1.160	1.057	1.107	0.434	0.304	3.765	0.458	0.2839	0.342	0.506
		XXS	..	0.382	0.896	0.631	1.534	0.434	0.2346	5.214	0.2732	0.341	0.411	0.472
1½ 1.900			5S	0.065	1.770	2.461	0.375	0.497	0.463	1.274	1.067	0.1580	0.1663	0.649
			10S	0.109	1.682	2.222	0.613	0.497	0.440	2.085	0.962	0.2469	0.2599	0.634
	40	Std	40S	0.145	1.610	2.036	0.799	0.497	0.421	2.718	0.882	0.310	0.326	0.623
	80	XS	80S	0.200	1.500	1.767	1.068	0.497	0.393	3.631	0.765	0.391	0.412	0.605
	160		..	0.281	1.338	1.406	1.429	0.497	0.350	4.859	0.608	0.483	0.508	0.581
		XXS	..	0.400	1.100	0.950	1.885	0.497	0.288	6.408	0.412	0.568	0.598	0.549
2 2.375			5S	0.065	2.245	3.96	0.472	0.622	0.588	1.604	1.716	0.315	0.2652	0.817
			10S	0.109	2.157	3.65	0.776	0.622	0.565	2.638	1.582	0.499	0.420	0.802
	40	Std	40S	0.154	2.067	3.36	1.075	0.622	0.541	3.653	1.455	0.666	0.561	0.787
	80	XS	80S	0.218	1.939	2.953	1.477	0.622	0.508	5.022	1.280	0.868	0.731	0.766
	160		..	0.343	1.689	2.240	2.190	0.622	0.442	7.444	0.971	1.163	0.979	0.729
		XXS	..	0.436	1.503	1.774	2.656	0.622	0.393	9.029	0.769	1.312	1.104	0.703
2½ 2.875			5S	0.083	2.709	5.76	0.728	0.753	0.709	2.475	2.499	0.710	0.494	0.988
			10S	0.120	2.635	5.45	1.039	0.753	0.690	3.531	2.361	0.988	0.687	0.975
	40	Std	40S	0.203	2.469	4.79	1.704	0.753	0.646	5.793	2.076	1.530	1.064	0.947
	80	XS	80S	0.276	2.323	4.24	2.254	0.753	0.608	7.661	1.837	1.925	1.339	0.924
	160		..	0.375	2.125	3.55	2.945	0.753	0.556	10.01	1.535	2.353	1.637	0.894
		XXS	..	0.552	1.771	2.464	4.03	0.753	0.464	13.70	1.067	2.872	1.998	0.844
3 3.500			5S	0.083	3.334	8.73	0.891	0.916	0.873	3.03	3.78	1.301	0.744	1.208
			10S	0.120	3.260	8.35	1.274	0.916	0.853	4.33	3.61	1.822	1.041	1.196
	40	Std	40S	0.216	3.068	7.39	2.228	0.916	0.803	7.58	3.20	3.02	1.724	1.164
	80	XS	80S	0.300	2.900	6.61	3.02	0.916	0.759	10.25	2.864	3.90	2.226	1.136
	160		..	0.437	2.626	5.42	4.21	0.916	0.687	14.32	2.348	5.03	2.876	1.094
		XXS	..	0.600	2.300	4.15	5.47	0.916	0.602	18.58	1.801	5.99	3.43	1.047
3½ 4.000			5S	0.083	3.834	11.55	1.021	1.047	1.004	3.47	5.01	1.960	0.980	1.385
			10S	0.120	3.760	11.10	1.463	1.047	0.984	4.97	4.81	2.756	1.378	1.372
	40	Std	40S	0.226	3.548	9.89	2.680	1.047	0.929	9.11	4.28	4.79	2.394	1.337
	80	XS	80S	0.318	3.364	8.89	3.68	1.047	0.881	12.51	3.85	6.28	3.14	1.307
4 4.500			5S	0.083	4.334	14.75	1.152	1.178	1.135	3.92	6.40	2.811	1.249	1.562
			10S	0.120	4.260	14.25	1.651	1.178	1.115	5.61	6.17	3.96	1.762	1.549
	40	Std	40S	0.237	4.026	12.73	3.17	1.178	1.054	10.79	5.51	7.23	3.21	1.510
	80	XS	80S	0.337	3.826	11.50	4.41	1.178	1.002	14.98	4.98	9.61	4.27	1.477
	120		..	0.437	3.626	10.33	5.58	1.178	0.949	18.96	4.48	11.65	5.18	1.445
	160		..	0.531	3.438	9.28	6.62	1.178	0.900	22.51	4.02	13.27	5.90	1.416
		XXS	..	0.674	3.152	7.80	8.10	1.178	0.825	27.54	3.38	15.29	6.79	1.374

Table 3-3 (Cont'd).

Nominal pipe size, O.D., in.	Schedule number a	b	c	Wall thickness, in.	I.D., in.	Inside area, sq in.	Metal area, sq in.	Sq ft outside surface, per ft	Sq ft inside surface, per ft	Weight per ft, lb	Weight of water per ft, lb	Moment of inertia, in.^{4}	Section modulus, in.^{3}	Radius gyration, in.
5 5.563			5S	0.109	5.345	22.44	1.868	1.456	1.399	6.35	9.73	6.95	2.498	1.929
			10S	0.134	5.295	22.02	2.285	1.456	1.386	7.77	9.53	8.43	3.03	1.920
	40	Std	40S	0.258	5.047	20.01	4.30	1.456	1.321	14.62	8.66	15.17	5.45	1.878
	80	XS	80S	0.375	4.813	18.19	6.11	1.456	1.260	20.78	7.89	20.68	7.43	1.839
	120			0.500	4.563	16.35	7.95	1.456	1.195	27.04	7.09	25.74	9.25	1.799
	160			0.625	4.313	14.61	9.70	1.456	1.129	32.96	6.33	30.0	10.80	1.760
		XXS		0.750	4.063	12.97	11.34	1.456	1.064	38.55	5.62	33.6	12.10	1.722
6 6.625			5S	0.109	6.407	32.2	2.231	1.734	1.677	5.37	13.98	11.85	3.58	2.304
			10S	0.134	6.357	31.7	2.733	1.734	1.664	9.29	13.74	14.40	4.35	2.295
	40	Std	40S	0.280	6.065	28.89	5.58	1.734	1.588	18.97	12.51	28.14	8.50	2.245
	80	XS	80S	0.432	5.761	26.07	8.40	1.734	1.508	28.57	11.29	40.5	12.23	2.195
	120			0.562	5.501	23.77	10.70	1.734	1.440	36.39	10.30	49.6	14.98	2.153
	160			0.718	5.189	21.15	13.33	1.734	1.358	45.30	9.16	59.0	17.81	2.104
		XXS		0.864	4.897	18.83	15.64	1.734	1.282	53.16	8.17	66.3	20.03	2.060
8 8.625			5S	0.109	8.407	55.5	2.916	2.258	2.201	9.91	24.07	26.45	6.13	3.01
			10S	0.148	8.329	54.5	3.94	2.258	2.180	13.40	23.59	35.4	8.21	3.00
	20			0.250	8.125	51.8	6.58	2.258	2.127	22.36	22.48	57.7	13.39	2.962
	30			0.277	8.071	51.2	7.26	2.258	2.113	24.70	22.18	63.4	14.69	2.953
	40	Std	40S	0.322	7.981	50.0	8.40	2.258	2.089	28.55	21.69	72.5	16.81	2.938
	60			0.406	7.813	47.9	10.48	2.258	2.045	35.64	20.79	88.8	20.58	2.909
	80	XS	80S	0.500	7.625	45.7	12.76	2.258	1.996	43.39	19.80	105.7	24.52	2.878
	100			0.593	7.439	43.5	14.96	2.258	1.948	50.87	18.84	121.4	28.14	2.847
	120			0.718	7.189	40.6	17.84	2.258	1.882	60.63	17.60	140.6	32.6	2.807
	140			0.812	7.001	38.5	19.93	2.258	1.833	67.76	16.69	153.8	35.7	2.777
		XXS		0.875	6.875	37.1	21.30	2.258	1.800	72.42	16.09	162.0	37.6	2.757
	160			0.906	6.813	36.5	21.97	2.258	1.784	74.69	15.80	165.9	38.5	2.748
10 10.760			5S	0.134	10.482	86.3	4.52	2.815	2.744	15.15	37.4	63.7	11.85	3.75
			10S	0.165	10.420	85.3	5.49	2.815	2.728	18.70	36.9	76.9	14.30	3.74
	20			0.250	10.250	82.5	8.26	2.815	2.683	28.04	35.8	113.7	21.16	3.71
				0.279	10.192	81.6	9.18	2.815	2.668	31.20	35.3	125.9	23.42	3.70
	30			0.307	10.136	80.7	10.07	2.815	2.654	34.24	35.0	137.5	25.57	3.69
	40	Std	40S	0.365	10.020	78.9	11.91	2.815	2.623	40.48	34.1	160.8	29.90	3.67
	60	XS	80S	0.500	9.750	74.7	16.10	2.815	2.553	54.74	32.3	212.0	39.4	3.63
	80			0.593	9.564	71.8	18.92	2.815	2.504	64.33	31.1	244.9	45.6	3.60
	100			0.718	9.314	68.1	22.63	2.815	2.438	76.93	29.5	286.2	53.2	3.56

Table 3-3 (Cont'd).

Nom. / O.D.	Sched. No.	Desig.	Wall	I.D.									
10 — 10.750	120		0.843	9.064	64.5	26.24	2.815	2.373	89.20	28.0	324	60.3	3.52
	140		1.000	8.750	60.1	30.6	2.815	2.291	104.13	26.1	368	68.4	3.47
	160		1.125	8.500	56.7	34.0	2.815	2.225	115.65	24.6	399	74.3	3.43
12 — 12.750		5S	0.165	12.420	121.2	6.52	3.34	3.25	19.56	52.5	129.2	20.27	4.45
		10S	0.180	12.390	120.6	7.11	3.34	3.24	24.20	52.2	140.5	22.03	4.44
	20		0.250	12.250	117.9	9.84	3.34	3.21	33.38	51.1	191.9	30.1	4.42
	30		0.330	12.090	114.8	12.88	3.34	3.17	43.77	49.7	248.5	39.0	4.39
		40S, Std	0.375	12.000	113.1	14.58	3.34	3.14	49.56	49.0	279.3	43.8	4.38
	40		0.406	11.938	111.9	15.74	3.34	3.13	53.53	48.5	300	47.1	4.37
		80S, XS	0.500	11.750	108.4	19.24	3.34	3.08	65.42	47.0	362	56.7	4.33
	60		0.562	11.626	106.2	21.52	3.34	3.04	73.16	46.0	401	62.8	4.31
	80		0.687	11.376	101.6	26.04	3.34	2.978	88.51	44.0	475	74.5	4.27
	100		0.843	11.064	96.1	31.5	3.34	2.897	107.20	41.6	562	88.1	4.22
	120		1.000	10.750	90.8	36.9	3.34	2.814	125.49	39.3	642	100.7	4.17
	140		1.125	10.500	86.6	41.1	3.34	2.749	139.68	37.5	701	109.9	4.13
	160		1.312	10.126	80.5	47.1	3.34	2.651	160.27	34.9	781	122.6	4.07
14 — 14.000	10		0.250	13.500	143.1	10.80	3.67	3.53	36.71	62.1	255.4	36.5	4.86
	20		0.312	13.376	140.5	13.42	3.67	3.50	45.68	60.9	314	44.9	4.84
	30	Std	0.375	13.250	137.9	16.05	3.67	3.47	54.57	59.7	373	53.3	4.82
	40		0.437	13.126	135.3	18.62	3.67	3.44	63.37	58.7	429	61.2	4.80
		XS	0.500	13.000	132.7	21.21	3.67	3.40	72.09	57.5	484	69.1	4.78
			0.562	12.876	130.2	23.73	3.67	3.37	80.66	56.5	537	76.7	4.76
	60		0.593	12.814	129.0	24.98	3.67	3.35	84.91	55.9	562	80.3	4.74
			0.625	12.750	127.7	26.26	3.67	3.34	89.28	55.3	589	84.1	4.73
			0.687	12.626	125.2	28.73	3.67	3.31	97.68	54.3	638	91.2	4.71
	80		0.750	12.500	122.7	31.2	3.67	3.27	106.13	53.2	687	98.2	4.69
			0.875	12.250	117.9	36.1	3.67	3.21	122.66	51.1	781	111.5	4.65
	100		0.937	12.126	115.5	38.5	3.67	3.17	130.73	50.0	825	117.8	4.63
	120		1.093	11.814	109.6	44.3	3.67	3.09	150.67	47.5	930	132.8	4.58
	140		1.250	11.500	103.9	50.1	3.67	3.01	170.22	45.0	1,017	146.8	4.53
	160		1.406	11.188	98.3	55.6	3.67	2.929	189.12	42.6	1,127	159.6	4.48
16 — 16.000	10		0.250	15.500	188.7	12.37	4.19	4.06	42.05	81.8	384	48.0	5.57
	20		0.312	15.376	185.7	15.38	4.19	4.03	52.36	80.5	473	59.2	5.55
	30	Std	0.375	15.250	182.6	18.41	4.19	3.99	62.58	79.1	562	70.3	5.53
			0.437	15.126	179.7	21.37	4.19	3.96	72.64	77.9	648	80.9	5.50
	40	XS	0.500	15.000	176.7	24.35	4.19	3.93	82.77	76.5	732	91.5	5.48
			0.562	14.876	173.8	27.26	4.19	3.89	92.66	75.4	813	106.6	5.46
			0.625	14.750	170.7	30.2	4.19	3.86	102.63	74.1	894	112.2	5.44
	60		0.656	14.688	169.4	31.6	4.19	3.85	107.50	73.4	933	116.6	5.43
			0.687	14.626	168.0	33.0	4.19	3.83	112.36	72.7	971	121.4	5.42

* See footnote at end of table.
† See footnote at end of table.

Table 3-3 (Cont'd).

Nominal pipe size, O.D., in.	Schedule number a	b	c	Wall thickness, in.	I.D., in.	Inside area, sq in.	Metal area, sq in.	Sq ft outside surface, per ft	Sq ft inside surface, per ft	Weight per ft, lb	Weight of water per ft, lb	Moment of inertia, in.⁴	Section modulus, in.³	Radius gyration, in.
16 / 16.000				0.750	14.500	165.1	35.9	4.19	3.80	122.15	71.5	1,047	130.9	5.40
	80			0.842	14.314	160.9	40.1	4.19	3.75	136.46	69.7	1,157	144.6	5.37
				0.875	14.250	159.5	41.6	4.19	3.73	141.35	69.1	1,193	154.1	5.36
	100			1.031	13.938	152.6	48.5	4.19	3.65	164.83	66.1	1,365	170.6	5.30
	120			1.218	13.564	144.5	56.6	4.19	3.55	192.29	62.6	1,556	194.5	5.24
	140			1.437	13.126	135.3	65.7	4.19	3.44	223.50	58.6	1,760	220.0	5.17
	160			1.593	12.814	129.0	72.1	4.19	3.35	245.11	55.9	1,894	236.7	5.12
18 / 18.000	10			0.250	17.500	240.5	13.94	4.71	4.58	47.39	104.3	549	61.0	6.28
	20			0.312	17.376	237.1	17.34	4.71	4.55	59.03	102.8	678	75.5	6.25
		Std		0.375	17.250	233.7	20.76	4.71	4.52	70.59	101.2	807	89.6	6.23
	30			0.437	17.126	230.4	24.11	4.71	4.48	82.06	99.4	931	103.4	6.21
		XS		0.500	17.000	227.0	27.49	4.71	4.45	93.45	98.0	1,053	117.0	6.19
	40			0.562	16.876	223.7	30.8	4.71	4.42	104.75	97.0	1,172	130.2	6.17
				0.625	16.750	220.5	34.1	4.71	4.39	115.98	95.5	1,289	143.3	6.15
				0.687	16.626	217.1	37.4	4.71	4.35	127.03	94.1	1,403	156.3	6.13
	60			0.750	16.500	213.8	40.6	4.71	4.32	138.17	92.7	1,515	168.3	6.10
				0.875	16.250	207.4	47.1	4.71	4.25	160.04	89.9	1,731	192.8	6.06
	80			0.937	16.126	204.2	50.2	4.71	4.22	170.75	88.5	1,834	203.8	6.04
	100			1.156	15.688	193.3	61.2	4.71	4.11	207.96	83.7	2,180	242.2	5.97
	120			1.375	15.250	182.6	71.8	4.71	3.99	244.14	79.2	2,499	277.6	5.90
	140			1.562	14.876	173.8	80.7	4.71	3.89	274.23	75.3	2,750	306	5.84
	160			1.781	14.438	163.7	90.7	4.71	3.78	308.51	71.0	3,020	336	5.77
20 / 20.000	10			0.250	19.500	298.6	15.51	5.24	5.11	52.73	129.5	757	75.7	6.98
	20			0.312	19.376	294.9	19.30	5.24	5.07	65.40	128.1	935	93.5	6.96
		Std		0.375	19.250	291.0	23.12	5.24	5.04	78.60	126.0	1,114	111.4	6.94
	30			0.437	19.126	287.3	26.86	5.24	5.01	91.31	124.6	1,286	128.6	6.92
		XS		0.500	19.000	283.5	30.6	5.24	4.94	104.13	122.8	1,457	145.7	6.90
	40			0.562	18.876	279.8	34.3	5.24	4.93	116.67	121.3	1,624	162.4	6.88
				0.593	18.814	278.0	36.2	5.24	4.91	122.91	120.4	1,704	170.4	6.86
				0.625	18.750	276.1	38.0	5.24	4.88	129.33	119.7	1,787	178.7	6.85
	60			0.687	18.626	272.5	41.7	5.24	4.84	141.71	118.1	1,946	194.6	6.83
				0.750	18.500	268.8	45.4	5.24	4.81	154.20	116.5	2,105	210.5	6.81
				0.812	18.376	265.2	48.9	5.24	4.78	166.40	115.0	2,257	225.7	6.79
	80			0.875	18.250	261.6	52.6	5.24	4.70	178.73	113.4	2,409	240.9	6.77
				1.031	17.938	252.7	61.4	5.24	4.57	208.87	109.4	2,772	277.2	6.72
	100			1.281	17.438	238.8	75.3	5.24		256.10	103.4	3,320	332	6.63

Table 3-3 (Cont'd).

Nominal / OD	Sched.	Desig.	t	d	Inside area	Area of metal	Sq ft outside surf.	Sq ft inside surf.	Wt. pipe per ft	Wt. water per ft	Moment of inertia	Section modulus	Radius of gyration
20 / 20.000	120		1.500	17.000	227.0	87.2	5.24	4.45	296.37	98.3	3,760	376	6.56
	140		1.750	16.500	213.8	100.3	5.24	4.32	341.10	92.6	4,220	422	6.48
	160		1.968	16.064	202.7	111.5	5.24	4.21	379.01	87.9	4,590	459	6.41
24 / 24.000	10		0.250	23.500	434	18.65	6.28	6.15	63.41	188.0	1,316	109.6	8.40
			0.312	23.376	430	23.20	6.28	6.12	78.93	186.1	1,629	135.8	8.38
	20	Std	0.375	23.250	425	27.83	6.28	6.09	94.62	183.8	1,943	161.9	8.35
			0.437	23.126	420	32.4	6.28	6.05	109.97	182.1	2,246	187.4	8.33
		XS	0.500	23.000	415	36.9	6.28	6.02	125.49	180.1	2,550	212.5	8.31
	30		0.562	22.876	411	41.4	6.28	5.99	140.80	178.1	2,840	237.0	8.29
			0.625	22.750	406	45.9	6.28	5.96	156.03	176.2	3,140	261.4	8.27
	40		0.687	22.626	402	50.3	6.28	5.92	171.17	174.3	3,420	285.2	8.25
			0.750	22.500	398	54.8	6.28	5.89	186.24	172.4	3,710	309	8.22
	60		0.968	22.064	382	70.0	6.28	5.78	238.11	165.8	4,650	388	8.15
	80		1.218	21.564	365	87.2	6.28	5.65	296.36	158.3	5,670	473	8.07
	100		1.531	20.938	344	108.1	6.28	5.48	367.40	149.3	6,850	571	7.96
	120		1.812	20.376	326	126.3	6.28	5.33	429.39	141.4	7,830	652	7.87
	140		2.062	19.876	310	142.1	6.28	5.20	483.13	134.5	8,630	719	7.79
	160		2.343	19.314	293	159.4	6.28	5.06	541.94	127.0	9,460	788	7.70
30 / 30.000	10		0.312	29.376	678	29.1	7.85	7.69	98.93	293.8	3,210	214	10.50
	20		0.500	29.000	661	46.3	7.85	7.59	157.53	286.3	5,040	336	10.43
	30		0.625	28.750	649	57.6	7.85	7.53	196.08	281.5	6,220	415	10.39

* The ferritic stainless steels may be about 5 percent less, and the austenitic stainless steels about 2 percent greater than the values shown in this table, which are based on weights for carbon steel. The following formulas were used in the computation of the values shown in the table:

Weight of pipe per ft, lb $= 10.6802t(D - t)$
Weight of water per ft, lb $= 0.3405d^2$
Sq ft outside surface, per ft $= 0.2618D$
Sq ft inside surface, per ft $= 0.2618d$
Inside area, sq in $= 0.785d^2$
Area of metal, sq in $= 0.785(D^2 - d^2)$
Moment of inertia, in.^{4} $= 0.0491(D^4 - d^4)$
$= A_M R_g^2$

Section modulus, in.^{3} $= \dfrac{0.0982(D^4 - d^4)}{D}$

Radius of gyration, in. $= 0.25\sqrt{D^2 + d^2}$

A_M = Area of metal, sq in.
d = I.D., in.
D = O.D., in.
R_g = Radius of gyration, in.
t = Pipe wall thickness, in.

† a. ASA B36.10 steel-pipe schedule numbers.
b. ASA B36.10 steel-pipe nominal wall-thickness designations.
c. ASA B36.19 stainless-steel-pipe schedule numbers (SS is not an approved standard).

Scheduled Designations

Piping was, at one time, and in some cases still designated as: standard, extra strong, and double extra strong. There are no provisions for thin wall pipe and no intermediate standard thicknesses between the three schedules. The three schedules do not cover a range flexible enough to be economical without intermediate thicknesses. Table 3-3 lists piping as a function of the more recent Schedule Number which is given, approximately, by the following relationship.

$$\text{Schedule No.} = 1{,}000 \times P/S$$

Where P is the operating pressure (psig) and
S = allowable stress value (psi)

Example

Find the required schedule of ASTM A106, Grade B pipe, operating at 1,150 psig and 600°F. Table 3-2 lists the S value as 15,000 psi. Substituting 1,000 (1150/15,000) = 76.6. Use Schedule No. 80 as a rough estimate. However, to determine an accurate wall thickness, use the equation in the following paragraph.

Thickness of Pipe

The following notes, covering power pipe systems, have been abstracted from Sec. I of the Code for Pressure Piping (ASA B31.1).

For inspection purposes, the minimum thickness of pipe wall to be used for piping at different pressures and for temperatures not exceeding those for the various materials listed in Tables 3-1 and 3-2 can be determined by the formula.

$$P = \frac{(2S + yP)t_m - C}{D}$$

where t_m = minimum pipe wall thickness, (in.) allowable on inspection
  P = maximum internal service pressure, psig (plus water-hammer allowance in case of cast iron conveying liquids)
  D = O.D. of pipe, (in.)
  S = allowable stress in material (taken from Tables 3-1 and 3-2 at service temperature)
  y = a coefficient, values for which are listed in Table 3-4.
  C = allowance for threading, mechanical strength, and corrosion, (in.) with values of C listed in Table 3-5.

Table 3-4. Values of *y* (Interpolate for intermediate values) (ASA Code for Pressure Piping, 1963).

	Temperature, deg F					
	900 below	950	1000	1050	1100	1150 and above
Ferritic steels	0.4	0.5	0.7	0.7	0.7	0.7
Austenite steels	0.4	0.4	0.4	0.4	0.5	0.7

Table 3-5. Values of *C*. (ASA Code for Pressure Piping, 1963)

Type of pipe	*Value of C, in.*
Cast-iron pipe, centrifugally cast	0.14
Cast-iron, pit case	0.18
Threaded-steel, wrought-iron, or non-ferrous pipe	
3/8" and smaller	0.05
1/2" and larger	Depth of thread
Grooved-steel, wrought-iron or non-ferrous pipe	Depth of groove
Plain-end steel, wrought-iron or tube	
1 in. and smaller	0.05
1-1/4 in. and larger	0.065
Plain-end non-ferrous pipe or tube	0.000

Physical and Chemical Property of Pipe

The design of piping for operation above 750°F presents many problems not encountered at lower temperatures. For the properties of steel applicable to high temperature service (as well as ordinary services) for pipe fittings, bolting material, etc. as specified by the ASA code for pressure piping, (ASA B31.1), is given in Table 3-6. The ASTM standard for welded and seamless steel pipe and dimensions for welded wrought iron pipe are given in Table 3-7(a) and 3-7(b).

Some common configurations are shown in Figure 3-1, with a list of threaded pipe fittings and their dimensions included in Table 3-8. To

Table 3-6. Specifications for Tensile Strength of Pipe
(ASTM—July, 1963)

ASA desig-nation	ASTM desig-nation	Style of pipe	Tensile strength, min, psi	Scope
B36.1	A53	Welded and seamless steel	Welded: Bessemer, 50,000; O.H. 45,000. Seamless: Grade A, 48,000; Grade B, 60,000	*a*
B36.2	A72	Welded wrought iron	40,000	*b*
B36.3	A106	Lap-welded and seamless steel for high temperatures	Seamless: Grade A, 48,000; Grade B, 60,000	*c*
B36.4	A134	Electric-fusion-welded steel, sizes 16 in. and over	See ASTM Standard A134	*d*
B36.5	A135	Electric-resistance-welded steel	Grade A, 48,000. Grade B, 60,000	*e*
B36.9	A139	Electric-fusion-welded steel, sizes 4 in. and over	Grade A, 48,000. Grade B, 60,000	*f*
B36.11	A155	Electric-fusion-welded steel pipe for high-temperature high-pressure service	Varies with material. See Spec. ASTM A155	*g*
G8.7	A120	Black and hot-dipped gal-vanized welded and seam-less steel	Same as B36.1	*h*

a Commercial steel pipe for general uses, also for coiling, bending, flanging, and similar forming operations when so specified.

b Commercial wrought-iron pipe for general uses, also for coiling, bending, flanging, and other special purposes.

c Lap-welded and seamless steel pipe for high-temperature service. Suitable for bending, flanging, and similar forming operations.

d Cover pipe 16 in. diam and over in wall thicknesses up to ¾ in., fabricated from steel plates by electric-fusion welding.

e Pipe up to 30 in. intended for conveying liquids, gas, or vapor at temperatures below 450 F. Adapted for flanging, bending, and similar forming operations in Grade A class.

f Covers sizes 4 to <16 in. in wall thicknesses not over ⅜ in., fabricated from steel plates by electric-fusion welding. Intended for conveying liquids, gas, or vapor at temperatures below 450 F. Adapted for flanging and bending.

g Electric-fusion-welded steel pipe having an outside diameter of 18 in. and over for high-temperature and high-pressure service. Suitable for bending, flanging, corrugating, and similar forming operations. Welding in accordance with Par. U-68 of the ASME code for unfired pressure vessels.

h Commercial steel pipe for ordinary uses such as low-pressure steam, liquid, or gas lines. Not intended for coiling or close bending, nor for high-temperature service.

elaborate on the figure, the term "Riser" defines a vertically oriented length of the pipe and "Run" a horizontally oriented length of pipe when drawn on a plan view, that it is laid out or is shown in an elevation view.

Some of the more common symbols used in piping drawings are included in the LIST OF TERMS at the bottom of Figure 3-1. The term IPS (Iron Pipe Size) is used as an abbreviation for all standard pipe sizes and screwed pipe thread sizes.

Flanges are flat circular disks made of steel. They are made of both ferrite and austenite steels depending on service delineated by pressure and temperature. They are attached to pipe in a number of different ways as shown in Figure 3-2. As a means of joining one flange with another, that is, connecting two pipes, holes are drilled through the flange plates for bolting. These holes are drilled using standard tem-

Table 3-7a. Dimension of Welded and Seamless Steel Pipe
(ASTM "Steel Piping Materials": July, 1963)

Nominal pipe size, in.	Outside diam, in.	Nominal wall thickness, in., for stated schedule numbers									
		10	20	30	40	60	80	100	120	140	16
⅛	0.405	....	....	....	0.068	....	0.095				
¼	0.540	....	....	....	0.088	....	0.119				
⅜	0.675	....	....	....	0.091	....	0.126				
½	0.840	....	....	....	0.109	....	0.147	....	....		
¾	1.050	....	....	....	0.113	....	0.154	....	....		
1	1.315	....	....	....	0.133	....	0.179	....	....		
1¼	1.660	....	....	....	0.140	....	0.191	....	....		
1½	1.900	....	....	....	0.145	....	0.200	....	....		
2	2.375	....	....	....	0.154	....	0.218	....	....	....	0.3
2½	2.875	....	....	....	0.203	....	0.276	....	....	....	0.3
3	3.5	....	....	....	0.216	....	0.300	....	....	....	0.4
3½	4.0	....	....	....	0.226	....	0.318				
4	4.5	....	....	....	0.237	....	0.337	...	0.437	....	0.5
5	5.563	....	....	....	0.258	....	0.375	...	0.500	....	0.6
6	6.625	....	....	....	0.280	....	0.432	....	0.562	....	0.7
8	8.625	....	0.250	0.277	0.322	0.406	0.500	0.593	0.718	0.812	0.9
10	10.75	....	0.250	0.307	0.365	0.500	0.593	0.718	0.843	1.000	1.1
12	12.75	....	0.250	0.330	0.406	0.562	0.687	0.843	1.000	1.125	1.3
14 O.D.	14.0	0.250	0.312	0.375	0.437	0.593	0.750	0.937	1.062	1.250	1.4
16 O.D.	16.0	0.250	0.312	0.375	0.500	0.656	0.843	1.031	1.218	1.437	1.5
18 O.D.	18.0	0.250	0.312	0.437	0.562	0.718	0.937	1.156	1.343	1.562	1.7
20 O.D.	20.0	0.250	0.375	0.500	0.593	0.812	1.031	1.250	1.500	1.750	1.9
24 O.D.	24.0	0.250	0.375	0.562	0.687	0.937	1.218	1.500	1.750	2.062	2.3

The schedule numbers are approximate values of $1,000P/S$ (see p. S-206 for the symbols).   Thic nesses include a mill tolerance of 12.5 percent.   Thicknesses in black type in schedules 30 and 40 agi with those of standard-weight pipe, those in schedules 60 and 80 with extra-strong pipe.

Table 3-7b. Dimensions of Welded Wrought-iron Pipe
(ASTM "Steel Piping Materials": July, 1963)

Nominal pipe size, in.	Outside diam, in.	Nominal wall thickness, in.	
		Standard weight pipe	Extra-strong pipe
¼	0.540	0.090	0.122
⅜	0.675	0.093	0.129
½	0.840	0.111	0.151
¾	1.050	0.115	0.157
1	1.315	0.136	0.183
1¼	1.660	0.143	0.195
1½	1.900	0.148	0.204
2	2.375	0.158	0.223
2½	2.875	0.208	0.282
3	3.5	0.221	0.306
3½	4.0	0.231	0.325
4	4.5	0.242	0.344
5	5.563	0.263	0.383
6	6.625	0.286	0.441
8	8.625	0.329	0.510
10	10.75	0.372	0.510
12	12.75	0.383	0.510

Wrought-iron pipe contains about 3 percent of slag, 0.5 percent C, and other impurities.   It is more resistant to corrosion than is steel pipe.

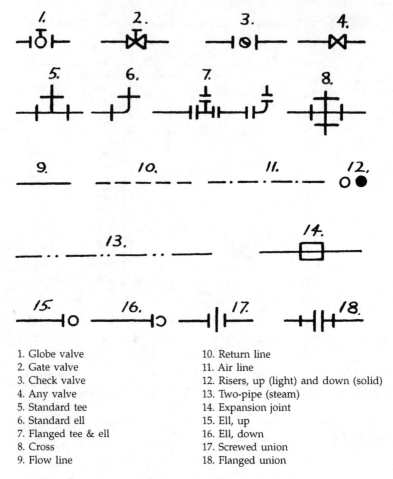

1. Globe valve
2. Gate valve
3. Check valve
4. Any valve
5. Standard tee
6. Standard ell
7. Flanged tee & ell
8. Cross
9. Flow line

10. Return line
11. Air line
12. Risers, up (light) and down (solid)
13. Two-pipe (steam)
14. Expansion joint
15. Ell, up
16. Ell, down
17. Screwed union
18. Flanged union

Figure 3-1. Standard Pipe Symbols.

plates. The diameter of the bolt circle and the number of holes and their diameters vary with different flange sizes. A selection of flanged fittings shown in Figure 3-2 include: welded flange, screwed flange, ring joint, and raised face, etc. Template dimensions for drilling various size flanges are shown in Table 3-9. Figure 3-3 shows two typical piping layouts including union, elbows, tees, and valves. Table 3-8 specifies the accepted dimensions for their fittings.

Welded Pipe Connections

Basically, welded pipe is used extensively for many steam and condensate applications, generally for high pressures and temperatures in-

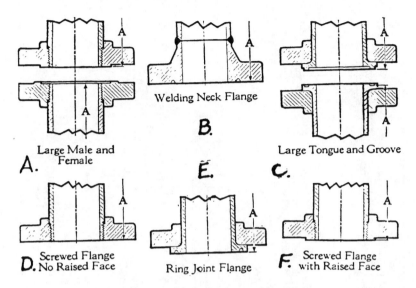

Figure 3-2. Types of Flanges.

stead of conventional pipe fittings. By combining both welding and bending in installations, many complicated pipe configurations can be assembled simply and economically. Several methods of welding pipe connections are shown in Figure 3-4.

Pipe Bends

Pipe bends are, for the most part, large radius elbows. They are fabricated with pipe bending machinery rather than cast or forged as are conventional fittings. Pipe bends have many uses in modern steam installations primarily to provide flexibility in piping assembly. They compensate for expansion and contraction as well as reducing the internal resistance to flow experienced in pipe fittings. Bends may have many different shapes and forms with numerous arrangements for outlets and connections. They can be made in all sizes of steel pipe. Pipe bends are shown in Figure 3-5. A list of dimensions for typical configurations is given in Table 3-10.

EXPANSION JOINTS

Provisions must be made for steam and condensate piping to expand from its cold dimensions to its hot dimensions. To do this, pipe

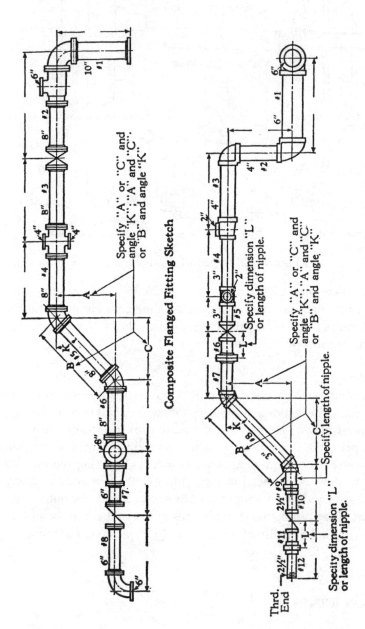

Figure 3-3. Screwed Fitting Illustration for a Typical Piping Layout

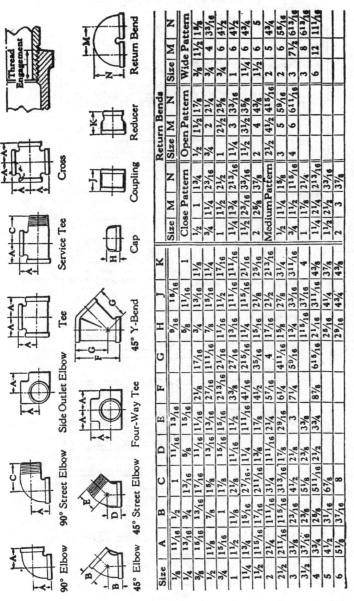

Figure fitting types: 90° Elbow, 90° Street Elbow, Side Outlet Elbow, Tee, Service Tee, Cross, Return Bend, 45° Elbow, 45° Street Elbow, Four-Way Tee, 45° Y-Bend, Reducer, Coupling, Cap, Thread Engagement.

Size	A	B	C	D	E	F	G	H	J	K
1/8	11/16	1/2	1	11/16	13/16			9/16	15/16	1
1/4	13/16	3/4	13/16	5/8	15/16			5/8	11/16	1
3/8	15/16	13/16	17/16	11/16	1 1/16	2 1/8	17/16	3/4	13/16	1 1/8
1/2	1 1/8	7/8	1 5/8	13/16	1 3/16	2 7/16	1 11/16	7/8	15/16	1 1/4
3/4	1 5/16	1	1 7/8	15/16	1 5/16	2 13/16	2 1/16	1 1/16	1 1/2	1 7/16
1	1 1/2	1 1/8	2 1/8	1 1/16	1/2	3 3/8	2 7/16	1 3/16	1 13/16	1 11/16
1 1/4	1 3/4	1 5/16	2 7/16	1 1/4	1 11/16	4 1/16	2 15/16	1 3/16	1 15/16	1 11/16
1 1/2	1 15/16	1 7/16	2 11/16	1 3/8	1 7/8	4 1/2	3 5/16	1 1/4	2 1/8	2 1/16
2	2 1/4	1 11/16	3 1/4	1 11/16	2 1/4	5 7/16	4	1 7/16	2 1/2	2 13/16
2 1/2	2 11/16	1 15/16	3 13/16	1 7/8	2 9/16	6 1/4	4 11/16	1 5/8	2 7/8	3 1/4
3	3 1/8	2 3/16	4 1/2	2 1/8	3	7 1/4	5 9/16	1 3/4	3 3/8	3 11/16
3 1/2	3 3/16	2 3/8	5 1/8	2 3/8	3 3/8			1 15/16	3 11/16	4
4	3 3/4	2 5/8	5 11/16	2 1/2	3 3/4	8 7/8	6 15/16	1 15/16	3 11/16	4 3/8
5	4 1/2	3 1/16	6 7/8					2 5/16	4 1/4	3 7/8
6	5 1/8	3 7/16	8					2 5/8	4 3/4	4 3/8

Return Bends

	Size	M	N	Size	M	N	Size	M	N
	Close Pattern			Open Pattern			Wide Pattern		
	1/2	1	1 3/4	1/2	1 1/2	1 7/8	3/8	1 1/2	1 1/16
	3/4	1 1/4	2 3/16	3/4	2	2 1/4	3/4	4	3 3/16
	1	1 1/2	2 1/2	1	2 1/2	2 5/8	3/4	6	4 1/2
	1 1/4	1 3/4	2 13/16	1 1/4	3	3 3/16	1	6	4 1/2
	1 1/2	2 3/16	3 3/16	1 1/2	4	4 3/8	1 1/4	6	4 3/4
	2	2 5/8	3 7/8	2	4	4 3/8	1 1/2	6	5
	Medium Pattern			2 1/2	4 1/2	4 15/16	2	5	4 3/4
	1/2	1 1/4	1 5/8	3	5	5 9/16	2	6	5 5/16
	3/4	1 1/2	1 15/16	4	6	6 11/16	3	7 1/2	6 13/16
	1	1 7/8	2 1/4				3	8	6 13/16
	1 1/4	2 1/4	3 1/16				6	12	11 1/16
	1 1/2	2 1/2	3 3/16						
	2	3	3 7/8						

Table 3-8. Dimensions of Pipe Fittings

Standard fittings. The table gives the dimensions in inches of the most currently used malleable iron straight-size fittings and reducers. Plain fittings have same center-to-end dimensions as corresponding banded fittings.

Table 3-9a. Templates for Drilling, American Standard Steel Pipe Flanges and Flanged Fittings* (ASA Spec. B16.5, 1961) (All dimensions in inches)

Nominal pipe size	400 lb standard					600 lb standard					900 lb standard					1,500 lb standard				
	Outside diam of flange	Thickness of flange, minimum	Diam of bolt circle	Number of bolts	Size of bolts	Outside diam of flange	Thickness of flange, minimum	Diam of bolt circle	Number of bolts	Size of bolts	Outside diam of flange	Thickness of flange, minimum	Diam of bolt circle	Number of bolts	Size of bolts	Outside diam of flange	Thickness of flange, minimum	Diam of bolt circle	Number of bolts	Size of bolts
1/2	For sizes below 4 in. use dimensions of 600 lb fittings					3 3/4	9/16	2 5/8	4	1/2	For sizes below 3 in. use dimensions of 1,500 lb fittings					4 3/4	7/8	3 1/4	4	3/4
3/4						4 5/8	5/8	3 1/4	4	5/8						5 1/8	1	3 1/2	4	3/4
1						4 7/8	11/16	3 1/2	4	5/8						5 7/8	1 1/8	4	4	7/8
1 1/4	......	......	......	......	......	5 1/4	13/16	3 7/8	4	5/8						6 1/4	1 1/8	4 3/8	4	7/8
1 1/2	......	......	......	......	......	6 1/8	7/8	4 1/2	4	3/4	......	......	......	......	......	7	1 1/4	4 7/8	4	1
2	......	......	......	......	......	6 1/2	1	5	8	5/8	......	......	......	......	......	8 1/2	1 1/2	6 1/2	8	7/8
2 1/2	......	......	......	......	......	7 1/2	1 1/8	5 7/8	8	3/4	......	......	......	......	......	9 5/8	1 5/8	7 1/2	8	1
3	......	......	......	......	......	8 1/4	1 1/4	6 5/8	8	3/4	9 1/2	1 1/2	7 1/2	8	7/8	10 1/2	1 7/8	8	8	1 1/8
3 1/2	......	......	......	......	......	9	1 3/8	7 1/4	8	7/8										
4	10	1 3/8	7 7/8	8	7/8	10 3/4	1 1/2	8 1/2	8	7/8	11 1/2	1 3/4	9 1/4	8	1 1/8	12 1/4	2 1/8	9 1/2	8	1 1/4
5	11	1 1/2	9 1/4	8	7/8	13	1 3/4	10 1/2	8	1	13 3/4	2	11	8	1 1/4	14 3/4	2 7/8	11 1/2	8	1 1/2
6	12 1/2	1 5/8	10 5/8	12	7/8	14	1 7/8	11 1/2	12	1	15	2 3/16	12 1/2	12	1 1/8	15 1/2	3 1/4	12 1/2	12	1 3/8
8	15	1 7/8	13	12	1	16 1/2	2 3/16	13 3/4	12	1 1/8	18 1/2	2 1/2	15 1/2	12	1 3/8	19	3 5/8	15 1/2	12	1 5/8
10	17 1/2	2 1/8	15 1/4	16	1 1/8	20	2 1/2	17	16	1 1/4	21 1/2	2 3/4	18 1/2	16	1 3/8	23	4 1/4	19	12	1 7/8
12	20 1/2	2 1/4	17 3/4	16	1 1/4	22	2 5/8	19 1/4	20	1 1/4	24	3 1/8	21	20	1 3/8	26 1/2	4 7/8	22 1/2	16	2
14 O.D.	23	2 3/8	20 1/4	20	1 1/4	23 3/4	2 3/4	20 3/4	20	1 3/8	25 1/4	3 3/8	22	20	1 1/2	29 1/2	5 1/4	25	16	2 1/4
16 O.D.	25 1/2	2 1/2	22 1/2	20	1 3/8	27	3	23 3/4	20	1 1/2	27 3/4	3 1/2	24 1/4	20	1 5/8	32 1/2	5 3/4	27 3/4	16	2 1/2
18 O.D.	28	2 5/8	24 3/4	24	1 3/8	29 1/4	3 1/4	25 3/4	20	1 5/8	31	4	27	20	1 7/8	36	6 3/8	30 1/2	16	2 3/4
20 O.D.	30 1/2	2 3/4	27	24	1 1/2	32	3 1/2	28 1/2	24	1 5/8	33 3/4	4 1/4	29 1/2	20	2	38 3/4	7	32 3/4	16	3
24 O.D.	36	3	32	24	1 3/4	37	4	33	24	1 7/8	41	5 1/2	35 1/2	20	2 1/2	46	8	39	16	3 1/2

Table 3-9b. Dimensions of American Standard Companion Flanges* (ASA Spec. B16.5, 1961) (All dimensions in inches)

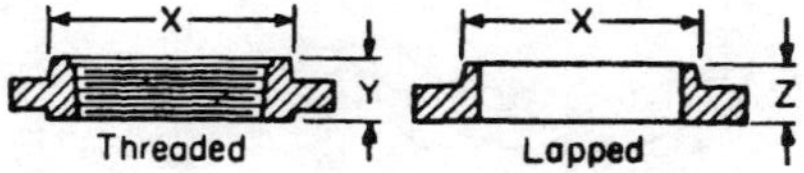

Nom pipe size	150 lb			300 lb			400 lb			600 lb			900 lb			1,500 lb		
	X	Y	Z	X	Y	Z	X	Y	Z	X	Y	Z	X	Y	Z	X	Y	Z
½	1 3/16	⅝	⅝	1½	⅞	⅞	For sizes below 4 in., use dimensions of the 600 lb flanges			1½	⅞	⅞	For sizes below 3 in., use dimensions of 1,500 lb flanges			1½	1¼	1¼
¾	1½	⅝	⅝	1⅞	1	1				1⅞	1	1				1¾	1⅜	1⅜
1	1 15/16	11/16	11/16	2⅛	1 1/16	1 1/16				2⅛	1 1/16	1 1/16				2 1/16	1⅝	1⅝
1¼	2 5/16	13/16	13/16	2½	1 1/16	1 1/16	....	.....	.....	2½	1⅛	1⅛				2½	1⅝	1⅝
1½	2 9/16	⅞	⅞	2¾	1 3/16	1 3/16	....	.....	.....	2¾	1¼	1¼	....	...	....	2¾	1¾	1¾
2	3 1/16	1	1	3 5/16	1 5/16	1 5/16	....	.....	.....	3 5/16	1 7/16	1 7/16	....	...	....	4⅛	2¼	2¼
2½	3 9/16	1⅛	1⅛	3 15/16	1½	1½	....	.....	.....	3 15/16	1⅝	1⅝	....	...	....	4⅞	2½	2½
3	4¼	1 3/16	1 3/16	4⅝	1 11/16	1 11/16	....	.....	.....	4⅝	1 13/16	1 13/16	5	2⅛	2⅛	5¼	2⅞	2⅞
3½	4 13/16	1¼	1¼	5¼	1¾	1¾	....	.....	.....	5¼	1 15/16	1 15/16						
4	5 5/16	1 5/16	1 5/16	5¾	1⅞	1⅞	5¾	2	2	6	2⅛	2⅛	6¼	2¾	2¾	6⅜	3 9/16	3 9/16
5	6 7/16	1 7/16	1 7/16	7	2	2	7	2⅛	2⅛	7 7/16	2⅜	2⅜	7½	3⅛	3⅛	7¾	4⅛	4⅛
6	7 9/16	1 9/16	1 9/16	8⅛	2 1/16	2 1/16	8⅛	2¼	2¼	8¾	2⅝	2⅝	9¼	3⅜	3⅜	9	4 11/16	4 11/16
8	9 11/16	1¾	1¾	10¼	2 7/16	2 7/16	10¼	2 11/16	2 11/16	10¾	3	3	11¾	4	4½	11½	5⅝	5⅝
10	12	1 15/16	1 15/16	12⅝	2⅝	3¾	12⅝	2⅞	4	13½	3⅜	4⅜	14½	4¼	5	14½	6¼	7
12	14⅜	2 3/16	2 3/16	14¾	2⅞	4	14¾	3⅛	4¼	15¾	3⅝	4⅝	16½	4⅝	5⅝	17¾	7⅛	8⅝
14 O.D.	15¾	2¼	3⅛	16¾	3	4⅜	16¾	3 5/16	4⅝	17	3 11/16	5	17¾	5⅛	6⅛	19½	.....	9½
16 O.D.	18	2½	3 7/16	19	3¼	4¾	19	3 11/16	5	19½	4 3/16	5½	20	5¼	6½	21¾	.....	10¼
18 O.D.	19⅞	2 11/16	3 15/16	21	3½	5⅛	21	3⅞	5⅜	21½	4⅝	6	22¼	6	7½	23½	.....	10⅞
20 O.D.	22	2⅞	4 1/16	23⅛	3¾	5½	23⅛	4	5¾	24	5	6¼	24½	6½	8¼	25¼	.....	11½
24 O.D.	26⅛	3¼	4⅜	27⅝	4 3/16	6	27⅝	4½	6¼	28¼	5½	7¼	29½	8	10½	30	.....	13

* Other dimensions are given in Tables 31 and 32. Finished bore on lapped flange to be such as method of attachment of pipe requires.

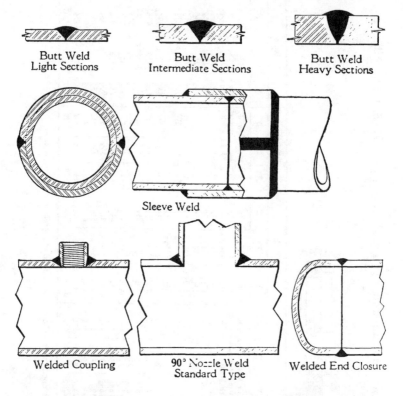

Figure 3-4. Methods of Welding Piping and Pipe Fittings.

lines can often expand several inches or feet depending upon their length and service temperature. One method of providing for this pipe dimensional increase is to use expansion joints at intervals along the length of the pipe. The Slip Joint type expansion joint is shown in Figure 3-6. The stuffing box usually contains a graphite impregnated calcium silicate fiber which is compressed with a gland to form a steam seal. It is good practice for these joints to leak steam ever so slightly to ensure that the gland is not so tight to inhibit expansion of the two parts. In other words, if the gland is tightened too much, the motion of the joint to compensate for the expansion of the pipe will not take place. Working pressures of expansion joints for saturated steam can go as high as 400 psi. The components of a joint are: the body, which are made of cast steel. Sleeves are made of machined and polished brass in joints 6 inches or smaller and from chromium plated steel in all joints 8 inches or larger.

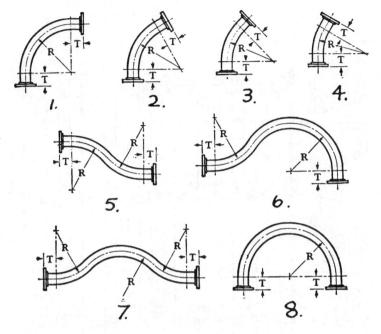

Figure 3-5. Types of Pipe Bends.

Stuffing boxes are proportioned to hold sufficient packing to ensure tight joints with a minimum of downtime for repacking.

Expansion joints are regularly furnished without packing. The packing used must meet the temperature and pressure requirement of the particular steam unit in which they are installed.

One side of an expansion joint is usually connected to a rigid base forming an anchor that permits only the protruding length of the pipe to expand into the fixed portion of the joint. Other pipe supports and anchors for direct burial steam pipe are discussed later in this chapter.

A table of Coefficients of Expansion for various piping materials lists them in inches of expansion per inch of length (coefficient A). And more importantly, it lists inches of expansion per 100 ft of pipe (coefficient B). Both components, coefficient A and coefficient B are given for various temperatures relative to an ambient temperature of 70°F. For example: in Table 3-10, 100 ft of carbon steel pipe will elongate (expand) 4.60 in. when elevated to a temperature of 600°F from a temperature of 70°F.

Table 3-10. Length of Pipe in Different Types of Bends.

Radius in Inches	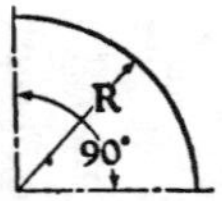 R 90°	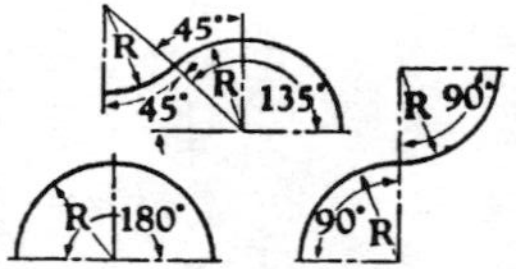 R 45° 45° R 135° R 90° R 180° 90° R	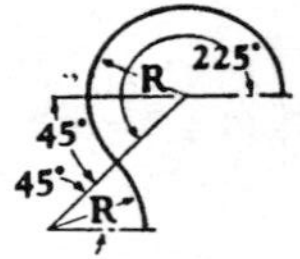 R 225° 45° 45° R	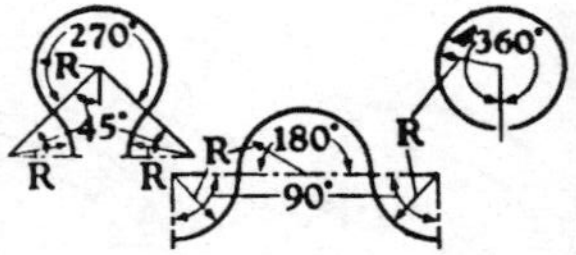 270° R 45° R R R 180° 90° R 360°	 270° 45° R 135°
5″	7¾″	15¾″	23½″	31½″	47¼″
6	9½	18¾	28¼	37¾	56½
7	11	22	33	44	66
8	12½	25¼	37¾	50¼	75½
9	14¼	28¼	42½	56½	84¾
10	15¾	31½	47¼	62¾	94¼
11	17¼	34½	51¾	69¼	103¾
12	18¾	37¾	56½	75½	113
13	20½	40¾	61¼	81¾	122½
14	22	44	66	88	132
15	23½	47¼	70¾	94¼	141½
20	31½	62¾	94¼	125¾	188½
25	39¼	78½	117¾	157	235½
30	47¼	94¼	141½	188½	282¾
35	55	110	165	220	330
40	62¾	125¾	188½	251¼	377
45	70¾	141½	212	282¾	424
50	78½	157	235½	314¼	471¼
60	94¼	188½	282¾	377	565½
70	110	220	330	439¾	659¾
80	125¾	251¼	377	502¾	754
90	141½	282¾	424	565½	848¼
100	157	314¼	471¼	628¼	942½
110	172¾	345½	518½	691¼	1036¾
120	188½	377	565½	754	1131
130	204¼	408½	612½	816¾	1225¼
140	220	439¾	659¾	879¾	1319½
150	235½	471¼	706¾	942½	1413¾

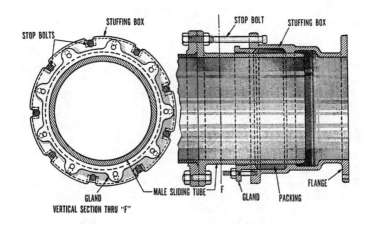

STUFFING BOX · STOP BOLT · STUFFING BOX · STOP BOLTS · GLAND · MALE SLIDING TUBE · F · GLAND · PACKING · FLANGE · VERTICAL SECTION THRU "F"

Figure 3-6. Slip Type Expansion Joint.

An equation for thermal expansion or elongation of pipe is

$$\text{Elongation} = \frac{\text{coef B} \times L}{100}$$

where coefficient B is from Table 3-11 and L = length of pipe at 70°F.

For example: a 700-ft length of Austenitic stainless steel pipe will expand at

$$E = \frac{7.5 \times 250 \, \text{ft}}{100} = 18.75 \text{ inches from its length of 70°F}$$

The standard traverse of a sleeve is 4 inches. However, 4 in, 6 in, and 8 in are available depending upon the diameter of the pipe.

Pipe Expansion

Figure 3-7 consists of two photos of steam lines under construction for direct burial. This type construction, direct burial, requires special consideration when providing for expansion. Figure 3-8 typifies a burial method for an underground steam line. Figure 3-9 illustrates a direct burial steam line anchor and line guide. Two wall penetration methods for low pressure direct burial pipe is shown in Figure 3-10. Figure 3-11

Table 3-11. Thermal-expansion Data.

(ASA Code for Pressure Piping)

A = Mean coefficient of thermal expansion × 10⁴, in. in. per in. per deg F } in going from 70 F to indicated temperature
B = Linear thermal expansion, in. per 100 ft

Temperature range: 70 F to

Material	Coefficient	70	200	300	400	500	600	700	800	900	1000	1100	1200	1300	1400
Carbon steel: carbon-moly steel low-chrome steels (through 3% Cr)	A	0	6.38	6.60	6.82	7.02	7.23	7.44	7.65	7.84	7.97	8.12	8.19	8.28	8.36
	B	0	0.99	1.82	2.70	3.62	4.60	5.63	6.70	7.81	8.89	10.04	11.10	12.22	13.34
Intermediate alloy steels 5 Cr Mo–9 Cr Mo	A	0	6.04	6.19	6.34	6.50	6.66	6.80	6.96	7.10	7.22	7.32	7.41	7.49	7.55
	B	0	0.94	1.71	2.50	3.35	4.24	5.14	6.10	7.07	8.06	9.05	10.00	11.06	12.05
Austenitic stainless steels	A	0	9.34	9.47	9.59	9.70	9.82	9.92	10.05	10.16	10.29	10.39	10.48	10.54	10.60
	B	0	1.46	2.61	3.80	5.01	6.24	7.50	8.80	10.12	11.48	12.84	14.20	15.56	16.92
Straight chromium stainless steels: 12 Cr, 17 Cr, and 27 Cr	A	0	5.50	5.66	5.81	5.96	6.13	6.26	6.39	6.52	6.63	6.72	6.78	6.85	6.90
	B	0	0.86	1.56	2.30	3.08	3.90	4.73	5.60	6.49	7.40	8.31	9.20	10.11	11.01
25 Cr–20 Ni	A	0	7.76	7.92	8.08	8.22	8.38	8.52	8.68	8.81	8.92	9.00	9.08	9.12	9.18
	B	0	1.21	2.18	3.20	4.24	5.33	6.44	7.60	8.78	9.95	11.12	12.31	13.46	14.65
Monel 67 Ni–30 Cu	A	0	7.84	8.02	8.20	8.40	8.58	8.78	8.96	9.16	9.34	9.52	9.70	9.88	10.04
	B	0	1.22	2.21	3.25	4.33	5.46	6.64	7.85	9.12	10.42	11.77	13.15	14.58	16.02
Monel 66 Ni–29 CuAl	A	0	7.48	7.68	7.90	8.09	8.30	8.50	8.70	8.90	9.10	9.30	9.50	9.70	9.89
	B	0	1.17	2.12	3.13	4.17	5.28	6.43	7.62	8.86	10.16	11.50	13.00	14.32	15.78
Aluminum	A	0	12.95	13.28	13.60	13.90	14.20								
	B	0	2.00	3.66	5.39	7.17	9.03								
Gray cast iron	A	0	5.75	5.93	6.10	6.28	6.47	6.65	6.83	7.00	7.19				
	B	0	0.90	1.64	2.42	3.24	4.11	5.03	5.98	6.97	8.02				
Bronze	A	0	10.03	10.12	10.23	10.32	10.44	10.52	10.62	10.72	10.80	10.90	11.00		
	B	0	1.56	2.79	4.05	5.33	6.64	7.95	9.30	10.68	12.05	13.47	14.92		
Brass	A	0	9.76	10.00	10.23	10.47	10.69	10.92	11.16	11.40	11.63	11.85	12.09		
	B	0	1.52	2.76	4.05	5.40	6.80	8.26	9.78	11.35	12.98	14.65	16.39		
Wrought iron	A	0	7.32	7.48	7.61	7.73	7.88	8.01	8.13	8.29	8.39				
	B	0	1.14	2.06	3.01	3.99	5.01	6.06	7.12	8.26	9.36				
Copper-nickel (70/30)	A	0	8.54	8.71	8.90										
	B	0	1.33	2.40	3.52										

shows a corrugated type expansion joint in a manhole. A more detailed illustration of a corrugated type joint shown in Figure 3-12. Corrugated joints, although requiring less maintenance, are not as good as the slip joint when the joint requires radial alignment. The corrugations do not lend themselves useful except for lateral alignment. Often an alignment support is used with the corrugated joint to insure radial alignment.

Figure 3-13 shows a stainless steel bellows type expansion joint for high pressure or temperature. The bellows joint and wall expansion joint shown in Figure 3-14 are typical expansion joints used in power plant steam lines from boiler houses indoors to the turbines outdoors.

Expansion joints, when located in tunnels, trenches, manholes, or above ground are usually fitted with a connection for draining condensate from the steam line. Steam traps are used to ensure that only condensate drains from the steam line. Traps are discussed in a separate chapter.

PIPE SUPPORTS

The code for pressure piping (ASA B31.1) is used as the standard of the designers for most types of supports and gives specific directions for their applications. A pipe support must have a strong rigid base and an adjustable feature to allow for expansion and maintain alignment of the pipe in specific directions. Several types of pipe supports are shown in Figures 3-15 through 3-17.

It is important to avoid excess friction caused by movement of the pipe in the support. They should be lubricated regularly as well as inspected both in the cold and hot conditions of the pipe. The designer should ensure that the support strength is sufficient to maintain alignment of the pipe before, during, and when expanded. Piping can be designed to expand from either or both ends of in the center of the pipe length under consideration.

Anchors are a special type of pipe support and are usually designed to resist movement of the pipe in any direction at the point where the anchor is installed. They must be securely fastened to structurally rigid locations either in a building, foundation, or frame and must be able to resist loads much more powerful than just the weight of the pipe and its contents. Large magnitude forces are exerted on anchors due to the thermal expansion loads placed on it between hot and cold pipe

**Figure 3-7. Buried Steam
Piping during Installation**
Note the anchor in the back-
ground of the lower picture.

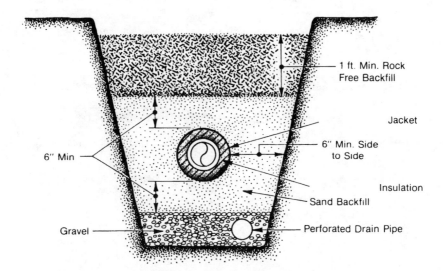

Figure 3-8. Backfill Detail on Insulated Steam Line.

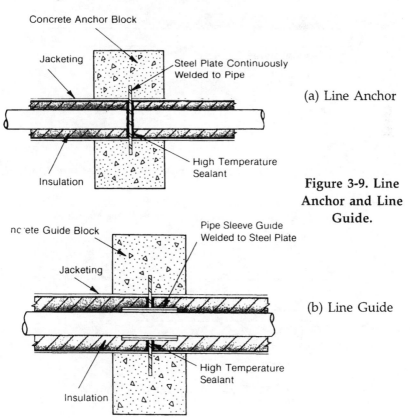

(a) Line Anchor

Figure 3-9. Line Anchor and Line Guide.

(b) Line Guide

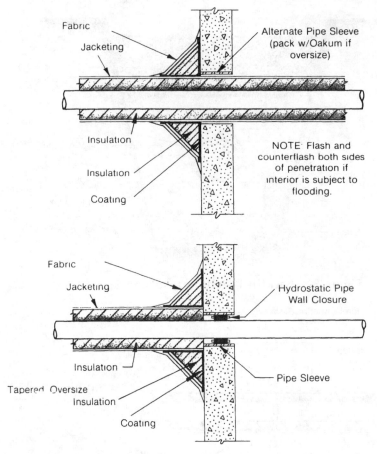

Figure 3-10. (top) Wall Penetrations; (bottom) Alternate Method.

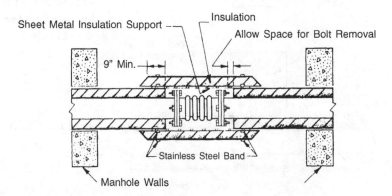

Figure 3-11. Expansion Joint in a Manhole.

INTERNAL SLEEVE
CENTERING RING

Figure 3-12. Corrugated Expansion Joint for Medium Pressures.

Figure 3-13. Bellows Expansion Joint for High Pressure Steam

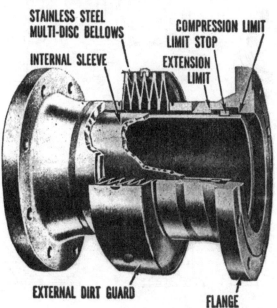

STAINLESS STEEL
MULTI-DISC BELLOWS

INTERNAL SLEEVE

COMPRESSION LIMIT
LIMIT STOP

EXTENSION
LIMIT

EXTERNAL DIRT GUARD

FLANGE

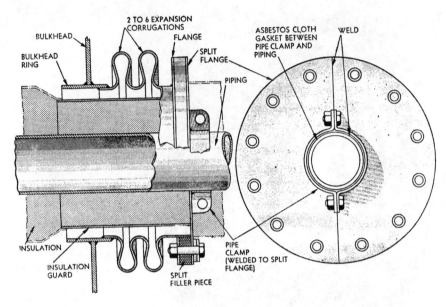

Figure 3-14. Wall Penetration for High Pressure Steam Pipe.

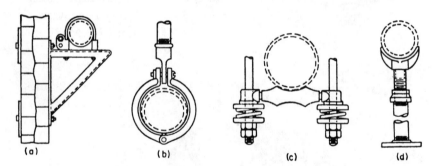

Figure 3-15. Typical Steam Pipe Support Methods.

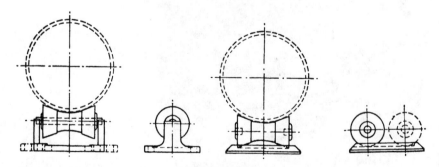

Figure 3-16. Steam Pipe Roller Supports.

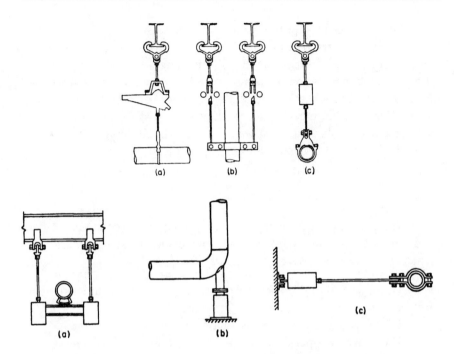

Figure 3-17. Spring Hangers and Braces for Steam Pipe.

conditions. Several methods of support are shown in Figure 3-15 (a) and Figure 3-9 (a). Figure 3-15 (a) is a typical above ground configuration, and Figure 3-9 (a) is for a direct burial application.

Table 3-12. Maximum Spacing of Pipe Supports at 750°F

Nominal pipe size, in.	1	1-1/2	2	3	4	6	8	10	12	14	16	18	20	24
Maximum span, ft.		7	9	10	12	14	17	19	22	23	25	27	28	30	32

*This tabulation assumes that concentrated loads, such as valves and flanges, are separately supported. Spacing is based on a combined bending and shear stress of 1,500 psi when pipe is filled with water; under this condition, sag in pipeline between supports will be approximately 0.1 in.

Several types of roller supports are shown in Fig, 3-16. Note that in Figure 3-16 (c) that a roller support plate is used. This must also be lubricated regularly with a high temperature lubricant.

Other types of supports shown in Figure 3-17 such as sway bars and spring hangers require only minimal lubrication and usually no other maintenance. When installing spring hangers, the cold spring locks shown on the installation drawings should be noted on the hanger itself. When the system is first started all the hot loads are noted and approved by the commissioning engineers, they should also be noted on the hanger. During subsequent start-ups and scheduled outages, the hangers should be checked to insure the piping is still in alignment.*

PACKING AND GASKETS

The primary purpose of packing is to seal joints of machinery and equipment against leakage. There are four principle types of joints with which packing is employed as follows:

1. Those which slide, such as piston rods, pistons, and expansion joints,
2. Those which rotate, such as shafts,
3. Those which operate helically and intermittently, such as valve stems; and
4. Those which are fixed, such as flanges and bonnets.

These moving joints offer the greatest difficulty, since it is necessary to seal against leakage without causing excessive friction, undue wear of the moving part, and early failure of the packing.

In the days of low pressure power plants, packing materials and packing spaces were given but little consideration. In many of the installations, packing spaces and methods had the appearance of being emergency measures added to the machinery to correct a situation which was giving trouble. The original packing materials which gave the best performance, were in the form of hemp rope and leather. Later, to these were added textile fibers and fabrics, asbestos fibers; then rubber, a combination of rubber and textiles; and later, metals. Today, packing can be chosen from almost endless variations of these basic materials.

Packing is inserted in stuffing boxes (Figure 3-18) which, generally speaking, consist of cavities so located around valve stems, rotating shafts, or reciprocating pump rods that they can be "stuffed" with a

---

*Expansion trams are used on many steam consuming machines.

packing material which is compressed as necessary and held in place by flanged and bolted or threaded gland bushings.

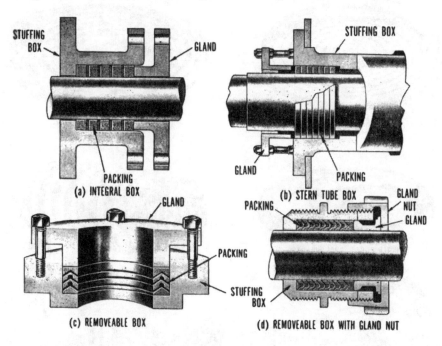

Figure 3-18. Packing and Packing Gland Arrangements.

Coils, rings, spirals, and molded rings (Figure 3-19) are the most common forms in which packing is prepared commercially. The materials shown are only a few of the many employed. For a long time high pressure asbestos rod packing was used almost exclusively for sealing moving steam joints (rods, valve stems, etc.). Its use for sealing the joints of high pressure, high temperature valve stems has been superseded by wire inserted square braided asbestos (up to 400 psi and 700°F) and by plastic non-metallic, asbestos packing encased in a braided wire covering (up to 650 psi and 850°F). The square braided asbestos packing is composed of 90 percent asbestos with brass or copper wire inserted yarns, and a high temperature lubricant. The plastic non-metallic, asbestos and wire jacket type is composed of a plastic core of asbestos fibers, graphite and a binder, encased in a braided monel wire jacket. These metal inserts or jackets tend to act as a bearing surface for the packing, cutting down friction and increasing the life of the packing.

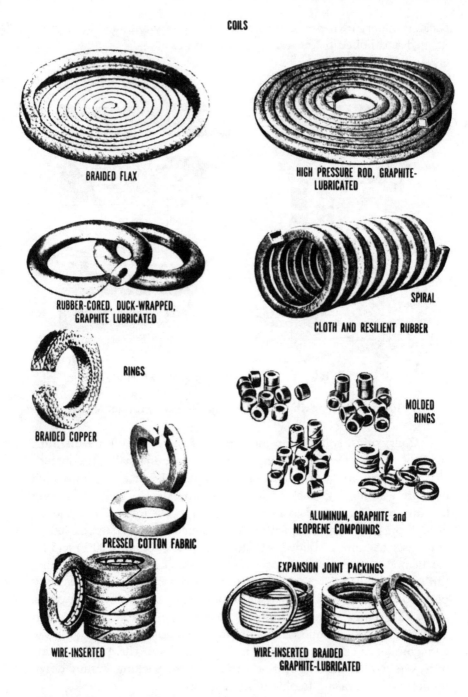

Figure 3-19. Assorted Types of Packing.

Gaskets

The sealing of fixed steam joints, until recent years, was performed in a satisfactory manner through the use of gaskets made from compressed asbestos sheet packing. This packing was composed of approximately 85 percent asbestos fibers and 15 percent rubber compounds. Asbestos is no longer used because of its health hazard; however, with higher temperatures and pressures in steam service, it has outgrown its usefulness as an effective gasket material anyway. For steam and condensate pipe the use of metallic or semi-metallic gaskets are considered necessary.

At the present time, there are three types of metallic or semimetallic gaskets in use in industrial and commercial plants:

1. **A flat ring or plain faced gasket** (Figure 3-20). These metal gaskets are made of monel metal or soft iron to specified shapes and sizes. A variation of this type is the ring gasket shown.

2. **A serrated-face gasket** also made from monel or soft iron (Figure 3-21). The raised serrations help to make a better seal at the piping flange joints and give the gasket some resiliency. A variation of the serrated gasket also shown is the expanding serrated gasket shown is the expanding serrated gasket (Figure 3-21). When this gasket is used, line pressure acts between the plates to force the serrated faces tighter against the adjoining flange. This is a relatively new type having limited service experience.

FLAT RING GASKET

3. **The semi-metallic, spiral-wound gasket,** (Figure 3-22), composed of alternate layers of plies of dove-tailed metal ribbon and strips of fiber spirally wound, ply upon ply,

FLAT FULL-FACE GASKET

Figure 3-20. Planed-face Metal Gasket.

**Figure 3-21. Serrated
Metal Gasket.**

until the desired diam-
eter is obtained. The
spiral wound gaskets
are so manufactured
that, with proper bolt
tension, the gasket will
compress to about
0.135" in thickness.

(a) SINGLE-PLATE TYPE

A property which
a gasket should possess
to maintain a tight joint
is resiliency. In tests, it
was found that the
plain-faced type gasket
depended entirely upon
the bolt stress for main-
taining tightness of the
joint; hence, when the
bolt stress was reduced

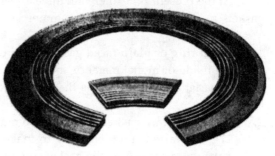

(b) EXPANDING (DOUBLE-PLATE) TYPE

below a certain amount through elongation of the bolts and through
sudden change in temperature conditions, leakage occurred from the
joint. The same was found to some extent with the ordinary serrated face
type gasket. The spiral wound, semi-metallic type gasket performed sat-
isfactorily, and the bolt stresses required to obtain tightness of the joint
were less than those required with the plain and serrated types of gasket.
Bolt stresses required for the expanding serrated type gasket should be
less.

Pipe insulation is covered in a separate chapter. It should be noted,
however, Occupational, Safety, and Health Act (OSHA) authorities are
enforcing strict regulations on the use of health hazardous materials that
were not considered dangerous to humans until ten to fifteen years ago.
In particular, asbestos is the prime offender attracting OSHA attention in
piping insulation. It is for the most part, being replaced by calcium silicate insulation and other insulation not containing asbestos.

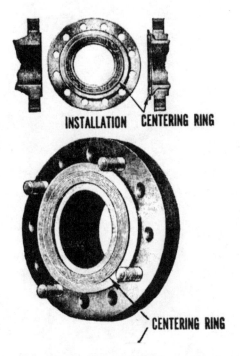

INSTALLATION   CENTERING RING

CENTERING RING

Figure 3-22. Metallic-fiber Spiral Wound Gasket.

CLOSURE

Chapter 2 has studied stem flow in pipe along with the economics and general techniques of steam distribution. Chapter 3 has studied the physical properties and dimensioned constraints of piping whether steam or condensate return, which the engineer can expect to encounter. A following chapter will study the specific problems operations engineers can expect in condensate return piping.

Chapter 4

Steam Piping Layout Excavations and Enclosures

S team distribution system construction falls into two categories: Factory prefabricated in lengths suitable for shipping, and then making final assembly in the field. Some field construction may also be used when short additional lengths of piping is needed to reach a new steam consumer, or to make a repair.

Figure 4-1 is a cross-section of a factory assembly. This type of steam pipe assembly is what would be used in a trench or buried in the ground. It consists of the pipe, preformed insulation, insulation cover, and steel spacers to accommodate an air space under the outer shell of the assembly. This outer shell is called a conduit. Until recently, the conduit has been constructed of steel. However, plastic conduits are being used in place of steel in many installations except where above-ground loads could damage or collapse the plastic conduit. Still where above-ground loads are low, they are useful, because they do not corrode or decompose as easily as steel.

Where the piping is to be used in a cold climate heat tracing should be installed to prevent freezing of the condensate or moisture collected in the system. This is particularly true of pipe installed above the freeze line. Tracing can be small diameter steam tubing wrapped around the steam and/or condensate pipe. This assembly is usually selected where there is another source of steam. In other applications, electrical heat tracing may be more desirable.

When piping is installed above ground, the construction is very similar to under ground assemblies. The main difference is that the conduit is usually corrugated aluminum sheet wrapped around the rigid preformed insulation. When using the light weight aluminum conduit,

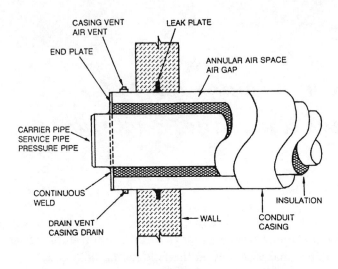

Figure 4-1. Conduit Systems Enclosures.

the supporting structure can be less costly. Another cost effective product is plastic insulation It serves well outdoors where other conduits may not be water proof. Plastic resists water penetration to the steam pipe thus maintaining its insulating qualities while preventing pipe corrosion.

Probably the most desirable pipe assembly is the one installed in a basement or sub-basement in a building or large interconnected industrial or commercial establishments. Typical cross-sections of steam conduits are shown in Figure 4-2. It closely resembles outdoor above-ground piping.

Tunnels constructed to accommodate steam and condensate may also be used to accommodate other utility services as well. Potable water, well insulated and grounded electrical cable. Some engineers may find it necessary, with community consent, to install drain water or sewage in the tunnels. This however is a violation of most community and state building regulations because of the proximity of sewer pipe and potable water.

Although, tunnels are expensive to construct they are easiest to own and operate. Where construction capital permits the building of tunnels will produce the best return on investment over other types of steam and condensate piping systems

Figure 4-5 is a cross section of a typical utility tunnel. It shows the hangers on the walls and the steam and condensate return lines. Electri-

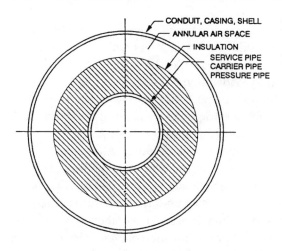

Figure 4-2. Conduit System with an Annular Air Space

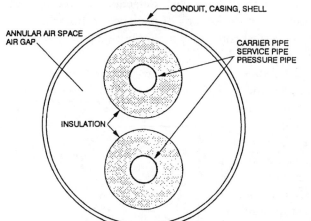

Figure 4-3. Conduit System with Two Carrier Pipes and an Air Space.

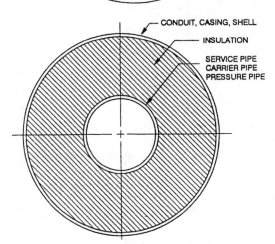

Figure 4-4. Conduit System with single Carrier Pipe and No Air Space.

cal conduits are in racks high up in the tunnel. Should the builder wish to install other utilities in the tunnel, it is recommended they be installed as low as possible in the tunnel except for potable water. The deck of the tunnel should be fitted with a wooden floor. The floor should be a checkerboard construction to offer the operations and maintenance men some degree of safety from burns and electric shock while in this wet environment. Non-metallic hard hats, rubber boots and rubber gloves should be worn by all persons whether inspecting or working in tunnels.

Only the above-ground system and the walk-through tunnel can be inspected and maintained easily. The shallow concrete trench system requires lifting of concrete lids, and other underground systems require excavation for repairs. Leak detection is easy in the above-ground systems, the walk-through tunnel system, and the concrete surface trench system. In other underground systems, leak detection is more difficult and often involves excavation of portions of a suspected branch.

The above-ground system consists of a distribution pipe, insulation surrounding the pipe, and a protective jacket that surrounds that insulation and which may have an integral vapor retarder. Figure 4-4 is a cross section of an above-ground conduit. In heating applications, the vapor retarder is not needed; however, a watertight jacket is required to keep storm water out of the insulation. The jacket material can be aluminum, stainless steel, plastic, a multi-layered fabric and organic cement, or a combination of these.

The structural supports for this system are typically wood, steel, or concrete columns. Pipe expansion and

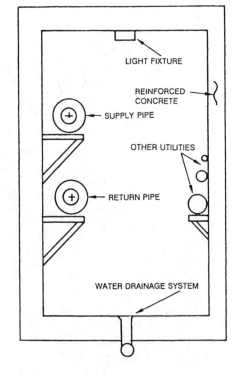

Figure 4-5. Walk-through Tunnel.

contraction is taken up in loops, elbows, and bends. Mechanical expansion joints may be used, but their maintenance must be considered. Welded joints have a higher level of integrity, but bolted joint assemblies may have lower installation costs.

The above-ground system usually has the lowest first cost and is the easiest to inspect and maintain; therefore, it has the lowest life cycle cost. It is frequently the system that all other systems are compared to when making a decision about which type of system to use. Drawbacks include poor aesthetics and the hazard of being struck by vehicles and equipment; also, the working medium can freeze in winter if circulation is stopped. To prevent this, electric or steam tracing is required.

Factory prefabricated systems (Figures 4-1 through 4-4) consist of a pipe that carries the working medium, insulation around that pipe, and a protective jacket surrounding the insulation This pipe is also called a carrier pipe, service pipe, or pressure pipe, and is designated as either air gap or not air gap. Systems with an air gap between the insulation and the casing are usually used when the working medium is above 250°F. The perimeter jacket is called a casing, shell, or conduit. The term conduit is often used to denote the entire factory prefabricated systems. These systems are designed so that the air space is continuous throughout the line from one manhole to the next. If water enters the conduit, it can be drained from the system. Once the water is drained and the required repairs made, the insulation can be dried in place by blowing air through the conduit and thus restoring it to its original thermal efficiency. Most underground systems will be flooded several times during their service life, even on reasonably dry sites.

A reliable water removal system should be used in the valve vaults and for systems that have an air space between the insulation and casing. Two schemes are used to ensure satisfactory insulation performance throughout the life of the system. One scheme uses insulation that can survive flooding. An annular air space around the insulation allows air to pass through the system to dry the insulation. Some closed cell, high-temperature insulations essentially do not absorb water, and little if any drying is required. Any of these systems must withstand earth loads and expansion and contraction forces due to temperature changes.

The second scheme used to retain insulation efficiency is to enclose the insulation in a waterproof envelope. Some manufacturers form the insulation envelope in each prefabricated section; others use watertight casing field joints. The first costs of these types of systems is more than that in the above-ground system but less that in the walk-through tunnel system.

Metal conduits are coated to protect against corrosion. In addition to the outside coating, cathodic protection of the conduit metal casing is used to extend the casing life in all but the most noncorrosive soils. Corrosion on the inside of a steel casing may develop but is of minimum consequence if water is kept out of the air gap.

Steam distribution systems, require careful design and construction to ensure the lowest maintenance and longest life. Proper slope, water drainage, flawless pipe and casing welds, proper supports, and anchors are essential for economic life. The designer must clearly delineate on the drawings and in the specification those parts of the system the conduit manufacturer must design. Unless a leak detection system is installed, these systems essentially cannot be inspected between vaults. If a problem develops in the system, excavation is required to repair it. Locating the excavation point may be difficult, even though some commercially available leak detection systems may narrow down the excavation area.

Products with nonmetallic casings are available, but, during design, carefully consider moisture presence and thermal expansion to avoid deterioration of the plastic. Not all plastic casings can tolerate the range of temperatures that can vary from the ground water temperature during wet site conditions and the carrier pipe with high temperatures. A site survey of installed systems is recommended to establish the overall performance and system life of a particular system or product with special attention given to the soil type and groundwater conditions.

The walk-through tunnel system (Figure 4-5) consists of a field-erected tunnel that is large enough for a person to walk through after the distribution pipes are in place. The top cover of the tunnel is covered with earth. The preferred construction material is reinforced concrete. Masonry units and metal preformed sections have been used with less success due to groundwater leakage and metal corrosion problems. Distribution pipes are supported from the tunnel wall or floor with conventional pipe supports. Walk-through tunnels generally cannot accommodate thermal expansion movement with pipe loops and, therefore, expansion joints are required. These are described in a later chapter. Because groundwater will penetrate the tunnel top and walls, a water drainage system must be provided. To ease inspection and maintenance, electric lights and electric service outlets can be provided. Tunnels usually have the highest first cost of all systems; however, they can have the lowest life cycle cost because (1) maintenance is easy, (2) construction errors can be corrected easily, and (3) they can be placed in areas where right-of-way is near the ground surface. In some cases, water or electrical utilities may be included in the tunnel, which further improves the economics and is an alternative in cold regions where it is necessary to protect water systems from freezing.

The shallow concrete surface trench system (Figure 4-6) is only partially buried. The floor is usually about 3 ft. below the surface grade and is only wide enough for the carrier pipes, insulation, clearance to allow for pipe movement, and possibly enough room to stand on the floor. Generally, the trench is at least as wide as it is deep. The floor and walls are usually cast-in-place reinforced concrete. Precast concrete floors and wall sections have not been successful because of the large number of oblique joints with non-standard sections required to follow the surface topography and to slope the floor to drain. Because this system controls the storm and groundwater entering it, the floor should slope toward a drainage point. The top is constructed of precast or cast-in-place reinforced concrete that protrudes slightly above grade to serve as a sidewalk.

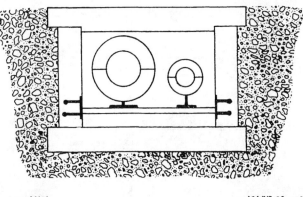
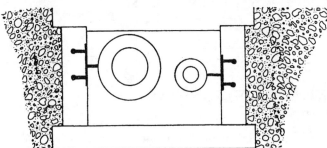

Figure 4-6. Shallow Concrete Trenches.

Pipes should be supported from cross beams attached to the side walls, because floor-mounted pipe supports tend to corrode and interfere with water drainage. This mounting system allows the distribution pipes to be assembled before lowering them on the pipe supports. These supports should be galvanized or oversized to account for corrosion. The carrier pipes, pipe supports, expansion loops and bends or expansion joints, and the insulation jacket are similar to those used in above-ground systems, with the exception of the pipe insulation. Insulation is covered with a metal or plastic jacket to protect it from storm water that can enter through the top cover joints. Inspection ports of about 12 in. diameter in the top covers, at key locations to permit inspection from outside. All replaceable elements such as valves, condensate pumps, steam traps, strainers, sump pumps, and meters should be located in valve vaults installed at strategically located points along the trench. Aesthetics are a large part in the design of trench systems. The first cost and life cycle cost of trenches are among the lowest for underground systems.

Deep buried trench systems (Figure 4-7) are used on sites where the groundwater elevation is lower than the system elevation. These systems are only large enough to contain the distribution system piping, the pipe insulation, and the pipe support system. The system is covered with earth and is essentially not, maintainable between valve vaults without

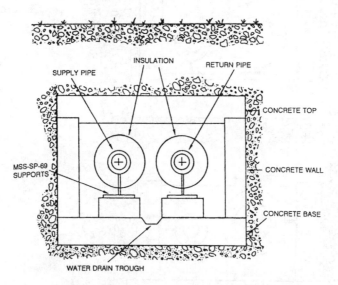

Figure 4-7. Deep Buried Trench/Tunnel

excavation. Therefore, select materials that will last for the intended life of the system should be used.

The shallow concrete surface trench system is covered with earth and sloped independent of the topography. Construction of this system is typically started in an excavated trench by pouring a cast-in-place concrete base that slopes so that groundwater can drain to the valve vaults. The slope must also be compatible with the pipe slope requirements of the distribution system. The concrete base has provisions for supports for the distribution pipes, the groundwater drainage system, and the mating surface for the top cover. The pipe supports, the distribution pipes, and the pipe insulation should be installed before installing the top cover. The cast-in-place concrete bottom should contain the groundwater drainage system, which may be a trough formed into the concrete bottom or a perforated drainage pipe either cast into the concrete base or located slightly above the base. Typically, the cover for the system is either cast-in-place concrete or preformed concrete or half-round clay tile sections. The top cover should mate to the bottom as tightly as possible to limit the entry of groundwater.

Poured insulation systems (Figure 4-8) usually encase the distribution system pipes in an envelope of insulating material, and the insulation envelope is covered with a thick layer of earth as required to match existing area condition. This system is used on sites where the groundwater is below the system elevation. Like other underground systems, the design must accommodate groundwater flood conditions. The insulation material must (1) serve as a structural member to support the distribution pipes and earth loads; (2) prevent ground water from entering the interior of the envelope; (3) allow the distribution pipes to expand and contract axially as the pipes change temperature; and (4) at elbows, expansion loops, and bends, the insulation must allow cavities to form around the pipe without significant distortion while still retaining the required structural load-carrying capacity. Special attention must be given to corrosion of metal parts and water penetration at anchors and structural supports that penetrate the insulation envelope. When the distribution pipes are hot, the heat loss tends to drive moisture out of the insulation as steam vapor.

Construction of this system is started by excavating a trench with a bottom slope that matches the desired slope of the distribution piping. The width of the bottom of the trench is the same as the insulation envelope. The distribution piping is assembled in the trench and supported

Figure 4-8. Poured Insulation Envelope Covered with Earth.

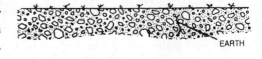

by the anchors and blocks that should be removed when the insulation is poured in place. The trench bottom and sides, wood, or sheets of plastic can be used as the form to hold the insulation. The insulation envelope is then covered with earth.

Insulating concrete (Figure 4-9) is a low-mass aggregate with additives in a mixture of cement. This insulation material has some structural properties to resist earth loads and to support the

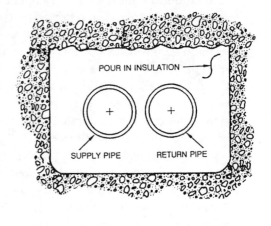

distribution piping. Insulating concrete may interfere with proper operation of rubber ring couplings; distribution pipe joints should be welded and a separating fluid agent placed around the pipes so that they can freely expand and contract axially.

Steel plate anchors are embedded in a thickened and reinforced base slab and welded to the distribution pipe. The design of anchor

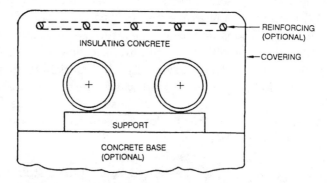

Figure 4-9. Insulated Concrete Poured Envelope System Installed above Ground.

points, loops, and bends is similar to other welded systems. Voids built into the corners of elbows, loops, and bends allow for pipe movement. Internal drains and vents remove construction water and other unexpected water, and also act as leak detectors.

Hydrocarbon powders (Figure 4-10) are thermoplastic materials composed of low-mass aggregate with asphalt binders. This insulation is poured around the pipe and its supports, which are fastened to a concrete base. A continuous, waterproof membrane helps keep water out of this system. Weld distribution pipe joints are used because the insulation material interferes with mechanical joints. Special care must be given to the design for lateral pipe movement at elbows, loops, and bends, because this material hardens as the number of expansion cycles increases and limits the amount of pipe movement that can be tolerated.

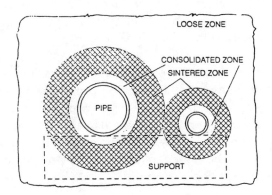

Figure 4-10. Poured Hydrocarbon Insulation Envelope System.

Hydrophobic powders are treated to make them water repellent. As a result, water will not dampen the powder, which may prevent water from entering the insulation envelope. Hydrophobic powders are installed in the same way as hydrocarbon powder.

THERMAL CONSIDERATIONS

Thermal Design Conditions

Three thermal design conditions must be met to ensure satisfactory system performance:

1. The "normal" condition used for the life cycle cost analysis determines appropriate insulation thickness. Average values for the temperatures, burial depths, and thermal properties of the materials are used for design. If the thermal properties of the insulating material are expected to degrade over the useful life of the system, appropriate allowances should be made in the analysis.

2. Maximum heat transfer rate determines the load on the central plant due to the distribution system. It also determines the temperature drop that determines the delivered temperature to the consumer. For this calculation, the thermal conductivity of each component must be taken at its maximum value, and the temperatures must be assumed to take on their extreme values, that would result in the greatest temperature difference between the carrier medium and the soil or air. The burial depth will normally be at its lowest point for this calculation.

3. During operation, none of the thermal capabilities of the materials may exceed design conditions. To satisfy this objective each of the components and the surrounding environment must be examined to determine if thermally induced damage is possible. A heat transfer analysis may be necessary in some cases. The conditions of these analyses must be chosen to represent the worse case conditions for the component being examined.

For example, in assessing the suitability of a coating material for a metallic conduit, assume that the thermal insulation is saturated, that the soil moisture is at its lowest probable level, and that the burial depth is maximum. These conditions, combined with the highest anticipated pipe and soil temperatures, give the highest conduit surface temperature to which the coating could be exposed.

Heat transfer in buried systems is influenced by the thermal conductivity of the soil and by the depth of burial, particularly when the insulation has low thermal resistance. Soil thermal conductivity changes significantly with moisture content; for example, the soil thermal conductivity ranges from 0.083 Btu/h-ft-F for dry soil to 1.25 Btu/h-ft-F for wet soil conditions.

Thermal Properties

Generally, the designer must rely on manufacturers' specifications and handbook data to obtain proper insulation design data.

Insulation provides thermal resistance against heat loss in steam distribution systems. Thermal properties and other characteristics of insulation normally used in thermal distribution systems are given in Table 4-1.

If an analysis of the soil is available or can be done, the thermal conductivity of the soil can be estimated. The thermal conductivity factors in Chapter 6, Tables 6-1 and 6-2, may be used as an approximation where detailed information on the soil is not known. Because dry soil is rare in the United States, a low moisture content should be assumed only where it can be validated values of 0.8 to 1 Btu/h-ft-°F are commonly used where soil moisture content is unknown.

Table 4-1. Comparison of Commonly Used Insulations in Underground Piping Systems.

Item	Calcium Silicate	Urethane Foam	Cellular Glass	Preformed Glass Fiber	Loose Glass Fiber	Insulating Concrete	Hydrocarbon Powder	Hydrophobic Powder
Thermal conductivity, Btu/h•ft•°F								
at 100°F	0.028	0.013	0.033	0.022	0.027	0.050 to 0.3330	.050 to 0.067	0.044 to 0.067
at 200°F	0.031	0.014	0.039	0.025	—	—	—	0.050 to 0.068
at 300°F	0.034	—	0.046	0.028	—	—	—	0.055 to 0.071
at 400°F	0.038	—	0.053	—	—	—	—	—
Density, lb/ft³	10 to 14	1.5 to 4	9 to 10	3 to 7	2 to 11	20 to 30	45 to 55	35 to 65
Maximum temperature, °F	1200	260	800	370 to 500	1000	1800	300 to 450	500
Compressive strength, psi	75 to 165	20 to 50	100	5	0	100 to 150	70 to 85	70 to 85
at compression	5%	5%	—	—	—	—	5%	5%
Moisture absorption	Great	Slight	Nil	Great	Great	Moderate	Moderate	Nil
Effect on k factor	Large	Slight	Slight	Large	Large	Slight	Large	Slight
Resistance to boiling	Good	Poor	Fair	Good	Good	Good	Poor	Poor
Recovery-drying	Good	Poor	Good	Good	Good	Good	Poor	Poor
Resistance to abrasion	Fair	Fair	Fair	Poor	Nil	Good	Poor	Good
Resistance to vibration	Fair	Good	Poor	Good	Poor	Fair	Fair	Good
Stability: shrink, shock	Good	Good	Fair	Excellent	Excellent	Good	Good	Good
Combustible	No	Yes	No	Yes	No	No	Yes	Yes

1. The descriptive terms in this table are only approximations, since any insulation and the conditions under which it is used will vary.

2. The thermal conductivities for these materials and for urethane foam used in the trench are for d– material. Since foam may absorb groundwater, and water vapor may eventually enter the interstices between powder particles, the ultimate thermal conductivity values are difficult to estimate.

Chapter 5

Steam Distribution System Protection

Piping systems require a substantial amount of protection. Some of the potential hazards include:

- Corrosion
- Water hammer
- Freezing
- Structural damage
- Personnel Safety
    - Burns
    - Suffocation
    - Drowning
    - Electrocution
    - Noise

The following discusses each of these hazards.

CORROSION

Corrosion can result in internal damage to steam and condensate piping. Impingement and overly acid condensate in steam or return lines can do havoc in a piping network. This occurs regardless of external corrosion. Steam and condensate pipelines can be protected by treating the water in the boilers from which the steam emits. Chemicals are used to control pH and carbon dioxide. There are two common destructive chemicals and compounds that can cause condensate in piping, and internal corrosion. They are oxygen and carbon dioxide. Volatile amines and phosphate are used to prevent corrosion by neutralizing the acid in the condensate. Some mechanical devices are also used to prevent collection of carbon dioxide in the steam and condensate systems. These are: deaerators and air venting steam traps. Impingement of condensate in the steam or steam carried into condensate piping from steam consuming machinery and steam and condensate piping protection can be prevented. See Figure 5-1. It can be best treated by configuring a nozzle in pipelines where they enter condensate return lines (see Figure 5-2). This sketch shows how moisture impingement on the bottom of condensate lines can be corrected. Experience in caring for piping will dictate the cost effectiveness of the level of corrosion protection best suited to your plant.

In addition to chemistry and mechanics, electrolytic corrosion can also be troublesome. It is controlled in several ways, one way is through cathodic protection.

Cathodic Protection

Corrosion of metals often takes place in the presence of liquids which act as a bridge between dissimilar metals or between two portions

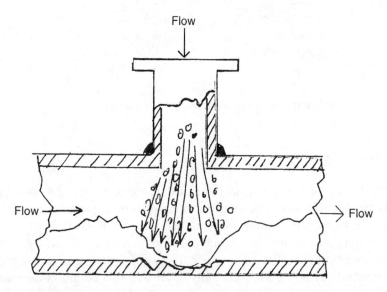

Figure 5-1. Moisture and Particle Impingement.

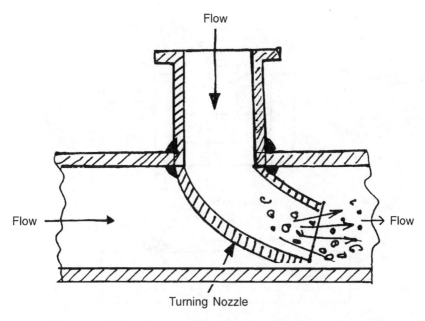

Figure 5-2. Moisture and Particle Impingement Protection.

of metal, thus allowing an electric current to flow and it is this current which causes the damage. For example, when in contact with soil or water, the corroding surface forms galvanic (electric) cells, these being identical in principle with the ordinary dry cell battery used in many appliances. See Figure 5-3 a. At the more reactive points of the metal surface, atoms of the metal are converted into positively charged ions by loss of electrons and at these points (anodes) the current flow is from the metal to the electrolyte (Figure 5-3b) and the metal is lost from the surface at all such points.

The difference in potential which forms the electro-chemical cells and corrode the metal are caused by one, or a combination of, the following factors:

Different metals in the same electrolyte such as cooper and steel, cast iron and steel, etc.;

A single metal whose internal structure has very small dissimilarities;

The same or different metals exposed to mixed environments, e.g., metal in contact with both sand and mud;

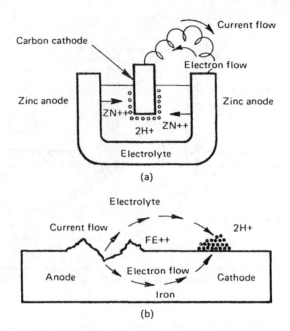

Figure 5-3. Principle of Cathodic Protection.

When current flows from the electrolyte to the metal (the cathodes), there is no metal loss and the general name, 'Cathodic Protection' is given to the methods used in such systems.

Prevention of Corrosion

To prevent corrosion arising from electro-chemical reaction, current must be prevented from leaving the metal and this is done by making the metal surfaces cathodic in relation to an auxiliary electrode—this is cathodic protection, where electrodes (anodes) are introduced into the electrolyte (liquids, e.g., water or soil/water mixture) to cause a current to flow from the latter into the entire metal surface, which then becomes cathodic, no current flows from it, metal loss to the electrolyte ceases and corrosion is prevented. Two methods are employed to produce a current flow from the electrolyte into the metal requiring protection, i.e.

(a)   Using self-energizing 'galvanic' electrodes where the anodes are magnesium, zinc, or aluminum alloys which are slowly consumed in the process of producing the protection current (sacrificial method).

(b)   Impressing a reverse current through the electrolyte from auxiliary anodes by use of a low voltage d.c. source to feed auxiliary anodes of various alternative materials which impress a current into the electrolyte and on to the metal surface of the piping. The anodes, therefore, act as conductors and not as energy sources.

The second method is generally used with large installations requiring large currents and eliminates the need for anode replacement.

Application To Pipelines

Cast-iron and steel pipes require cathodic protection when buried underground or when under water and to supplement external protective coatings. Protection can be applied with galvanic anodes (generally magnesium) or with impressed current and silicon iron anodes. The choice is governed by ground resistivity, whether other pipelines are buried nearby and the availability of a d.c. supply or other power source. Underwater pipes are generally serviced with impressed current.

Typical diagrammatic layouts of galvanic and impressed cathodic protection systems are shown schematically in Figure 5-3 and Figure 5-4. Types of anodes used are made from magnesium which is normally the first choice for pipelines, zinc used in low resistivity applications, and also aluminum, silicon-iron, and platinized titanium.

WATER HAMMER

With a better understanding of the nature and severity of the water hammer problem, we can avoid its destructive forces. This greater understanding should also help with the introduction of more preventative measures into system designs and installations, which will help provide maximum safety for personnel, lower maintenance cost and reduce system downtime.

Where Water Hammer Occurs

Water hammer can occur in any water supply line, hot or cold. Its effects can be even more pronounced in heterogeneous or bi-phase sys-

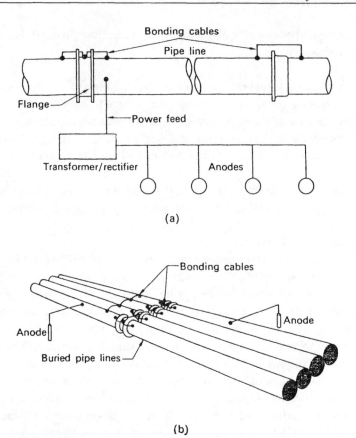

Figure 5-4. Diagrammatic Arrangements of Cathodic Protection; (a) Impressed System; (b) Galvanic System.

tems. Bi-phase systems carry water in two states, as a liquid and as a gas. Such a condition exists in a steam system where condensate coexist with live or flash steam: in heat exchangers, tracer lines, steam mains, condensate return lines and in some cases, pump discharge lines

Effects of Water Hammer

Water hammer has a tremendous and dangerous force that can collapse floats and thermostatic elements, over stress gauges, bend mechanisms, crack trap bodies, rupture fittings and heat exchange equipment and even expand piping. Over a period of time, this repeated stress of the pipe will weaken it to the point of rupture. Water hammer is not always accompanied by noise. Some types of water hammer, resulting from localized abrupt pressure drops, are never heard. The consequences, however, may be severe.

Conditions Causing Water Hammer

Three conditions have been identified that can cause the violent reactions known as water hammer. These conditions are hydraulic shock, thermal shock and differential shock.

Hydraulic Shock

Visualize what happens at home when a faucet is open. A solid shaft of water is moving through the pipes from the point where it enters the house to the faucet. This could be 100 pounds of water moving at 10 feet per second, about 7 miles per hour. When the faucet is shut suddenly, it is like a solid object hitting the valve This shock wave is similar to a hammer hitting a piece of steel. The shock pressure wave of about 600 psi is reflected back and forth from end to end until the energy is dissipated. Similar action can take place in the suction or discharge piping of a pump when the pump starts and stops if check valves are in the line. Slow closure of the valve or faucet and slow-closing check valves along with water hammer arrestors are solutions to these problems. If the column of water is slowed before it is stopped, its momentum is reduced gradually and, therefore, damaging water hammer will not be produced.

Water hammer arrestors, if correctly sized, placed and maintained, will reduce water hammer by providing a controlled expansion chamber in the system. As the forward motion of the water column in the pipe is stopped by the valve, a portion of the reversing column is forced into the water hammer arrestor. The water chamber of the arrestor expands at a rate controlled by the pressure chamber and gradually slows the column, preventing hydraulic shock. If a check valve is used in a system without an arrestor, excessive pressure may be exerted on the system when the reversing water column is violently stopped by the check valve. Equipment damage can occur.

Thermal Shock

In a bi-phase system steam bubbles may become trapped in pools of condensate in a flooded main, branch or tracer line, as well as in heat exchanger tubing and pumped condensate lines. Since condensate temperature is almost always below saturation, the steam will immediately collapse.

One pound of steam at 0 psig occupies 1,600 times the volume of water at atmospheric conditions. This ratio drops proportionately as the pressure increases. When the steam collapses, water is accelerated into the resulting vacuum from all directions. This happens when a steam trap discharges relatively high-pressure flashing condensate into a pump discharge line.

Another cause of water hammer is lack of proper drainage ahead of a steam control valve. When the valve opens, a slug of condensate will enter the equipment at a high velocity, producing water hammer when it impinges on the inside of the pipe. In addition to this, the mixing of the steam that follows with the relatively cool condensate will produce water hammer from thermal shock. This condition can be corrected by dropping the supply riser as shown in Figure 5-5.

Water hammer can also occur in steam mains, condensate return lines and heat exchange equipment where steam entrapment can take place (Figure 5-6). A coil constructed and installed as shown here, except with just a steam trap at the outlet, permits steam from the control valve to be directed through the center tube(s) first. Steam then gets into the return header before the top and bottom tubes are filled with steam.

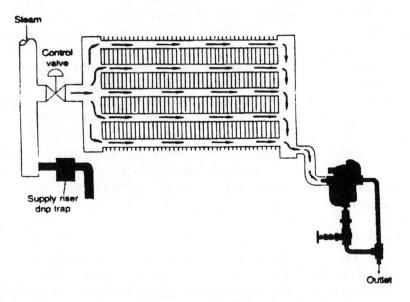

Figure 5-5. Steam Trap and a Condensate Controller Connected to a Space Heater.

Consequently, these top and bottom tubes are fed with steam from both ends. Waves of condensate are moved toward each other from both ends, and steam can be trapped in the waves.

Water hammer results from the collapse of this trapped steam. The localized sudden reduction in pressure caused by the collapse of the steam bubbles has a tendency to erode the pipe and tube interiors. Oxide layers that otherwise would resist further corrosion are removed, resulting in accelerated corrosion.

One means of overcoming this problem is to install a condensate controller, which maintains a positive pressure differential across all the tubes (Figure 5-5). A condensate controller provides a specialized purge line, which assures a positive flow through the coil at all times.

Differential Shock

Differential shock, like thermal shock, occurs in bi-phase systems. It can occur whenever steam and condensate flow in the same line but at different velocities, such as in condensate return lines. In bi-phase systems, velocity of the steam is often 10 times the velocity of the liquid. If condensate waves rise and fill a pipe, a seal is formed with the pressure of the steam behind it (Figure 5-6). Since the steam cannot flow through the condensate seal, pressure drops on the downstream side. The condensate seal now becomes a "piston" accelerated downstream by this pressure differential. As it is driven downstream it picks up more liquid, which adds to the existing mass of the slug, and the velocity increases. If this slug of condensate gains high enough momentum and is then required to change direction, for example at a tee, elbow or valve, great damage can result. Since a bi-phase mixture is possible in most condensate return lines, correct sizing becomes essential.

Condensate normally flows at the bottom of a return line. It flows because of the pitch in the pipe and also because of the higher velocity flash steam above it, dragging it along. The flash steam moves at a higher velocity because it moves by differential pressure.

Flash steam occurring in return lines, due to the discharge of steam traps, creates a pressure in the return line. This pressure pushes the flash steam at relatively high velocities toward the condensate receiver, where it is vented. Condensing of some of the flash steam, due to heat loss, contributes to this pressure difference and amplifies the velocity. Because the flash steam moves faster than the condensate, it makes waves. As long as these waves are not high enough to touch the top of the pipe and

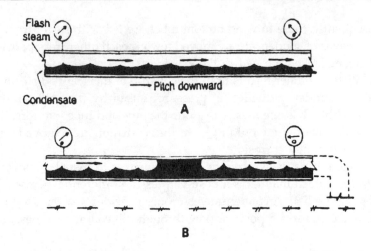

Figure 5-6. Illustration of How Water Hammer Happens.

so do not close off the flash steam's passageway, all is well.

Damaging water hammer similar to that just described is experienced also when elevated heat exchange equipment is drained with a long vertical drop to a trap. Condensing of steam downstream of the slug produces a drop in pressure differential across the slug. This pressure differential, together with gravity, accelerates the slug downward. It does not take much of this strong force to cause damage to the piping system. Back-venting the trap to the top of the vertical drop can correct this problem (Figure 5-7). Since condensation is what produces the acceleration, uninsulated pipes and their appurtenances would be expected to suffer greater damage than those with insulation.

To control differential shock, the condensate seal must be prevented from forming in a bi-phase system. Steam mains must be properly pitched, condensate lines must be sized and pitched correctly, and long vertical drops to traps must be back-vented. The length of lines to traps should be minimized, and the pipes may have to be insulated to prevent water hammer.

Steam main drainage is one of the most common applications for steam traps. It is important that water is removed from steam mains as quickly as possible, for reasons of safety and to permit greater plant efficiency. A build-up of water can lead to water hammer, capable of fracturing pipes and fittings. When carried into the steam spaces of heat exchangers, it simply adds to the thickness of the condensate film which reduces heat transfer. Inadequate drainage leads to leaking joints, and is

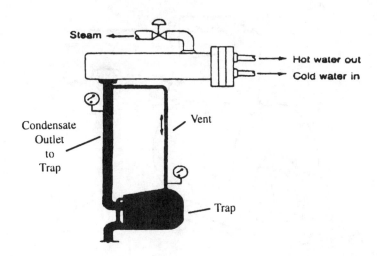

Figure 5-7. Condensate Controller Connected to a Heater with a Back Vent.

a potential cause of wiredrawing of control valve seats, along with other system damages.

Water Hammer Velocities

Water hammer occurs when a slug of water, pushed by steam pressure along a pipe instead of draining away at the low points, is suddenly stopped by impact on a valve or fitting such as a pipe bend or tee. The velocities which such slugs of water can achieve are not often appreciated. They can be much higher than the normal steam velocity in the pipe, especially when the water hammer is occurring at start-up.

When these velocities are destroyed, the kinetic energy in the water is converted into pressure energy and a pressure shock is applied to the obstruction. In mild cases, there is noise and perhaps movement of the pipe. More severe cases lead to fracture of the pipe or fittings, and consequently escape of live steam at the fracture. Water hammer is avoided completely if steps are taken to ensure that water is drained away before it accumulates in sufficient quantity to be picked up by the steam. Careful consideration of steam main drainage can avoid damage to the steam main and possible injury or even loss of life. It offers a better alternative than an acceptance of water hammer and an attempt to contain it by choice of materials, or pressure ratings, of equipment.

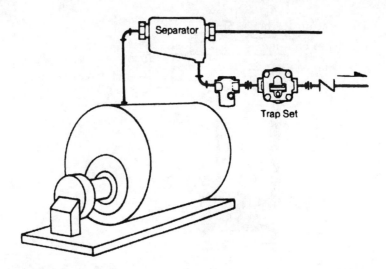

Figure 5-8. Steam Separator Mounted on a Boiler with a Steam Trap.

Efficient Steam Main Drainage

Proper drainage of lines (see Figure 5-9), and some care in start-up methods, not only prevent damage by water hammer, but help improve steam quality, so that equipment output can be maximized and maintenance of control valves reduced. The use of oversized steam traps giving very generous "safety factors" does not necessarily ensure safe and effective steam main drainage during start-up. A number of points must be kept in mind, for a satisfactory installation during start-up and operation of steam lines:

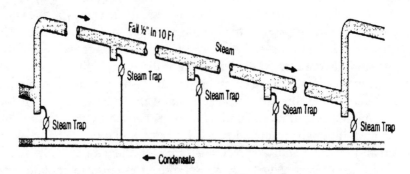

Figure 5-9. Schematic of a Draining Steam Main.

1)   The heat-up method employed,

2)   Provision of suitable collecting legs or reservoirs for condensate,

3)   Provision of a minimum pressure differential across the steam trap,

4)   Choice of steam trap type and size,

5)   Proper trap installation.

Heat-up Method

The choice of steam trap depends on the heat-up method adopted to bring the steam main up to full pressure and temperature. The two most usual methods are (a) supervised start-up and (b), automatic start-up.

A) Supervised Start-up

In this case, at each drain point in the steam system, a manual drain valve is fitted, by-passing the steam trap, and discharging to atmosphere. These drain valves are opened fully before any steam is admitted to the system. When the heated condensate has been discharged and as the pressure in the main begins to rise, the valves are closed. The condensate formed under operating conditions is then discharged through the traps. Clearly, the traps need only be sized to handle the losses from the lines under operating conditions, given in Tables 5-1 and 5-2.

This heat-up procedure most often used in large installations where start-up of the system is an infrequent, perhaps even an annual, occurrence. Large heating systems, and chemical processing plants are typical examples.

B) Automatic Start-up

One traditional method of achieving automatic start-up is simply to allow the steam boiler to be fired and brought up to pressure with the main steam stop valve wide open. Thus the steam main and branch lines come up to pressure and temperature without supervision, and the steam traps are relied on to automatically discharge the condensate as it is formed.

This method is generally confined to small installations that are regularly and frequently shut down and started up again. For example, the boilers in many laundry and dry-cleaning plants are often shut down at night and restarted the next morning.

In anything but the smallest plants, the flow of steam from the

Table 5-1. Warm-up Load in Pounds of Steam per 100 ft of Steam Main. Ambient temperature 70°F based on sch. 40 pipe to 250 psi, sch. 80 above 250 except sch. 120, 5" and larger above 800 psi.

Steam Pressure (psi)	2"	2-1/2"	3"	4"	5"	6"	8"	10"	12"	14"	16"	18"	20"	24"	0°F Correction 24" Factor[†]
0	6.2	9.7	12.8	18.2	24.6	31.9	48	68	90	107	140	176	207	208	1.50
5	6.9	11.0	14.4	20.4	27.7	35.9	48	77	101	120	157	198	233	324	1.44
10	7.5	11.8	15.5	22.0	29.9	38.8	58	83	109	130	169	213	251	350	1.41
20	8.4	13.4	17.5	24.9	33.8	43.9	66	93	124	146	191	241	284	396	1.37
40	9.9	15.8	20.6	29.3	39.7	51.6	78	110	145	172	225	284	334	465	1.32
60	11.0	17.5	22.9	32.6	44.2	57.3	86	122	162	192	250	316	372	518	1.29
80	12.0	19.0	24.9	35.3	47.9	62.1	93	132	175	208	271	342	403	581	1.27
100	12.8	20.3	26.6	37.8	51.2	66.5	100	142	188	222	290	366	431	600	1.26
125	13.7	21.7	28.4	40.4	54.8	71.1	107	152	200	238	310	391	461	642	1.25
150	14.5	23.0	30.0	42.8	58.0	75.2	113	160	212	251	328	414	487	679	1.24
175	15.3	24.2	31.7	45.1	61.2	79.4	119	169	224	265	347	437	514	716	1.23
200	16.0	25.3	33.1	47.1	63.8	82.8	125	177	234	277	362	456	537	748	1.22
250	17.2	27.3	35.8	50.8	68.9	89.4	134	191	252	299	390	492	579	807	1.21
300	25.0	38.3	51.3	74.8	104.0	142.7	217	322	443	531	682	854	1045	1182	1.20
400	27.8	42.6	57.1	83.2	115.7	158.7	241	358	493	590	759	971	1163	1650	1.18
500	30.2	46.3	62.1	90.5	125.7	172.6	262	389	535	642	825	1033	1263	1793	1.17
600	32.7	50.1	67.1	97.9	136.0	186.6	284	421	579	694	893	1118	1387	1939	1.16
800	37.7	57.6	77.0	112.5	203.2	273.5	455	670	943	1132	1445	1835	2227	3227	1.156
1000	42.1	64.2	86.0	125.5	226.7	305.2	508	748	1052	1263	1612	2047	2485	3601	1.147
1200	46.9	71.5	95.7	139.7	252.5	340.0	566	833	1172	1407	1796	2280	2767	4010	1.140
1400	51.9	79.2	106.0	154.7	279.6	376.3	626	922	1297	1558	1968	2524	3084	4440	1.135
1600	57.3	87.4	117.0	170.8	308.6	415.4	692	1018	1432	1720	2194	2786	3382	4901	1.130
1750	61.8	94.3	126.1	184.2	332.8	448.0	746	1098	1544	1855	2367	3006	3648	5285	1.128
1800	63.4	96.7	129.4	188.9	341.3	459.3	764	1125	1584	1902	2427	3082	3741	5420	1.127

[†]For outdoor temperature of 0°F multiply load value in table for each size by correction factor shown.

Table 5-2. Running Load in Pounds per Hour per 100 ft of Insulated Steam Main. Ambient temperature 70°F—Insulation 80% efficient load due to radiation and convection for saturated steam.

Steam Pressure (psi)	Main Size														0°F Correction Factor[†]
	2"	2-1/2"	3"	4"	5"	6"	8"	10"	12"	14"	16"	18"	20"	24"	
10	6	7	9	11	13	16	20	24	29	32	36	39	44	53	1·58
30	8	9	11	14	17	20	26	32	38	42	48	51	57	68	1·53
60	10	12	14	18	24	27	33	41	49	54	62	67	74	89	1·45
100	12	15	18	22	28	33	41	51	61	67	77	83	93	111	1·41
125	13	16	20	24	30	36	45	56	66	73	84	90	101	121	1·39
175	16	19	23	26	33	38	53	66	78	86	98	107	119	142	1·38
250	18	22	27	34	42	50	62	77	92	101	116	126	140	161	1·36
300	20	25	30	37	46	54	68	85	101	111	126	138	154	184	1·35
400	23	28	34	43	53	63	80	99	118	130	148	162	180	216	1·33
500	27	33	39	49	61	73	91	114	135	148	170	185	206	246	1·32
600	30	37	44	55	61	82	103	128	152	167	191	201	232	277	1·31
800	36·2	44	53	69	85	101	131	164	194	214	244	274	305	365	1·30
1000	43·1	52	63	82	101	120	156	195	281	254	290	326	383	435	1·27
1200	51·0	62	75	97	119	142	185	230	274	301	343	386	430	515	1·26
1400	60·4	73	89	114	141	188	219	273	324	356	407	457	509	610	1·25
1600	69·8	85	103	132	163	195	253	315	375	412	470	528	588	704	1·22
1750	76·6	93	113	145	179	213	278	346	411	452	518	580	845	773	1·22
1800	79·2	96	117	150	185	221	288	358	425	467	534	600	867	800	1·21

†For outdoor temperature of 0°F, multiply load value in table for each main size by correction factor.

boiler into the cold pipes at start-up, while the boiler pressure is still low can lead to excessive carry-over of boiler water with the steam. Such carry-over can be enough to overload separators in the steam off-take, where these are fitted. Very careful start-up is needed if water hammer is to be avoided.

For these reasons, modern practice calls for an automatic valve to be fitted in the steam supply line, arranged so that the valve stays closed until a reasonable pressure is attained in the boiler. The valve can then be made to open over a timed period so that steam is admitted slowly into the distribution pipework. The pressure within the boiler may be climbing at a fast rate, but the slow opening valve protects the piping network.

Where these valves are used, the time available to warm up the pipework will be known, and set on the valve control. In other cases it is necessary to know the details of the boiler start-up procedure so that the time can be estimated. Boilers started from cold are often fired for a short time and then shut off while temperatures equalize. The boilers are protected from undue stress by these short bursts of firing, which extend the warm-up time and reduce the rate at which condensation in the mains is to be discharged at the traps. Whichever of these automatic start-up methods is adopted, steam will only flow into the pipes, and discharge air from the air vents, if it is at a pressure above atmospheric. The temperature of the condensate will at first be quite low, probably well below 212°F. This means that the steam traps, which are usually sized for the hot condensate condition, will have a greater than normal capacity at this time. The pressure within the pipe will be a little more than atmospheric. If the traps can be located 28" below the main, then a hydraulic head of 1 psi is available in addition to this line pressure.

Collecting Legs

Sizing the drain traps requires estimation of the amount of condensate produced in bringing the main up to working temperature, and allowing for the time available for this warm up, determining the hourly rate of flow through the traps. With an estimate of the steam line pressure giving differential across the trap, suitable sized traps can be chosen. Warm up loads for the most common pipe sizes and pressures appear in Table 5-1. The most rapid rate of condensation occurs when the mains are cold and the pressure of the airsteam mixture within them is still low, perhaps only 1 psi. The mains may have been heated to some temperature less than the corresponding saturation temperature by this time, so the amount of condensate will not have reached the values shown in Table 5-1. Further, although some of the condensate is passing through the trap, the remainder is filling the condensate collecting leg. The amount of condensate being formed and the pressure available to discharge it are both varying continually and at any given moment are indeterminate because of the many unknown variables. A compromise, based on experience, is necessary when sizing these traps.

In most cases it seems reasonable to take perhaps 2/3 of the total amount of condensate produced in bringing the pipe up to 212°F. Allow for a total steam/air pressure plus hydraulic head of 2 psi to push condensate through the traps, and then make the best possible estimate of the time needed to bring the mains to this temperature. The time will vary with the rate of admission of the steam end with the temperatures of the cold mains.

Start-up Condition Example

Consider a length of 6" main which is to carry steam at 250 psi. Drip points are located at 150 ft. intervals, with collecting legs 4" diameter × 28" long. The main is brought up to pressure from 50°F in 30 minutes. From Table 5-1,

Total Warm Up Load to 406°F	= 89.4 lb/100 ft
	= 134.1 lb/150 ft
Condensate Load up to 212°F	= 31.9 lb/100 ft
	= 47.9 lb/1500 ft

Two-thirds of this load is about 31.9 lb of water. A collecting pocket 4" dia. × 28" long will hold about 12.68 lb of water:

Allow for 47.9 lb - 12.68 lb	= 35 lb of water
Temperature rise to 250 psi	= 406°F - 50°F = 356°F
Temperature rise to 0 psi	= 212°F - 50°F = 162°F
Time to reach 212°F	= 162/356 × 30 minutes
	= 13 minutes
Trap discharge rate	= 35 lb/13 min = 162 lb/hr

Capacity of a 1/2" thermostatic disc trap at 2 psi differential is about 255 lb/hr and this trap would have ample capacity in this case.

Draining Steam Mains

Note from the example that in most cases other than large distribution mains 1/2" thermodynamic traps have ample capacity. For shorter lengths between drip points, and far smaller diameter pipes, the 1/2" low capacity TD trap more than meets even start-up loads, but on larger mains it may be worth fitting parallel 1/2" traps as in Figure 5-9. Low pressure mains are best drained using float and thermostatic traps, and these traps can also be used at higher pressures.

Condensate Collecting Legs

Condensate collecting legs should be the same diameter as the main, up to 4" size. Larger pipes can have collecting legs 2 or 3 sizes smaller than the main but not less than 4". The length of the pockets used with automatic start-up is usually 28" or more, to give 1 psi hydraulic head. With supervised start-up the length of the legs can be 1-1/2" diameter but not less than 8".

The spacing between the drainage points is often greater than is desirable. On a long horizontal run (or rather one with a fall in the direction of the flow of about 1/2" in 10 feet or 1/250) drain points should be provided at intervals of 150 to 300 feet. Longer lengths should be split up by additional drain points. Any natural collecting points in the systems, such as at the foot of any riser, should also be drained.

A very long run laid with a fall in this way may become so low that at intervals it must be elevated with a riser The foot of each of these "relay points" also requires a collecting pocket and steam trap. Sometimes the ground contours are such that the steam main can only be run uphill. This will mean the drain points should be at closer intervals, say 50 ft apart, and the size of the main increased. The lower steam velocity then allows the condensate to drain in the opposite direction to the steam flow.

Air Venting

Air venting of steam mains is of paramount importance and is far too often overlooked. Steam entering the pipes tends to push the air already there in front of it as would a piston. Automatic air vents, fitted on top of trees at the terminal points of the main and the larger branches, will allow discharge of this air. Absence of air vents means that the air will pass through the steam traps (where it may well slow down the discharge of condensate) or through the steam-using equipment itself.

Over Pressure Protection

Safety Valves, Relief Valves, and Safety-Relief Valves

These valves protect the entire plant from over pressure operation and their reliability is most important for equipment and personnel safety. They are the subject of mandatory regulations and insurance acceptance in most countries and standards exist dealing with their design and use, requiring careful calculation based on established practice and known data. Confusion is sometimes caused by loose interchange of the terms 'safety valve' and 'relief valve,' suggesting that they are one and the same thing. There is a difference. A safety valve is automatically actuated by the static upstream pressure and is used for gas or vapor (compressible fluids). It has rapid full opening action, to give immediate protection by release of pressure.

A relief valve automatically operates in a similar manner to a safety valve, but it opens in proportion to the increase in pressure over the opening pressure and it is used for liquids (non-compressible fluids). To complicate matters further, a safety-relief valve is automatically actuated in the same manner as safety valves and relief valves and, dependent upon the application, can be used either for gas and vapors or for liquids. The majority of these types usually function by means of a spring-loaded disc, an increasing pressure compressing the spring (or springs), rifting the disc and allowing gas or liquid to pass from the pressurized zone (Figure 5-10). The valve is adjusted to suit relief conditions by means of precompression of the spring using an adjusting screw on top of the valve. Some designs have an external combination of lever and balance weight, the latter being adjustable so that the relieving point can be set at varying distances from the valve stem to suit the release pressure.

Another design is the dead weight open discharge valve, where the blowdown pressure is adjusted by cast iron disc weights; increasing or decreasing the number of weights alters the blow-down pressure accordingly. This valve does not require springs and it occupies a small amount of space. Blowdown pressure, which is the difference between the release pressure and the slightly less pressure of the system at which the valve closes, is regulated in safety valves by a single ring, known as the blowdown ring (Figure 5-10) which when moved upwards, increases blowdown; i.e., decreases the reseating pressure; lowering the ring increases reseat pressure. For ordinary type valves, blow-off pressures are set a minimum of 10 percent above the operating pressure of the system.

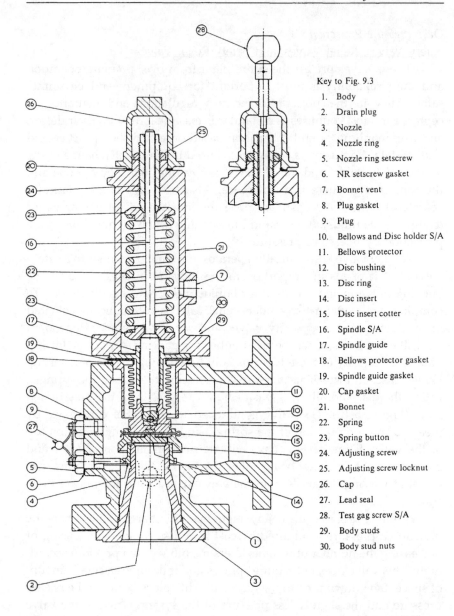

Key to Fig. 9.3

1. Body
2. Drain plug
3. Nozzle
4. Nozzle ring
5. Nozzle ring setscrew
6. NR setscrew gasket
7. Bonnet vent
8. Plug gasket
9. Plug
10. Bellows and Disc holder S/A
11. Bellows protector
12. Disc bushing
13. Disc ring
14. Disc insert
15. Disc insert cotter
16. Spindle S/A
17. Spindle guide
18. Bellows protector gasket
19. Spindle guide gasket
20. Cap gasket
21. Bonnet
22. Spring
23. Spring button
24. Adjusting screw
25. Adjusting screw locknut
26. Cap
27. Lead seal
28. Test gag screw S/A
29. Body studs
30. Body stud nuts

Figure 5-10. Safety Valve with a Nozzle Ring and Disc Ring for Blowdown Control.

Back pressure falls into two classes, constant and variable. When the valve is required to discharge into a system containing constant back pressure, it must be set to lift at a pressure less than the design pressure but more than the operating pressure of the equipment. It will then stay open until the pressure reaches a preset pressure below the operating pressure. The lift of the ordinary safety valve must be at least 1/24 of the valve seat bore.

Other design considerations are as follows:

(a)   High lift safety valve where the lift is at least 1/2 of the value of the seat bore.

(b)   Full lift safety valve, the fully lifted valve head presenting a net discharge are equal to the seat bore.

(c)   Nozzle type valve which is an automatic pressure relieving valve of the full lift type operating by the static upstream pressure of the valve and has rapid full opening or 'pop' action characteristics.

(d)   Pilot operated safety or relief valve is indirectly operated by initial press release from a pilot valve. This valve gives maximum 'full bore' discharge, and is installed for protection of boilers and superheaters, vapor or gas pressure vessels, and allied piping systems.

(e)   Torsion bar safety valve desired because of the difficulties in procuring helical springs with stability and accuracy at extremes of pressure, the loading being achieved by means of twin torsion bars giving the necessary accuracy and stability.

(f)   Electrically assisted safety valve—is generally used for the overall protection of an entire system (principally related to steam boilers). It can be considered as a powered easing lever, the valve lifts and closes instantaneously, and there is no 'simmering' or wire-drawing and it thus relieves all other valves in the system from damage by excessive 'simmering.' Failure of electrical supply allows the valve to continue as an ordinary safety valve.

Like the safety valve, ordinary relief valves must have a lift of at least 1/24 of the valve seat bore, and they are also designed for full lift and pilot operation. In the relief valve, the area presented to over-pressure is the same whether the valve is open or closed and a continuous increase of pressure results in a gradual lifting of the disc to the fully open position.

GROUP CLASSIFICATION

Safety, relief, safety-relief valves can be conveniently arranged into the three groups:

(a) For use with saturated and superheated steam where temperature is related to pressure. These valves require an extremely exacting standard of design and manufacture and cover a wide range of capacities and types.

(b) Simpler types of valves for use with compressed air, clean gases, and clean liquids and having an extensive range of capacities, and suited to conditions from vacuum to medium pressures.

(c) Valves in special materials for use in the chemical and process industries and oil refining where the main difficulty is corrosion. Valves are generally assessed on orifice ratings rather than inlet bore, and relieving pressures rang as high as 690 bar (10,000 lb/in^2) with maximum temperatures in the region of 540°C. Materials of manufacture include cast iron, bronze, various grades of cast steels and stainless steels, bar stock and forgings for the smaller sizes, etc., and specially hardened seats and resilient seats (principally in PTFE) can be fitted to meet special conditions. Springs are in various grades of steels, including stainless steel and tungsten alloy.

When located in corrosive atmospheres, the valve spring requires particular attention, as pitting can take place, leading to reduction in strength of spring material which, because of its nature, is limited in range.

Pressure rating, of springs are identified by standard color codes and valves are fitted either with outside springs or inside springs according to service conditions. The latter require to be fitted when noxious gases or fluids have to be piped away to a point where no harm can be done, and also for protection of the spring against damage. The outside spring can be adopted where no danger is possible to operating personnel, or when spring operating temperatures are so high that advantage must be taken of atmospheric cooling to maintain a reasonably low and constant spring temperature.

Valves can be supplied with various top fitments, such as a dome to make the valve fluid tight on the discharge side, this being particularly suited if the valve is to be used for hydraulic relief, an easing lever and padlock or lead seal, the former being for release pressure checking purposes and the padlock or seal for prevention of unauthorized interference with the lever. The lift gear can be arranged with air-operating lift for remote opening of the valve.

Other variations of valves are angle type and jacketed valve for viscous fluid service.

It is important to note that proper valve functioning will be nullified if attention is not paid to correct sizing of exhaust piping. This should be calculated so that the bore pressure falls within the prescribed limits, the pipe should be as straight as practicable, the use of right angle bends should be avoided, and the pipe must be well anchored against thrust, and allowance for expansion should be made if above-ambient temperatures are involved. Also, if conditions are such that long inlet piping requires to be fitted, then the safety valve itself must be firmly anchored to avoid possible damage from vibration arising from pulsation.

DRIP PAN ELBOWS OR
EXPANSION CHAMBERS AND SILENCERS

A drip pan elbow is a form of slip joint fitted to the outlet of a safety valve where the discharge has to be piped away from source. For optimum safety valve performance, the discharge piping should be supported independently of the valve. An arrangement is shown in Figure 5-11.

The design is such that strains caused by pipe extension do not react on the valve, and a variety of different designs exist. Silencers are recommended to solve any noise problems arising from discharge of steam, air, or gas from safety valves.

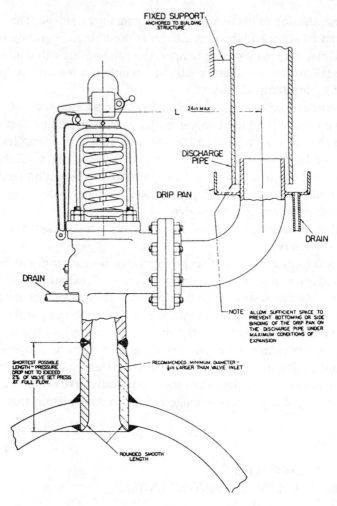

Figure 5-11. Safety Valve Assembly Equipped with a Drip Pan and Steam Vent Stack.

Dump or Unloading Valve

A dual purpose valve is at all times a conventional spring-loaded safety-relief valve, but serves a further purpose as a depressurizing unit when, under emergency conditions of fire or some mechanical failure, it will rapidly depressurize to a safe condition a vessel or number of vessels, from a single control point.

The same valve can be used as a 'drop-out' valve under normal operating conditions to release contents from a vessel during periodic shutdown.

This type of arrangement is widely used as a safety precaution in chemical plant and refinery processing, and can be operated by hand or by remote control by fitting an actuator (steam or air operating) or by use of a pilot control unit.

FREEZE PROTECTION

Freeze protection of above-ground steam piping, a pipe buried shallow below ground can be exposed to freezing temperatures. The chapter on steam traps explains how piping is protected against freezing using steam in inclement weather.

Many engineers choose to use electrical tracing on steam or condensate piping. It is easier to install and operate and maintain than steam tracing and most times a more effective freeze protection. It is, however, more costly to operate but is less expensive to install. Since both use only a minimum quantity of energy when considering total steam or electrical load of a plant. The energy used in freeze protection is minimal, it operates on very low voltage utilizing a.c. electrical power with fail safe. Therefore personnel are not in danger of electrocution. Burns to personnel are discussed in Personnel Safety.

STRUCTURAL DAMAGE

The most common cause of extreme structure damage happens with overhead steam lines being hit by cranes, trucks and ships, etc. The best form of prevention is to erect a sign large enough for all to see stating its height of the steam line is above the ground surface. Training in use of machinery should include an effort to ensure drivers and operators are aware of the hazards of overhead steam lines.

PERSONNEL SAFETY

There are many areas of concern in maintaining the safety of personnel when working on steam distribution systems. One way of helping to prevent personnel injury is to ensure there is a co-worker or supervisor present outside the space. This attendant can provide assistance if necessary. He should be in contact with the person in the enclosure at all times by voice or two-way radio. The five hazards are:

- Burns
- Suffocation
- Drowning
- Electrocution
- Noise

Burns

As much of a steam distribution system as possible should be thermally insulated. In areas where exposure of steam piping or equipment is necessary for the equipment's performance, the exposed surface should be covered with a cage. This prevents contact of personnel's body parts with the heated surface. Cages are available from steam specialty equipment suppliers. They come in a variety of sizes, shapes and configurations. They usually are uninsulated but where extremely high steam temperatures exist it is advisable to wrap the cage itself with insulation to prevent more burns or discomfort to personnel coming in contact with them.

Another source of burns is where a quantity steam leaking or dripping from piping or steam equipment. Often times the discharge is part of the operation of the machine but nevertheless, can harm personnel. These areas should be covered in a way to prevent accidental contact with personnel.

Also steam emitting from a steam leak may not be detected visually but can be located and identified from the sound (hissing) at the point of leakage. To further identify the leak a glove or a cloth can be used to locate the leak. The cloth will flitter when it enters the steam flow. It should be noted that steam is invisible. The moisture carried in it is what is seen. Therefore, in the case of superheated steam where there is no moisture, the emissions are invisible. Superheated steam can cause severe burns to an unsuspecting person since its temperature often is in excess of 1000°.

Suffocation

Suffocation occurs when a person is not getting sufficient oxygen. If a person is in a confined space and the oxygen in the air is too low then suffocation can occur. In many cases as a person is suffocating he/she may pass out and be unable to do anything to help themselves. Often, suffocation can occur without the victim experiencing any breathing difficulty. There may be no warning before it is too late and death occurs.

Another cause of suffocation can occur from toxic gases in the space where the technician is located. These gases attack the blood or lung tissue rendering them ineffective in absorbing or transferring oxygen. Again the victim will die unless rushed to a medical facility. Other toxic gases attack humans in many ways. Some of the gases are present in our environment. Suffocation is most likely to occur in manhole, tunnels, and large trenches.

To provide a minimum level of safety portable toxic gas meters can be used to examine the atmosphere in enclosed spaces. These meters will identify a large number of typical gases found in industrial plants, and will measure their level of toxicity. They will also measure the oxygen content of the gas in the space. Very often only part of the enclosure may be hazardous. Therefore, it is important to examine the space while passing through it. First aid equipment and personnel familiar will EPR should be present at the site where the tunnel is located.

Drowning

Drowning is most likely to occur when a workman enters a manhole or other enclosure unassisted by a co-worker outside the space to render help as necessary. The man entering the space should have a rope or harness around his waist. This can be used to remove the worker from the space in the event a drowning hazard arises. Ropes can often be connected to a worker's tool belt. This safety arrangement can assist any of the persons in the tunnel. The major cause of drowning usually occurs when a man enters an enclosure unassisted

Electrocution

Electrocution can occur causing various levels of injury. For more information consult an electrical equipment handbook, first-aid or medical textbook. The caution against electrocution to steam system personnel is the responsibility of the plant engineer or his designated employer. It is recommended that safety training sessions including electrical safety be mandatory for your steam system personnel on a regular basis.

Noise Pollution

Noise problems are not confined to any particular industry or process and, in so far as pipework is concerned it can occur in atmospheric vents, blowdowns, safety and relief valve exit piping, intake and exhausts from engines, etc.

Noise is difficult to define but it can be considered as unwanted sound or undesired disturbance, and it may be intermittent, erratic or in a continuous steady state depending upon the source. Also, it can often be a measure of mechanical inefficiency. The accepted unit of measurement is that dimensionless ratio, the decibel.

Excessive noise, like other forms of pollution, can result in deterioration of community relations and damage the corporate image of a company. Unfortunately, vulnerability to this can occur during start-up on a new plant. It can also interrupt departmental communication, prevent coordinated operations thus causing decreased production, increased tension among operators, and in many cases it is directly responsible for higher accident rates.

Noise is the subject of legislation in many countries, and in recent years compensation—sometimes with legislative action—has shown a sharp downward trend, principally due to increasing hearing injuries to personnel.

Silencers

The silencer is universally used in pipework to eliminate or at least to reduce noise to an acceptable level and designs are grouped into three basic categories.

   (a)   Snubber type silencer (low reactive),
   (b)   Absorptive type silencer (dissipative),
   (c)   Combinations of (a) and (b).

Various configurations used in industry are illustrated in Figure 5-12 (a-g), all these types being suitable for vertical or horizontal installation. Combinations combine features from both the 'low reactive' and dissipative (or absorptive) types shown in the illustrations.

Major uses occur in vent and blowdown pipe lines, and in pressure reduction regulating systems. It is important to note that when selecting a silencer, one should not attempt to comply with all possible design parameters as it is seldom, if ever, that all parameters need to be met.

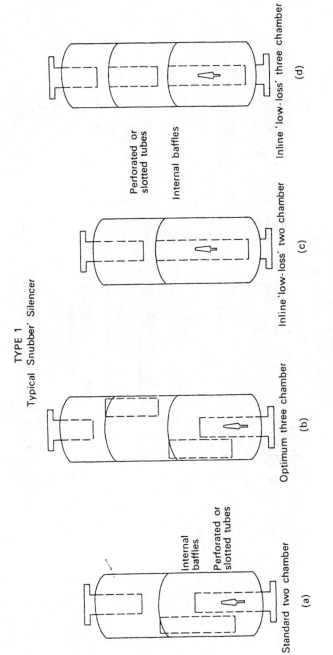

Figure 5-12. A Selection of Silencers.

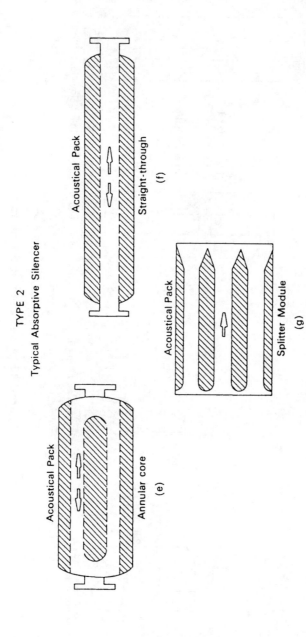

TYPE 2

Typical Absorptive Silencer

Figure 5-12. A Selection of Silencers (Continued).

The essential parameters of silencer design are:
    (a)   maximum noise reduction (attenuation)
    (b)   virtually no pressure drop
    (c)   minimum size and weight
    (d)   durable long life and maintenance free
    (e)   lowest possible cost

Separators

Common types of mechanical separators are the cyclone, mesh pad or filter types, gravity settling or sonic types, and electrostatic. All use the principle of impingement in that particles colliding with other particles, or with surfaces, adhere after collision, and the use of centrifugal force is employed to some degree in separation. This is maximized in, for example, the cyclone. Gravity force is also employed at least to carry off the separated product.

The equipment employed can, depending upon volume and conditions, be large and occupy considerable space volume, or small enough to be a pipe line fitting not much larger than a normal sized valve.

The subject is a specialized one which lies largely in the hands of the specialist manufacturers, and consultation with them is advisable when a problem arises. Steam separators as with other specialty equipment are produced by a large number of companies and distributed by an even greater number of reliable shops. Sorting out what distributors have products that suit your needs can be a monumental job for a small steam producer and distributing firms. It is suggested you obtain referrals from your counterparts at other sites, steam specialty shops or whatever other sources engineering know-how can provide.

Chapter 6

Thermal Insulation

The preservation of heat by means of insulating materials has always been a problem posing many difficulties in the solution. In fact, at one time the user paid too much attention to first cost and tended to neglect the question of economic cost. It is true that the concept of determining an economic insulation has long been appreciated. But it is also true to say that the lengthy and laborious calculations necessary to establish heat loss through varying insulation thicknesses over an extensive range of pipe sizes, through a wide temperature range, meant that economic calculations were by no means always carried out.

Through the years, various methods were used or attempted, to arrive at a basis on which an accurate guarantee could be made. An outdated system of specification was to use the 'efficiency' method—i.e.,

$$\frac{\text{Heat saved by insulation}}{\text{Heat lost by bare pipe}} \times 100$$

but this has no bearing on modern insulation problems. For example, "efficiency" on the above basis can increase by say 12 percent, but the heat loss can decrease by 80 to 90 percent.

After much trial and error, the effective basis for an accurate guarantee was found to be that of conductivity of the materials. Extensive work has been carried out in this respect by manufacturers, users, and such organizations as the National Laboratories. Loss of heat is invariably accompanied by loss of valuable fuel such as oil, coal and natural gas. These in themselves have high social and industrial value. They can be used to manufacture literally thousands of useful products.

133

If a 5-in. steam pipe having a surface temperature of 200°F is uninsulated for a length of approximately 200 ft., it can waste 9 lb of coal per hour—or approximately 35 tons per year. If the same uninsulated, 200 ft. pipe had a surface temperature of 360°F the corresponding waste of coal would be about 70 tons per year.

Until recently, the only really quick way of attaining a reasonably economic insulation thickness was by acceptance and use of rather arbitrary thicknesses specified in the National Laboratories' Standards.

These standards, together with work by Research Laboratories and Institutes and specialized handbooks, such as that issued by The Engineering Equipment Manufacturers Association, have gone a considerable way towards clarifying and greatly speeding up the effective design work so necessary in this field of study.

The large users of thermal insulation in the chemical and petroleum industries have also produced their own standards and specifications which are kept up to date as new conditions and new insulation products arise. Perhaps the major step forward not only in speeding up the work of the designer, but also in applying a quick and accurate method to the calculation of economic thicknesses, is the tabulation method. This was prepared by leading insulation manufacturers in conjunction with the National Laboratories. This tabulation method avoids generalizations and assumptions necessary in algebraic solutions by the publication of comprehensive data for thermal conductivity, surface coefficients, temperatures of surfaces, heat losses, and insulation thicknesses for a complete range of pipe sizes and a large temperature range.

INSULATION COSTING

All that need be known are the evaluation period, the cost of heat and the applied cost of a range of insulation thicknesses, together with their respective heat losses. Tables 6A, 6B, and 6C show a selection of pipe insulations and their respective economic thicknesses.

The evaluation period is determined by the anticipated life of the plant in operating hours.

Cost of heat must be taken at the point of application of the insulation and should include cost proportions of the boilerhouse and other heat producing units. Capital and recurring costs, efficiency of boilers, fuel costs and the cost of the distribution system are continuously influenced by fuel cost alterations.

Applied cost of insulation is usually available from insulation contractors or within companies employing their own laborers, on the basis of materials, thicknesses, pipe sizes, locations, required finish and site conditions. This information should be precise, with proper accounting taken of change from single to multi-layer application.

Economic heat loss is obtainable from the data in Tables 6A, 6B and 6C. An example calculation of heat loss from an underground steam distribution system is shown in a technical report in Appendix F.

INSULATING MATERIALS—GENERAL REQUIREMENTS

Material Categories:
1)   Fibers, which include the mineral wools such as rock, slag, glass, ceramic and hair felt.

2)   Granular, such as Calcium Silicate, Magnesium, diatomaceous earth and cork.

3)   Cellular, such as foamed glass, polyurethane and polystyrene.

4)   Metallic, such as Aluminum foil.

Thermal insulating materials (including protection) should have:

1)   Resistance to attack by chemicals with which they may come into contact. If this is not practical, then the insulation should be provided with a resistant coating or jacket.

2)   Resistance to moisture sufficient that they do not deteriorate under wet conditions. This is particularly important when operating in the open.

3)   Resistance to vibration, mechanical shock and abrasion or, as most insulations by their nature are mechanically weak, at least protection against damage by same.

4)   Characteristics which allow them to be formed, as required, to effectively insulate awkward pipe and fitting shapes.

Cost obviously enters into the matter. An insulation may fulfill all requirements but at a high initial cost. This is where economic costing is important. Consideration also has to be given to the cost of maintenance if a chosen insulation requires frequent replacement. Sometimes a more expensive first cost for the insulation may be the more economic.

HEALTH HAZARDS
ASSOCIATED WITH THERMAL INSULATION

Precautions are required in handling and application of insulation materials, as hazards may be present which could have a serious effect on health.

Asbestos fiber was at one time a principal ingredient in several insulations. Now, other types of fibers are being introduced which are much less harmful, with many insulations now specified as 'asbestos-free'.

Since this book does not deal with waste removal, procedures used in safe asbestos removal is beyond the scope of this book. However, it is recommended that a skilled asbestos removal contractor be retained to insure that the plant is in compliance with government statutes regarding asbestos.

INSULATING MATERIALS

Material Categories
1)   Fiber including: mineral wool, glass, ceramic and felt.
2)   Calcium silicate, magnesium and diatomaceous earth.
3)   Fiber glass, polyurethane and polystyrene

Insulating materials should have:

1)   Resistance to deterioration from chemicals used in the plant,
2)   Moisture resistance,
3)   Resistance to damage caused by vibration, mechanical or hydraulic shock and abrasion.

MATERIAL DESCRIPTION

The physical, chemical and thermal properties of most insulation used in steam and condensate systems are listed in Table 6-1. Table 6-2 contains heat loss data for Mineral Fiber, Calcium Silicate and Cellular Glass.

INSULATION FINISHES AND PROTECTIVE COATINGS

Finish and protective coatings are of importance to the efficiency and the elimination of moisture penetration, and mechanical damage and corrosion. Finishes differ according to operating conditions, types of insulation and the physical shape of the insulation.

Some types of insulation finishes and their method of application are as follows:

1) Wire netting is employed as a key for air drying cement. Air drying cements are the most common type of finish used with calcium silicate. These coatings dry to a hard finish and are durable in tunnels, water proof trenches and indoors. Although moisture resistant air drying cements are not suitable for outdoor use or underground conditions or where flooding occurs.

2) Metal bands and wire are used for holding rigid sections of insulation in place. Their spacing depends upon the size of pipe and the thickness of the insulation. Good judgment dictates its spacing along the pipe's length. Metal bands and wire are available and commonly used in stainless steel, aluminum and galvanized steel.

3) Sheet metal can be used on all insulating material that have cylindrical shapes, and can be cut and fitted over insulated valves and flanges, etc. It can be installed so that it provides a waterproof covering over the insulation. It is readily available in galvanized steel, aluminum and stainless steel. First cost is high but this is compensated for in low maintenance costs. Also, Aluminum sheet is light in weight and is an obvious choice for most outdoor use.

4) Protective tapes that are self adhering come in a variety of widths. They can be wrapped about pipe insulation and also to many insu-

lated pipe fittings. They have a low first cost but are not as durable as sheet metal or self setting cements. Self sealing tape is cost effective for indoor conditions where moisture content of the air is low. Since it is basically a plastic material it is susceptible to damage from certain chemicals also it should not be used where mechanical damage is prevalent.

5)   Another protective coating is mastic. It can be applied by spray, trowel or brush and is an excellent moisture retardant. It can be used on insulation for indoors, outdoors, in tunnels, trenches and if heavily applied can be an effective sealant for underground use.

6)   Blankets are formed by stuffing a heavy grade of cotton canvas with a granular or fibrous insulating material. The filling is sealed in the blanket by sewing the blanket closed. These are usually factory assembled and fitted in the field by fastening them to valves, flanges or other pipe fitting. When these are properly manufactured and fitted they make an excellent heat seal. They are used where the insulation requires frequent removal.

When writing an insulation specification choose the material with the lowest thermal conductivity within the framework of the lowest life cycle cost, reliability and safety.

Figures 6-1 to 6-13 inclusive are diagrams of how to fit the various types and shapes of rigid insulation, blankets and other insulation and covers. Although, the diagrams of insulation assembly are not intended to be all inclusive, with a little imagination most configurations can be dealt with effectively.

See also Figure 3-7, Figure 3-8, and Figure 4-1 to Figure 4-9.

It is important not to apply insulation that is beyond the maximum or minimum limits presented in Table 6-1.

Table 6-1. Physical and Thermal Properties of Common Insulation.

Class	Material	Temperature range min/max °C	Resistance to			Forms available			
			Moisture	Fire	Mech. damage	Rigid	Plastic	Loose fill	Flexible
Fibrous	Glass fibre	−185° to 540° flex −185° to 510° rigid	Good	Excellent	Poor	Yes	—	Yes	Blanket rolls
	Rock wool	Up to 600° rigid Up to 750° flex	Excellent	Excellent	Fair	Yes	Yes	—	Blanket rolls
	Slag wool	Up to 600°	Good	Excellent	Fair	Yes	—	—	Blanket rolls
	Ceramic	1260°	Moderate	Excellent	Fair	Yes	Sprayed	Yes	Blanket rolls
Granulated	Calcium silicate	200° to 1000°	Good	Good	Fair	Yes	—	—	—
	Magnesia	Up to 315°	Poor	Fairly good	Poor	Yes	Yes	—	—
	Diatomaceous earth	850°/1000°	Poor	Fairly good	Poor	Yes	Yes	Yes	—
	Cork	−155° to 90°	Good	Poor	Moderate	Yes	—	—	—
Cellular	Foamed glass	−240° to 425°	Excellent	Excellent	Good	Yes	—	—	—
	Polyurethane	−240° to +110°	Good	Good	Fair	Yes	Sprayed and dispensed	—	—
	Isocyanurate foam	−240° to +110°	Good	Good	Fair	Yes	Sprayed and dispensed	—	—
	Polystyrene	−240° to +75°	Good	Good	Fair	Yes	Sprayed and dispensed	—	Yes
Metallic	Aluminium foil	Up to 600°	Excellent	Excellent	Poor	—	—	—	Yes

Table 6-2. Insulation Heat Loss Data. (A) Mineral Fiber.

Nominal Pipe Size, in.		MINERAL FIBER (Fiberglass and Rock Wool) Process Temperature, °F									
		150	250	350	450	550	650	750	850	950	1050
½	Thickness	1	1½	2	2½	3	3½	4	4	4½	5½
	Heat loss	8	16	24	33	43	54	66	84	100	114
	Surface temperature	72	75	76	78	79	81	82	86	87	87
1	Thickness	1	1½	2	2½	3½	4	4	4½	5	5½
	Heat loss	11	21	30	41	49	61	79	96	114	135
	Surface temperature	73	76	78	80	79	81	84	86	88	89
1½	Thickness	1	2	2½	3	4	4	4	5½	5½	6
	Heat loss	14	22	33	45	54	73	94	103	128	152
	Surface temperature	73	74	77	79	79	82	86	84	88	90
2	Thickness	1½	2	3	3½	4	4	4	5½	6	6
	Heat loss	13	25	34	47	61	81	105	114	137	168
	Surface temperature	71.	75	75	77	79	83	87	85	87	91
3	Thickness	1½	2½	3½	4	4	4½	4½	6	6½	7
	Heat loss	16	28	39	54	75	94	122	133	154	184
	Surface temperature	72	74	75	77	81	83	87	86	87	90
4	Thickness	1½	3	4	4	4	5	5½	6	7	7½
	Heat loss	19	29	42	63	88	102	126	152	174	206
	Surface temperature	72	73	74	78	82	86	85	87	88	90
6	Thickness	2	3	4	4	4½	5	5½	6½	7½	8
	Heat loss	21	38	54	81	104	130	159	181	208	246
	Surface temperature	71	74	75	79	82	84	87	88	89	91
8	Thickness	2	3½	4	4	5	5	5½	7	8	8½
	Heat loss	26	42	65	97	116	155	189	204	234	277
	Surface temperature	71	73	76	80	81	86	89	88	89	92

10	Thickness	2	3½	4	4	5	5½	5½	7½	8½	9
	Heat loss	32	50	77	115	136	170	220	226	259	307
	Surface temperature	72	74	77	81	82	85	90	87	89	91
12	Thickness	2	3½	4	4	5	5½	5½	7½	8½	9½
	Heat loss	36	57	87	131	154	192	249	253	290	331
	Surface temperature	72	74	77	82	82	86	91	88	89	91
14	Thickness	2	3½	4	4	5	5½	6½	7½	9	9½
	Heat loss	40	61	94	141	165	206	236	271	297	352
	Surface temperature	72	74	77	82	83	86	87	89	89	91
16	Thickness	2½	3½	4	4	5½	5½	7	8	9	10
	Heat loss	37	68	105	157	171	228	247	284	326	372
	Surface temperature	71	74	78	83	82	87	86	88	89	91
18	Thickness	2½	3½	4	4	5½	5½	7	8	9	10
	Heat loss	41	75	115	173	187	250	270	310	354	404
	Surface temperature	71	74	78	83	83	87	87	88	90	91
20	Thickness	2½	3½	4	4	5½	5½	7	8	9	10
	Heat loss	45	82	126	189	204	272	292	335	383	436
	Surface temperature	71	75	78	83	83	87	87	89	90	92
24	Thickness	2½	4	4	4	5½	6	7½	8	9	10
	Heat loss	53	86	147	221	237	295	320	386	439	498
	Surface temperature	71	74	78	83	83	86	86	89	91	93
30	Thickness	2½	4	4	4	5½	6½	7½	8½	10	10
	Heat loss	65	105	179	268	286	332	383	439	481	591
	Surface temperature	71	74	79	84	84	85	87	89	89	94
36	Thickness	2½	4	4	4	5½	7	8	9	10	10
	Heat loss	77	123	211	316	335	364	422	486	556	683
	Surface temperature	71	74	79	84	84	84	86	88	90	94
Flat	Thickness	2	3½	4	4½	5½	8½	9½	10	10	10
	Heat loss	10	14	20	27	31	27	31	38	47	58
	Surface temperature	72	74	77	80	82	80	82	85	89	93

Table 6-2. Insulation Heat Loss Data. (B) Calcium Silicate.

	CALCIUM			SILICATE						
	Process Temperature, °F									
	150	250	350	450	550	650	750	850	950	1050
	1	1½	2	2½	3	3½	4	4	4	4
	13	24	34	42	53	63	75	90	108	128
	75	78	80	81	82	83	84	87	91	94
	1	2	2½	3	3½	4	4	4	4	4
	16	26	38	49	60	72	89	109	130	154
	76	76	79	80	82	83	86	90	94	98
	1½	2½	3	3½	4	4	4	4	5	5
	17	29	42	54	68	86	106	128	139	164
	73	75	78	80	81	85	88	92	91	94
	1½	2½	3	3½	4	4½	5	5½	6	6
	19	32	47	61	75	90	106	123	142	167
	74	76	79	81	82	84	85	87	88	91
	2	3	3½	4	4½	5	5½	6	6	6
	21	37	54	71	87	105	123	143	71	202
	73	75	78	80	82	84	85	87	90	94
	2	3	4	4	4½	5	5½	6	6½	7
	25	43	58	82	101	121	142	164	187	213
	70	76	77	81	83	85	87	89	90	92
	2	3½	4	4	4½	5	5½	6	7	8
	33	51	75	105	129	153	178	205	224	245
	74	75	79	83	85	87	89	91	91	91
	2½	3½	4	4½	5	5	6	7	8	8½
	35	62	90	117	144	183	200	220	243	277
	73	76	79	82	85	89	89	89	90	92

2½	4	4	4	5	5½	6	7½	8½	9
41	66	106	149	168	200	233	243	269	306
73	75	80	85	86	88	90	89	89	91
2½	4	4	4	5	5½	7	8	8½	9½
47	75	121	170	191	266	236	262	300	330
73	76	81	86	86	89	88	88	90	91
2½	4	4	4	5	5½	7	8	9	9½
51	81	130	183	205	242	252	262	308	352
73	76	81	86	87	89	88	88	89	91
3	4	4	4	5½	6½	7½	8	9	10
50	90	144	204	211	237	265	307	338	372
72	76	82	87	85	86	87	89	90	91
3	4	4	4	5½	6½	7½	8½	9	10
55	99	159	225	232	259	289	320	367	403
73	76	82	87	86	87	87	88	90	91
3	4	4	4	5½	6½	7½	8½	9½	10
60	108	174	245	252	281	312	346	381	435
73	77	82	87	86	87	88	89	90	92
3	4	4	4	5½	6½	7½	8½	9½	10
71	127	203	287	293	325	360	397	437	497
73	77	82	88	87	88	88	89	90	93
3	4	4	4	5½	7	8	9	10	10
86	154	247	349	353	368	409	452	498	589
73	77	83	88	87	87	88	89	90	94
2½	4	4	4	6½	7½	8	9	10	10
119	181	291	410	359	406	475	524	576	681
74	77	83	89	84	86	88	89	91	94
2½	3½	4	5½	6½	7½	8½	9½	10	10
12	20	28	29	33	36	39	43	49	58
73	77	81	81	83	84	85	87	89	93

Table 6-2. Insulation Heat Loss Data. (C) Cellular Glass.

CELLULAR GLASS						
Process Temperature, °F						
150	250	350	450	550	650	750
1½	1½	2	2½	3	3½	4
9	23	34	48	62	78	92
70	76	78	82	83	85	84
1½	2	2½	3	3½	4	4
12	25	38	52	68	86	112
71	75	7̃	79	81	83	88
1½	2½	3	4	4	4	4
15	28	44	56	79	105	137
72	75	77	78	82	87	92
1½	2½	3	4	4	4	4½
17	31	47	61	84	113	140
72	74	77	78	82	86	89
1½	3	3½	4	4	4½	5
22	35	54	75	105	132	161
73	74	77	79	84	86	89
2	3	4	4	4	4½	5
22	41	59	87	122	150	185
71	74	76	80	85	87	90
2	3½	4	4	4½	5½	6
30	48	74	111	144	171	212
72	74	77	82	85	86	89
2½	3½	4	4	5	5½	6½
30	58	90	134	161	203	238
71	74	78	83	84	87	89
2½	4	4	4	5½	5½	7
37	63	106	159	178	238	264
71	74	79	84	84	87	88
2½	4	4	4	5½	5½	7½
42	71	121	181	201	269	284
71	74	79	85	84	90	88
2½	4	4	4	5½	5½	8
47	79	134	199	219	293	293
72	74	80	85	85	91	87
2½	4	4	4	5½	5½	8
53	88	149	222	242	325	322
72	75	80	86	86	91	88
2½	4	4	4	5½	5½	8
59	96	164	245	266	356	351
72	75	80	86	86	92	88

Table 6-2. Insulation Heat Loss Data. (C) Cellular Glass (Cont'd).

2½	4	4	4½	5½	5½	8
64	105	179	243	289	387	379
72	75	81	84	86	92	88
2½	4	4	5	5½	5½	8
76	123	209	260	336	449	436
72	75	81	83	87	93	89
2½	4	4	5½	5½	5½	8
93	150	254	290	405	542	521
72	75	81	82	87	93	90
2½	4	4	5½	5½	5½	8
110	176	229	340	474	635	606
73	76	81	82	88	94	90
2½	4	4	5½	5½	7½	8½
11	17	29	31	44	43	50
73	76	83	84	90	90	93

Figure 6-1. Pipe with Insulation Provided for Expansion.

6 mm (1/4 in.) thick asbestos-free cement finish.
All curves not less than 25 mm (1 in.) radius.

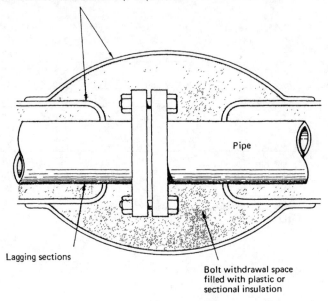

Pipe

Lagging sections

Bolt withdrawal space
filled with plastic or
sectional insulation

Figure 6-2. Typical Insulation over Pipe Flange.

Bolt withdrawal space
stemmed with plastic insulation

6 mm (¼ in) thick
asbestos-free cement finish

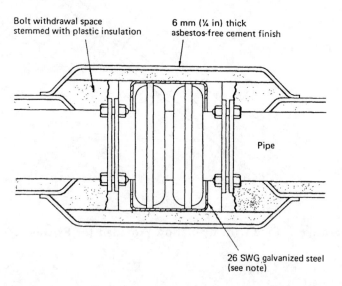

Pipe

26 SWG galvanized steel
(see note)

Figure 6-3. Lagging over Bellows Expansion Joint.

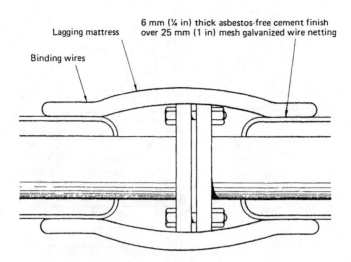

Lagging mattress

6 mm (¼ in) thick asbestos-free cement finish
over 25 mm (1 in) mesh galvanized wire netting

Binding wires

Figure 6-4. Removable Lagging over a Pipe Flange.

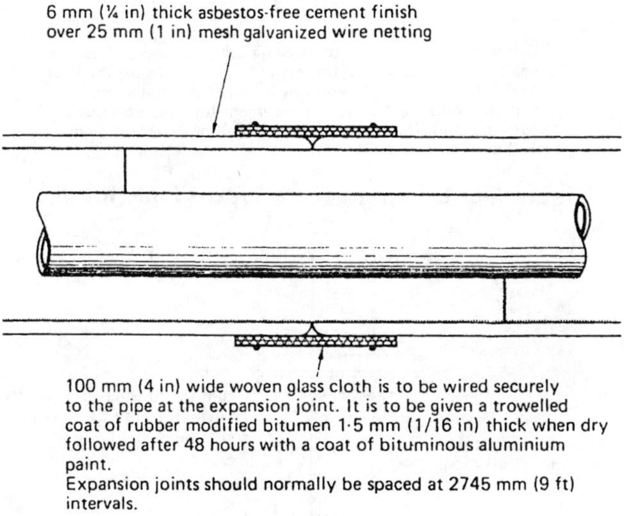

6 mm (¼ in) thick asbestos-free cement finish
over 25 mm (1 in) mesh galvanized wire netting

100 mm (4 in) wide woven glass cloth is to be wired securely
to the pipe at the expansion joint. It is to be given a trowelled
coat of rubber modified bitumen 1·5 mm (1/16 in) thick when dry
followed after 48 hours with a coat of bituminous aluminium
paint.
Expansion joints should normally be spaced at 2745 mm (9 ft)
intervals.

Figure 6-5. Pipe Showing Lagging Expansion Provision.

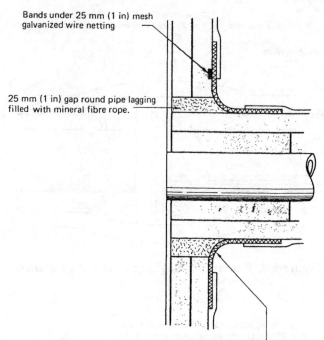

Bands under 25 mm (1 in) mesh
galvanized wire netting

25 mm (1 in) gap round pipe lagging
filled with mineral fibre rope.

Woven glass cloth wired to branch pipe and given a trowelled
coat of rubber modified bitumen (1·5 mm (1/16 in) thick when
dry) followed after 48 hours with a coat of bituminous
aluminium paint. Asbestos-free cement finish to be lapped
over the cloth for at least 50 mm (2 in), but no more than
100 mm (4 in)

Figure 6-6. Lagging over a Pipe—Tank Joint.

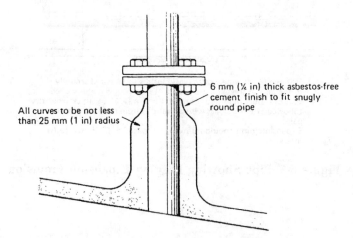

6 mm (¼ in) thick asbestos-free
cement finish to fit snugly
round pipe

All curves to be not less
than 25 mm (1 in) radius

Figure 6-7. Lagging over a Tank Top Piping Joint.

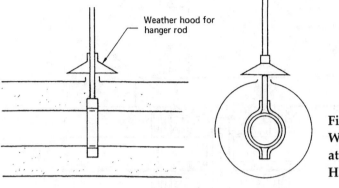

Figure 6-8. Weather Hood at Pipe Hanger.

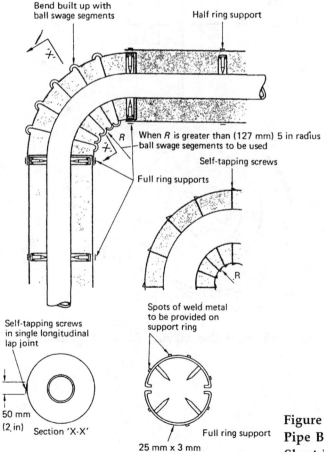

Figure 6-9. Insulated Pipe Bend with Sheet Metal Cover.

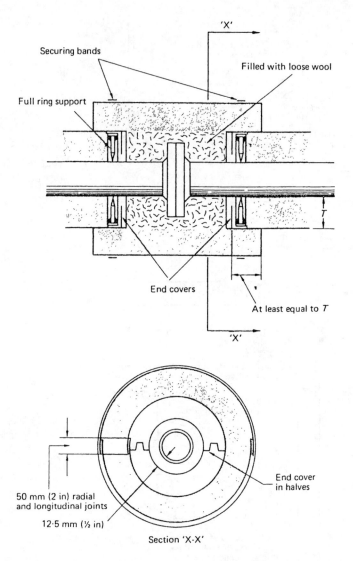

Figure 6-10. Flange with Insulation Having Radial Expansion Provision.

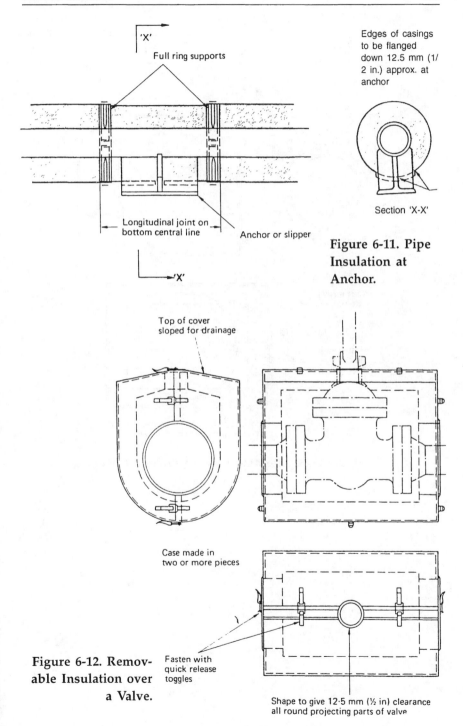

'X'

Full ring supports

Edges of casings
to be flanged
down 12.5 mm (1/
2 in.) approx. at
anchor

Section 'X-X'

Longitudinal joint on
bottom central line

Anchor or slipper

'X'

**Figure 6-11. Pipe
Insulation at
Anchor.**

Top of cover
sloped for drainage

Case made in
two or more pieces

**Figure 6-12. Removable Insulation over
a Valve.**

Fasten with
quick release
toggles

Shape to give 12·5 mm (½ in) clearance
all round projecting parts of valve

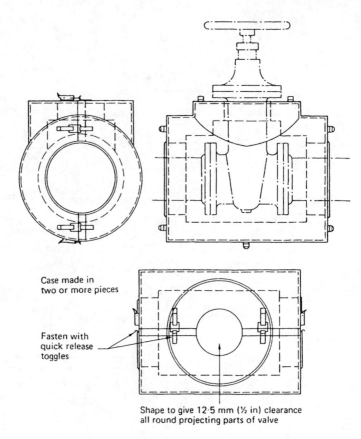

Case made in
two or more pieces

Fasten with
quick release
toggles

Shape to give 12·5 mm (½ in) clearance
all round projecting parts of valve

Figure 6-13. Alternate Valve Insulation Cover.

Chapter 7

Valves for Steam and Condensate Systems

VALVING

Gate Valves

*G*ate valves are in common use in the process industries for condi
tions requiring an 'on-off' service. This type valve is not practical
for throttling purposes because the disc can suffer 'wire drawing,
and vibration and chatter of the disc caused by flow.

Gate valves come in a wide range of sizes and permit full flow
when the bore of the valve is the same size as the pipe inside diameter.
They offer negligible resistance to flow when fully open and therefore
pressure drop is minimal. Large gate valves may require removal for
repair. It is usually more cost effective to scrap small gate valves when
damaged than to repair them. They should be replaced with new valves
meeting the same specification

Gate valves are suitable to steam or condensate service when a fully
opened or fully closed valve position is all that is required. The gate
valve cannot normally be used for throttling service.

Single Wedge Gate Valve

The *Single Wedge Valve* (Figure 7-1), usually fitted with a solid
wedge, can also be obtained with a flexible wedge to facilitate a better fit
when the valve is closed. It is less likely to be damaged when closing the
valve should foreign material be present. The solid wedge valve has no
internal moving parts to cause damage in services where vibration may
be present. Split wedge gate valves (Figure 7-2) have the wedge in male

153

and female sections which function in a similar manner to the solid wedge valve but give greater flexibility in the wedge action and ensure extended longevity. The *solid wedge tapered disc* offers maximum resistance to strains caused by pressure but, compared with other types, the tapered seat construction makes it more costly to repair.

Ball and Socket Valve

The *double disc ball and socket type valve* (Figure 7-3) has discs which rotate freely and independently and can therefore adjust to changes in seat angle, thus allowing good sealing and more wear resistance.

Globe Valves

This is a generic term for a class of valves where a disc is lifted or lowered from or on to a body seat, actuated by a screw type stem and

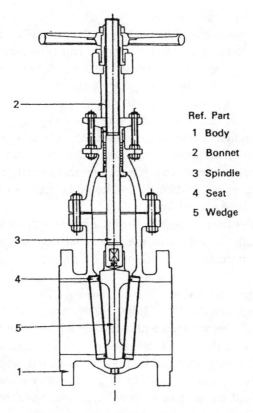

Ref. Part
1 Body
2 Bonnet
3 Spindle
4 Seat
5 Wedge

Figure 7-1. Single Disc Solid Wedge Gate Valve.

extensively used for flow regulation by throttling.

The principle types are: (a) the globe valve, (b) the oblique or 'Y' valve, (c) the angle valve and (d) the needle valve.

The *globe valve* (Figure 7-3 and Figure 7-5) as their name implies, have a bulbous shape and are fitted with a disc or plug which sits in an orifice perpendicular to the axis of flow. The seat is replaceable or ma-

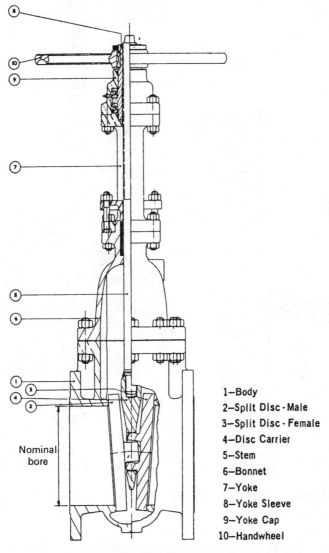

1—Body
2—Split Disc - Male
3—Split Disc - Female
4—Disc Carrier
5—Stem
6—Bonnet
7—Yoke
8—Yoke Sleeve
9—Yoke Cap
10—Handwheel

Nominal bore

Figure 7-2. Single Disc Ball and Socket Gate Valve.

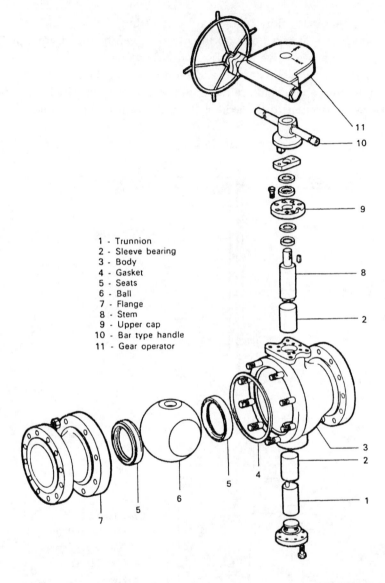

1 - Trunnion
2 - Sleeve bearing
3 - Body
4 - Gasket
5 - Seats
6 - Ball
7 - Flange
8 - Stem
9 - Upper cap
10 - Bar type handle
11 - Gear operator

Figure 7-3. Trunnion Mounted and Gear Operated Ball Valve.

chined integral with the body. The design has a greater pressure drop than the gate, plug or ball valves but it has very desirable throttling characteristics and positive shut-off.

Discs are manufactured from metal or a composition of fiber and binder. See (Figure 7-5a). The former fits into a taper seat, offers good

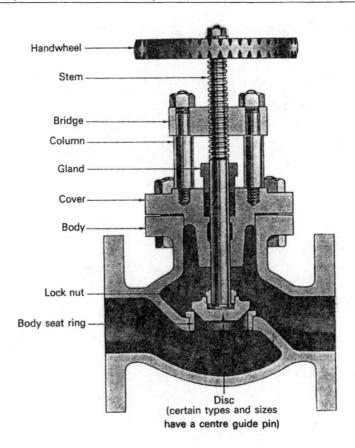

Handwheel

Stem

Bridge

Column

Gland

Cover

Body

Lock nut

Body seat ring

Disc
(certain types and sizes
have a centre guide pin)

Figure 7-4. Globe Valve

resistance to erosion and wire drawing and gives the best throttling results of the three types.

Composition discs are suitable for low pressure low temperature
condensate where there are no throttling requirements. They can be easily replaced and they protect the valve seat from damage and give positive shut-off. The valve is adaptable to many different condensate
services and less power is required to seat tightly (see Figure 7-5b).

The metal disc (Figure 7-5c) usually has a spherical or taper seat
and breaks down deposits which may form on the seat surface.

The normal size range for globe valves is one-quarter inch through
16 inches bore although they can be manufactured in sizes up to 30
inches bore with working pressures to 6000 psi and 1200°F.

**Figure 7-5a. Globe Valve Seat
with Metallic Disc.**

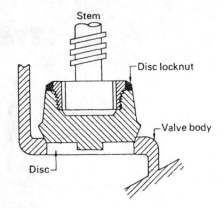

**Figure 7-5b. Globe Valve
Seat with Composition
Disc.**

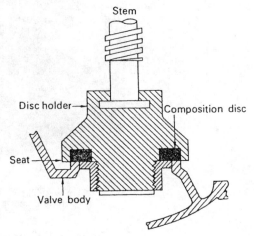

**Figure 7-5c. Glove Valve
with Tapered Seat and Disc.**

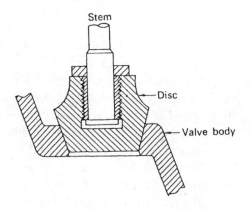

Y Valves

Oblique or 'Y' valves are a variation of the globe valve; the difference being that they have the stem and seat at approximately a 45 degree angle to the path of flow. The advantage of this design is that it has less pressure drop than the globe valve and has good throttling characteristics. They can have a renewable seat clamped between the two body parts, thus allowing quick replacement. This also, makes the valve suitable for conversion to an angle valve. Y valves are available in sizes from one-quarter inch to 10 inches bore, with maximum pressures to 6000 psi and temperatures to 1200°F.

Angle Valves

Angle valves are another version of the globe valve, with inlet and outlet ends at a 90 degree angle. Its use reduces the number of pipe fittings required in a piping system. It is usually in a one-piece body but can be provided in two parts similar to a 'Y' valve. Sizes and operating conditions are essentially as the 'Y' valve although a heavy duty type is also manufactured, suited to working pressures up to 6000 psi, with a size range of one-quarter inch to 6 inches bore.

Needle Valves

Needle valves are a type of globe valve used where fine control of flow is required. It is fitted with a tapered, needle-like plug that fits accurately into the valve seat. The stem usually has a fine thread for accurate flow control. The size ranges are one-eighth inch to 1 inch. They can be designed for service up to 6,000 psi, 1200°F.

General

All four types of valves; Globe, 'Y' Valve Angle and Needle valves are manufactured in a large material range including: cast iron, ductile iron, forged steel, cast steel, brass, bronze, and various grades of stainless steel, and corrosion resistant alloys.

The globe type valve and its variations are used extensively in industry, for throttling steam and/or condensate. They operate precisely under difficult conditions of pressure, temperature and corrosion. They can control a multitude of different fluids.

Check Valves

Automatic, and designed to prevent reversal of flow in piping systems, check valves are kept open by the pressure of the flowing liquid or gas and closed by back pressure or by weight of the checking mechanism. The valve is designed in various forms, such as the swing check, the tilting disc, the lift type, the ball or piston and the stop-check valve. They vary in size from one-quarter inch bore to 20 inches. However, they are seldom used in steam or condensate systems in sizes above 12 inches. Check valve selection is determined by the velocity of the conveyed fluid, the allowable pressure drop, the working pressure and the temperature. The three typical check valve designs described in what follows are illustrated in (Figures 7-6, 7-7 and 7-8).

Swing Check Valve

The *swing check valve* has a low pressure drop. It incorporates a hinged disc that opens when the inlet pressure of the fluid exceeds the pressure downstream of the valve. See (Figure 7-6). It is most suited to low velocities with infrequent flow changes. It is commonly used in pipe lines containing gravity flow or pumped liquids.

Composition discs can be used to prevent noise nuisance or where positive low pressure shut-off is required. The valve can also be equipped with an outside lever and weight for faster closure or to keep the valve open until a predetermined reverse pressure is reached.

Tilting Disc Check Valve

The *tilting disc check valve* is a low pressure-drop, straight-through design which provides rapid closure and is sometimes fitted with an external dashpot. The valve operates by the disc tilting at an increasing angle to the flow path as the flow rate increases. Pressure drop at low velocities is less than in the swing check valve, while at high velocities it is greater than in the swing check valve.

Lift Check Valve

The *lift check valve* (Figure 7-7) depends upon upward flow pressure raising a ball or disc. When the flow ceases or reverses, the ball or disc is forced back on its seat by back pressure and gravity.

Piston Check Valve

The *piston check valve* (Figure 7-8) is similar in principle to the ball or disc lift check valve operation except the piston is cushioned in its cylinder as it is forced upwards. This check valve should be considered for use if water hammer or hydraulic shock is a problem in the piping

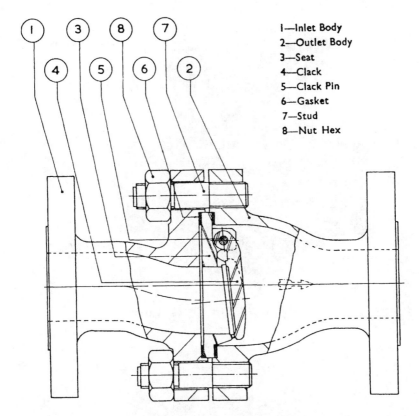

1—Inlet Body
2—Outlet Body
3—Seat
4—Clack
5—Clack Pin
6—Gasket
7—Stud
8—Nut Hex

Figure 7-6. Swing Check Valve.

network. This is particularly true if the water hammer is caused by another type check valve. See Chapter 5 for more information on water hammer and its prevention, and Chapter 10 for some specialty uses of check valves.

Stop Check Valve

The *stop-check valve* incorporates a spindle which can be operated in the same manner as an ordinary globe valve and can be adjusted for tight seating or restriction of valve lift. The spindle is not attached to the disc, but merely prevents it from reacting to a pressure drop across the valve. By adjusting the stem, the valve can be used for throttling the flow of condensate. Check valves can be manufactured of ferrous and nonferrous metals including: cast iron, forged steel, brass, bronze, manganese, other common alloys and stainless steel.

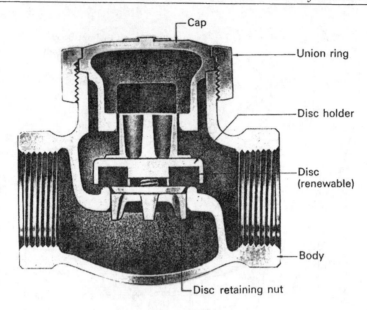

Figure 7-7. Lift Check Valve.

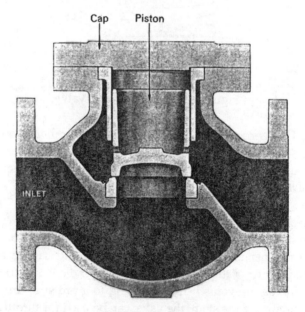

Figure 7-8. Piston Check Valve.

Foot Valves

Foot valves are a type of check valve, sometimes fitted with a strainer to prevent particles from passing through the valve and making their way from the tank into the condensate system. The valve is used to maintain sufficient head at the suction to permit automatic priming of the condensate make-up water pump.

Plug Valves

The last decade or so, has seen continuous development and refinement in plug type valves, to meet industrial requirements. It is used mainly as an on-off service valve. But adaptations to the shape of the inlet and outlet ports as well as changing the shape of the valve spindle it can be used as a throttle. The basis of operation is in turning a plug, either tapered or cylindrical, within a valve body containing a rectangular or circular through-port for the passage of the fluid.

The basic valve opens to full bore and is designed to insure there are no pockets where solids can gather, thus insuring easy cleaning and minimizing friction loss. It is quick operating; a quarter turn of the lever opens or closes the valve, fully.

The fitting of a cage within the valve and or modifications to the stem make them suitable for throttle control. See Automatic Process Control, Chapter 9.

Multiport Valves

The valves can be adapted to multiport construction for special uses in steam and condensate piping systems. Three-way, four-way and five-way valves are available in a variety of sizes but are seldom stock items in the larger sizes. See (Figure 7-12 and Figure 7-13) They are more often used in transport of the commodities that are being manufactured. But oftentimes, steam and condensate are mixed with the product or extracted from it. Precision valves that can accurately meter both steam and product are often needed in the manufacturing and process industries

By adopting the multiport design, important advantages are achieved in simplification of piping layout, elimination of pipe fittings and convenient operation. One multiport valve can replace up to four straight-through valves and there is less risk of product inter-mixing when it is not wanted. Care must be taken when using plug valves and positive displacement pumps. Damage to the pump, piping and equipment can result if the pump discharges against a closed valve.

Figure 7-9 indicates the various arrangements possible using the multiport principle. Study of these diagrams will suggest more economic layouts than can be obtained with a system of gate or globe valves. The multiport is also arranged in forms of 'Transflow' and 'Non-Transflow' as shown in (Figures 7-10a and 7-10b, respectively.

When the plug of a 'transflow' valve is rotated from one flow position to another, the second flow position starts to open before the first flow position is completely closed. This form is essential when momentary interruption of flow is not permissible—for example, on the outlet from a positive displacement pump.

When the plug of a non-transflow valve is rotated from one flow position to another, the first flow position is isolated before the second flow position starts to open. The midway position, however, is not intended to be used as a shut-off. This form is essential when momentary flow from one port to the other two is not permissible—for example, on the outlet of a measuring vessel or on certain types of tank manifolds. Another type of valve could permit mixing of a contaminated fluid with makeup condensate.

Multiport valves can only provide positive shut-off against full rated working pressure in the positive direction, that is, as illustrated in (Figures 7-11a and 7-11b, with the line pressure to hold the plug against the body port which is to be shut off from the higher pressure. For vacuum also, care should be taken to ensure that the conditions tend to hold the plug against the body port which is to be shut off.

Negative pressure as illustrated in (Figures 7-11c and 7-11d will not give a tight shut-off as the line pressure applied in the direction as shown tends to force the plug away from the body port which is to be shut off.

Multiport valves are designed for flow diversion and should not be expected to provide a positive shut-off in intermediate positions.

PLUG VALVE LUBRICATION

There are two basic types of plug valves: the lubricated and the non-lubricated, the latter being provided with plastic sleeves, linings or coatings, thus avoiding the need to lubricate. *Standard type lubricated plug valves* have a tapered plug, the large end of which is at the operating shank. Lubricant is fed into the shank and travels via a system of lubricant grooves, The lubricant film between plug and body ensures an effective seal. When the valve is open, the seating surfaces are protected

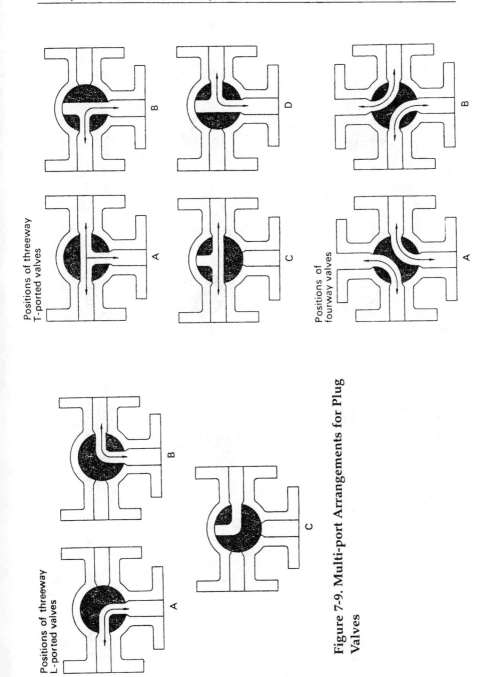

Figure 7-9. Multi-port Arrangements for Plug Valves

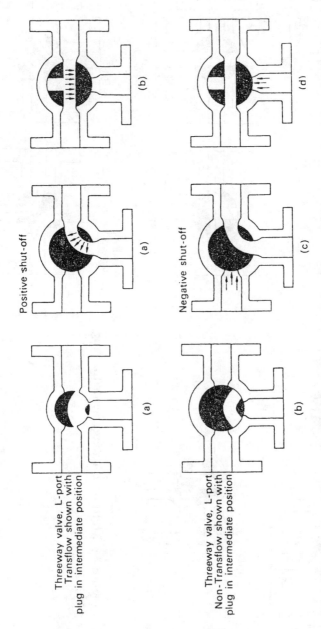

Figure 7-10. Transflow and Non-transflow
Arrangements for a Multi-port Valve.

Figure 7-11. Positive and Negative Shut-off of
Multi-port Valves.

from corrosion or erosion from the pipe line fluid. Valves can be lubricated under full line pressure and the design of grooves is such that loss of lubricant into the pipe line is prevented.

An important feature is that the lubricating system provides the means of building up a very high hydraulic pressure at the small end of the taper plug. This serves to ease the plug in its tapered seat, thus giving easy operation even though the valve may not have been operated for a long period of time.

A wide range of lubricants have been developed to suit many conditions of service; including steam and condensate systems. The correct choice ensures an inert barrier, resistant to the line fluid, including acids, solvents, foodstuffs and pharmaceuticals, hot gases and other high temperature applications with or without steam or condensate.

Inverted type lubricated plug valves are used in the larger sizes. In this case the plug is, in effect, reversed, the taper reducing to the shank with the base seated on a pressure screw set in the valve base. The lubrication is similar to the standard type valve.

NON-LUBRICATED PLUG VALVES

The principle of the sleeve type design is the introduction of a plastic sleeve such as TFE a fluorocarbon which completely surrounds the plug and is locked in place on the metal body of the valve. This results in a continuous covering of the primary sealant between seal and plug. The plastics used are durable and self-lubricating, have a low friction coefficient and are inert to all but a few chemicals.

Research and development has insured good thermal stability and that deformation of sleeves is minimized even under severe conditions of operation. The result is a valve of simplicity in construction, giving pressure-tight conditions and the sleeve overcomes the problems of galling and sticking.

A unique feature of this design prevents deformation of the sleeve into the flow passages and protect it against erosion from throttle effect when the valve is being opened or closed. The built in partial cover of the flow passage affords further protection from abrasion by removal of adhering contaminants from the plug as it rotates. Most important, they act as a barrier against line fluid under pressure getting behind the sleeve. This can cause damage to the valve and product.

SLEEVED VALVES

Sleeved valves are in the size range of one-quarter inch to 12 inches bore. Operating pressures should not exceed 300 psi saturated steam or condensate. They meet a wide range of conditions of corrosion in the process industries and with special sleeve designs, are used in nuclear power plants.

SPLIT PLUG VALVES

Split plug valves (Figure 7-12), were developed for a wide range of anti-corrosion, pressure and temperature conditions. They work on a similar principle to the double disc gate valve. When closed, the fluid pressure acts on both plug segments. The valve is designed to operate over long periods of service between outages. They are manufactured in various stainless steels, nickel, and other common alloys. Split plug valves are non-lubricated valves.

The lubricated plug valve is the usual choice of steam and condensate system designers, however, there is a trend to use coated plug valves at low pressure and temperature applications.

Plug valves in general are manufactured in addition to those already referred to, including cast iron, ductile iron, steel, stainless steels, bronze, nickel, etc., together with a range of plastics for sleeves, linings and coatings.

The principle advantages of plug valves are quick opening and closing, simplicity of construction and operation, a minimum of moving parts, minimum space to operate and adaptability to a wide range of operating conditions. In addition to steam and condensate it can transport acids and caustic materials. Generally, the valves operate well in corrosive conditions in the process industries.

BALL VALVES

Basically an adaptation of the plug valve, ball valves have been in use for many years. Recent developments in the technology of plastics and elastomers, together with improved and more economic methods of manufacture, have led to a much wider use of plug valves. This has

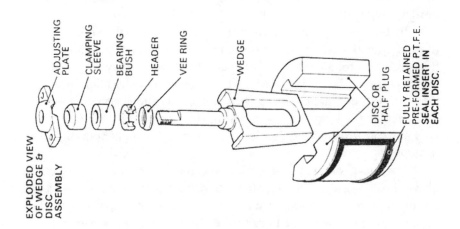

EXPLODED VIEW OF WEDGE & DISC ASSEMBLY

ADJUSTING PLATE

CLAMPING SLEEVE

BEARING BUSH

HEADER

VEE RING

WEDGE

DISC OR 'HALF' PLUG

FULLY RETAINED PRE-FORMED P.T.F.E. SEAL INSERT IN EACH DISC.

Figure 7-12. Split Check Valve.

resulted in a wide spread acceptance in the process and manufacturing industries. Like the gate and plug valves they are essentially for 'on-off' service, although they have throttling application dependent upon the design employed.

Instead of a plug, the valve is fitted with a ball having a hole through one axis which, when open, connects the inlet and outlet ports. It is straight-through in this position and a 90° turn achieves quick opening or closing. The design is compact, requires no lubrication, is simple in principle, gives tight shut-off with low torque and is easy to maintain. The circular ports ensure low pressure drop, its compactness makes it very suitable for manifold use. It gives positive sealing in both directions of flow.

The most common design allows a free floating ball to be pushed into a flexible seat by line pressure, but a more recent design uses a ball mounted on top and bottom trunnions to maintain constant seat loading at high line pressures and to reduce operating torque for easier operation. The use of trunnions eliminates the possibility of damage from vibration.

Valve variations are the full opening, reduced opening and venturi bores for top entry or end entry. The top entry is best suited for use with manifolds. Each valve in the manifold can be serviced without interference with the other valves.

The valves are manufactured in a variety of materials including cast iron, ductile iron, carbon steel and stainless steel and a variety of other ferrous and non ferrous materials. Various corrosion-resistant materials including monel, titanium, hastelloy, and plastics. The valve designs can be suitable for both high temperature and cryogenic purposes and when conditions of commercial vacuum are in common use.

The size range is from one-quarter inch to 20 inches for steam and condensate. Pressure for steam use is as high as 3000 psi.

SPECIALTY VALVES

Piston valves are seatless valves based on a sliding piston which uncovers cylinder ports. They are used extensively in hydraulic and pneumatic systems and other power equipment.

A type for wider use in services such as steam industrial process pipe systems consists of a piston operated by the valve spindle which moves through two non-metallic resilient packing rings separated by a ported lantern bush which fits inside the body. The tightness of the valve is dependent upon the fit of the pistons in the valve rings.

The design is such that any abrasion or erosion which may take place has no effect upon the tightness of the valve. This keeps flow resistance at a minimum. The spindle screw is external and stuffing boxes are not required up to sizes of 2-inch bore. In larger sizes, balanced pistons are adopted and a stuffing box is then necessary in the valve. Sizes range from one-half inch to 4-inch bores; working pressures of 1200 psi, maximum and its corresponding saturation temperature are acceptable for steam systems.

Valves for high vacuum conditions. Thermal systems are often subject to high vacuum and special vacuum breaker valves become part of a well designed system. This permits rapid pressure reduction and material cooling. Chapter 10, Steam Traps, covers the subject of steam system cool-down in more detail and vacuum breakers, steam traps and thermal tracing are covered more thoroughly. Since vacuum breakers can be used over a wide range of operation, good engineering practice dictates that they be designed to resist leakage, and thermal and dynamic shock. Working pressures are as high as 10,000 psi and the valves are precision made from a variety of steel and stainless steel.

Vacuum breakers automatically release air from water mains and steam lines and also allow air into a pipeline to prevent possible implosion or collapse of the line by differential pressure caused by the vacuum conditions.

Other specialty valves such as safety valves, relief valves and solenoid valves are covered under Chapter 9, Automatic Controls Metering.

VALVE MAINTENANCE

In the past, valve maintenance was done on site by plant maintenance personnel. The trend today, is to replace leaking valves during scheduled outages. Most plants will contract with valve repair companies to replace damaged or leaking valves with rebuilt or new ones. The valves removed and repaired are reused. Small and inexpensive valves are usually discarded rather than repaired and replaced when their discs or seats are damaged.

Rebuilding large or expensive globe valves will usually require that their valve disc and seat be machined or replaced. In both cases the parts are removed from the valve body. The discs and seats are precision machined on a grinder first, then reassembled in the valve body. They are finished by hand grinding to a precision fit.

Rebuilding gate valves usually requires replacement of the wedges and seating surfaces in the valve body. The seating surfaces are screwed into the body on large or expensive valves. See (Figure 7-1). It is seldom that the valve seat is an integral part of valve body and when this is the case the seat can sometimes be machined on a shaper. Skilled experienced machinists are required. Other parts needing repair are: stems, bonnets, flanges and packing glands.

Stem repair is simply manual polishing to remove scratches, deposits and packing deposits. Corrosion can be removed from the stems screw, either manually or on a lathe. Care must be exercised not to under cut into the threaded part of the stem. Corrosion can be removed from the threaded portion of the bonnet using a milling machine. The same care must be taken not to under cut the threads.

The bonnet is the top cover of the valve. It contains the stem, disc and packing gland. The packing gland prevents leakage around the stem. A bolted flange connects the bonnet to the valve body. Occasionally, the bonnet and body flanges can be damaged or corroded. The fitted surfaces of the flanges can be cleaned of corrosion and gasket material, or if necessary machined on a milling machine.

The flanges at the fluid entrance and outlet may require cleaning to remove corrosion and some of the remnants of the old gaskets. Should there be damage to the flanges on each side of valve, they can be repaired on a milling machine. Care should be taken not to remove more than 0.001 or 0.002 inches from the surfaces. Valves are designed to allow for minor reductions in their matching surfaces during repair. Repair of valve mating surfaces need not be necessary during every annual outage. With good planning and care in operation, valves may need complete overhaul about every five years.

A valve packing gland is detailed in (Figure 7-13). Repacking requires removal of the old packing material and replacement with new cut to fit packing rings. Packing rings are available in graphite reinforced with stainless steel wire. Cut joints in the formed to fit packing rings should be staggered when inserting them into the gland. The upper part of the gland is then used to compress the packing and make a tight seal around the stem.

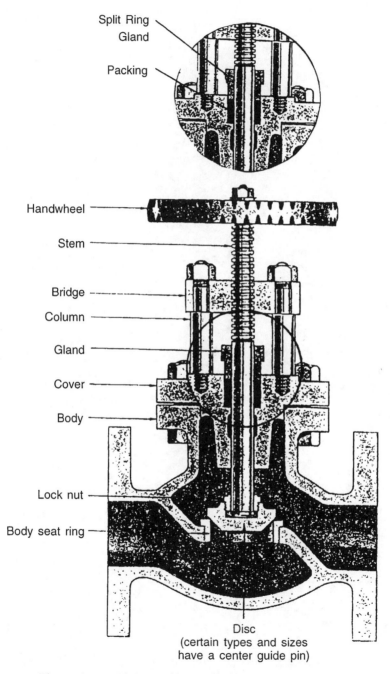

Figure 7-13. Globe Valve with Packing Gland Detail.

Chapter 8

Valve Drive Mechanisms

Rotation and the lever are the basic means used for moving a valve control element, whether it be disc, diaphragm, plug, ball or globe valve. Actuation is by a threaded or plain stem or spindle which either rotates or moves axially. It may have a combination of these actions. The stem extends from the control element to the outside of the valve and terminates in a hand-wheel, lever or wrench. This is but one of the many actuating devices designed to suit a variety of situations and conditions. Exceptions include check valves and foot valves, some types of safety valves, and regulating valves, where the operation is pressure dependent.

The following methods of stem operation are the most commonly used in steam and condensate system valves:

(a) Stem rotates only (rotating stems)—Non-rising stem gate valves; ball valves; plug and sleeve valves; quick opening rotating disc valves.

(b) Non-rotating stem with endwise movement (sliding stem), outside sliding with yoke: gate valves, globe, slide, piston and sleeve valves.

(c) Rotating stem with inside movement; rising stem gate valves, globe, angle, 'Y' and needle valves.

Other major features are the inside threaded stem with non-rising stem or with rising stem, and the outside threaded stem with rising stem. All of these particularly apply to gate valves (see Figure 8-1).

175

Globe valves are obtainable with either inside or outside threaded stem. Rising stem valves both turn and move axially The threaded stem of the outside stem gate valve moves axially only (see figure 8-2).

The choice of inside or outside thread is influenced by service requirements, particularly operating temperature and corrosion. Other

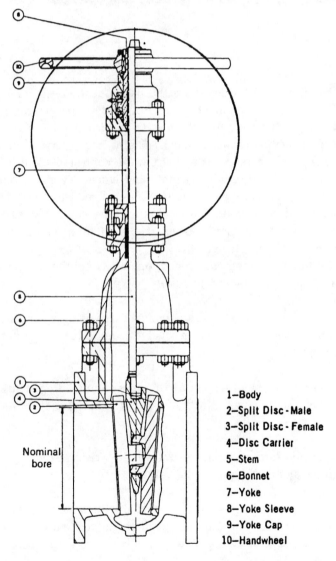

1—Body
2—Split Disc - Male
3—Split Disc - Female
4—Disc Carrier
5—Stem
6—Bonnet
7—Yoke
8—Yoke Sleeve
9—Yoke Cap
10—Handwheel

Nominal bore

Figure 8-1. Gate Valve with Outside Stem and Yoke OS&Y.

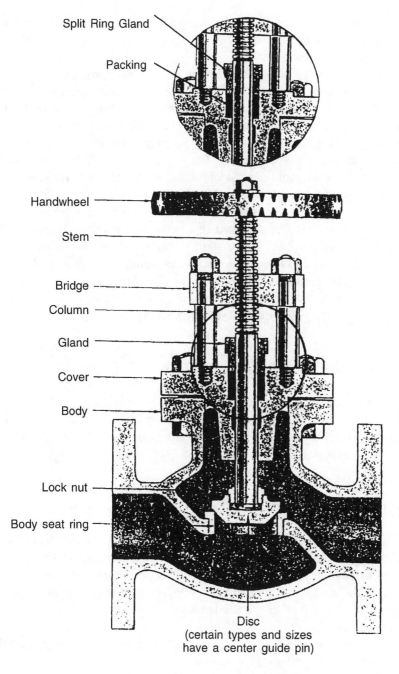

Split Ring Gland

Packing

Handwheel

Stem

Bridge

Column

Gland

Cover

Body

Lock nut

Body seat ring

Disc
(certain types and sizes
have a center guide pin)

Figure 8-2. Globe Valve with Packing Gland Detail.

considerations are space to accommodate the valve equipment and an operator. It is important to allow clearance for a rising stem valve if installed.

If the steam or condensate temperature is high, then it is prudent to install an outside threaded stem to obtain maximum atmospheric cooling conditions. An inside-threaded stem may seize or bind and thus become inoperable. In OS and Y valves, the thread is not in contact with the process fluid due to the intervening packing. The stem position indicates the extent of valve opening. These valves are suited to conditions of high temperature and corrosion, but sufficient clearance must be provided for the rising stem when opening the valve.

Inside threaded stems are exposed to steam and condensate temperatures. They may be exposed to corrosion conditions, particularly oxidation. Stainless steel is the stem material that is cost effective for most steam and condensate service. The rising stem gate and globe valves indicate positions of the wedge or disc. The yoke of these valves can be equipped with a scale to use in determining the amount of valve opening, and the resultant flow through the valve.

LEVER OPERATED SLIDING AND ROTATING STEMS

Quick opening service is obtained with sliding stem valves. They can be fitted in gate and globe valves. Rotary stems on quick opening valves only require a quarter turn from fully open to fully closed and are used with plug valves and ball valves.

EXTENSION SPINDLES, GEARING, ETC.

It is sometimes necessary to operate a valve from a position where a direct mounted valve hand-wheel or lever is not accessible. In such cases, extension spindles are frequently used. There are innumerable types of these and an arrangement can be assembled to fit most applications. Occasionally, gears are needed to accomplish the task of operating a remote valve.

Figure 8-3a to 8-3g and Figure 8-4h to 8-4m show diagrammatic arrangements of extension spindles to meet some of a multitude of situations that may arise.

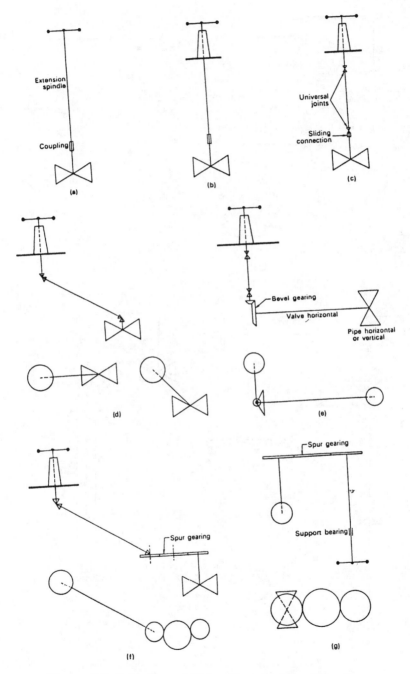

Figure 8-3. Selection of Valve Remote Operating Gear.

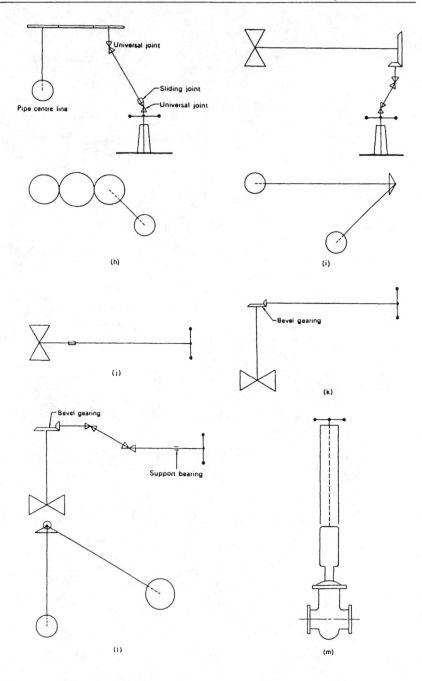

Figure 8-4. Additional Remote Valve Operating Gear.

FOR OPERATING POSITION ABOVE THE VALVE

Type (a)—Is a direct mounted spindle and hand wheel connected to the valve stem. The valve stem and the extension spindle are connected by a mechanical or pipe coupling.

Type (b)—Is similar to (a) but the extension spindle protrudes through a platform and carried in a standard floor column pedestal. The rigid extension spindle will only accommodate small amounts of movement of the steam or condensate piping and, where greater flexibility is required, use type (c).

Type (c) should be adopted as the use of universal joints takes up lateral movement and the sliding connection at one end of extension spindle permits vertical movement of the valve.

Type (d) is an arrangement to satisfy conditions where the valve is offset to the floor column in one or both directions but it will not accommodate axial movement. Types (e) and (f) are variations on type (c) where the valve operation is accomplished by bevel or spur gearing to overcome difficult locations and to provide, if desirable, a reduction ratio to ease the valve operation.

OPERATING POSITIONS BELOW THE VALVE

Type (g) is a spur gear arrangement and is of rigid construction. It also permits dual operation above the valve as well as below, by extending the hand-wheel shaft upwards beyond the spur gear.

Type (h) is similar to (g) but is arranged to take valve pipe movement by the introduction of universal and sliding joints.

Type (i) is a bevel gear arrangement particularly for valves mounted in vertical pipe lines, or where headroom in horizontal pipe installations precludes the use of spur gearing.

ARRANGEMENTS WITH OPERATING
POSITION ALONGSIDE THE VALVE

Type (j) is identical with type (a) except that the valve is mounted horizontally and is used where the standard length of valve spindle is inaccessible from an adjacent platform or floor. The arrangement can also be used for extending and operating a hand-wheel through a safety barrier or wall.

Type (k) is a bevel gear arrangement which can be used where the valve is either horizontally or vertically mounted in the piping, or where barriers or walls or overhead may interfere.

Type (1) is similar to type (e) except that the hand-wheel is mounted on a wall bracket instead of a floor column.

Type (m) extension spindle is used where (a) is inadequate. It is enclosed in a cylindrical conduit for conditions where valves are buried, partially buried or likely to be subjected to flood water. The valve yoke is equipped with a tubular extension to raise the operating mechanism to the desired height.

Chain wheels are used for overhead valves where safety regulations permit. These are fitted with chain guides and can also be obtained in spark proof material for safety in hazardous areas.

Other devices for hand operation include: single and 'T' shaped wrenches for 90° operated valves, ratchet wrenches for operation in confined spaces, bell-crank levers with chains for 90° valve operation.

VALVE INTERLOCKING DEVICES

Considerations of safety for the operator and plant, may require padlocking for prevention of unwanted operation at certain times. For this reason valves can be obtained that can be padlocked. Also valves are available with mechanical interlocks to manage safe operation. Some of these features are:

(a) Valves can be coupled to operate together from a single valve wheel.

(b) The valve coupling described in (a) can be provided with reduction gears to give the operator or power driver a mechanical advantage in operating the valve.

(c) Where a number of valves require to be interlocked or where valves in a system are not in close proximity, interlocking is achieved by fitting a metal hood over the top of the valve allowing only the insertion of a special wrench or key when the valve is in the fully open or fully closed position. Once the valve is operated it must be returned to its original position before the wrench or key can be removed.

(d)   For more complex arrangements of valve interlocking and also where valves are required to be interlocked with other equipment, special locks are fitted which can only operate by means of a coded key system. Interlocking systems of this type can be designed to ensure the correct sequential operation of any number and combination of valves.

(e)   Valve manufacturers can usually supply the fittings referred to in the above.

VALVE ACTUATORS

Power operated, valves are distinct from valves that are a part of an automatic control system. They can be remotely operated without a controller or input from a sensor. Although, power operators for valves can be part of an automatic control loop. That is discussed in Chapter 9. What is discussed here is the remote operation of a valve using a power drive to operate the valve. The power actuator is a precision mechanism for control of most types of valves. The advantages are that a valve or group of valves can be remotely operated by one control room operator. Also, one valve or a group of valves can be operated from a control point, other than the main control room. These types of remote control stations along with a centrally located automatic control system, improve plant efficiency and product quality control. Inaccessible valves can be readily operated. Appropriate action can be taken in response to an emergency.

There are two basic types of power actuators: the double acting power drives and the fail-safe drives. Fail-safe drives return to a predetermined open or close position on loss of power. Manual control of the valves may be desirable upon power failure. This can be provided for in the plans and specifications for a retrofit, or an original design, if desired.

Figure 8-5 shows a pneumatic actuator that is opened by air pressure and is closed by springs upon loss of air pressure. In this case air is used as the power supply. Other sources of power used in actuators are

used as the power supply. Other sources of power used in actuators are liquids and gases such as water or steam. Electrical power is frequently used. It can be a motor drive or a solenoid. Solenoids are discussed in more detail in Chapter 9.

The most usual method of mounting an actuator is directly onto the valve with a connection to the valve stem. They can also be remote mounted on a pedestal, wall or floor mounted with extension spindles and universal couplings in the manner shown in Figure 8-1a to 8-1m, as shown previously. Power drives a available for: gate, globe and the entire range of plug and ball valves. Power drives are available for valves in all sizes.

Many refinements can be provided, such a jilting: torque limit switches to ensure application of limited thrust in positive seating valves, provision for indicator switches, integral starters, solenoid valves for fast operation, and selector switches for three-position valve control.

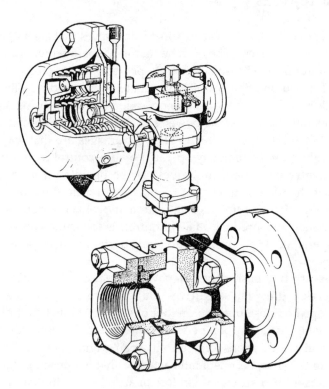

Figure 8-5. Diaphragm Operated Actuator Mounted on a Ball Valve.

Provisions can also be made for rising or non-rising stem valves.

Electric motors and equipment can be standard squirrel cage motors: weatherproof, totally enclosed, explosion proof, or certified flameproof to meet conditions of operation and satisfy the regulations of various authorities.

Actuators can be operated by push-button or control valve from positions located on the main control panel in the control room or from control stations or by fully automatic control. Instruments as described later in the Chapter 9, deal with automatic controls.

VALVE POSITIONERS

A positioner is an instrument generally equipped with characterizing cams and usually mounted on the valve actuator and operated by pneumatic or electro-pneumatic means. It is used for certain control valve applications where a high degree of accuracy and a positive response is required and is of particular importance for the following operating conditions:

(a) To maintain valve position when high pressure drop or friction creates unbalanced forces in the valve. Under these conditions, the controller output pressure signal would not be sufficient to provide linear positioning of the inner valve.

(b) In situations where high pressures need tight stem packing to prevent leakage. This causes high torsional load on the stem requiring heavy duty valve drives

(c) Where long transmission lines between the controller and control valve may produce unacceptable time lags.

(d) To allow two or more valves to be operated in sequence or in parallel with each other.

(e) To improve the response time of the valve.

Chapter 9

Controls and Metering

This is a large field and distinct from that of manually operated valves. Its growth has been rapid over the past few decades and it will undoubtedly continue to expand to meet the increasing needs of industries requiring finer accuracy for more complicated processes. Control in steam distribution systems may be for conditions of pressure, temperature and rate of flow. They may be for steam or condensate control. Sensors are used to detect deviations from a previously set value and send signals through a control loop to actuate the valve being controlled. In steam distribution and condensate systems, valves, pumps and tank levels are the devices that are usually controlled. Some devices found in steam distribution systems have self contained controls while others depend upon sensors to transmit a signal to a controller outside of the device.

SELF CONTAINED VALVES

These valves do not require an external power signal. They include backpressure regulating and pressure reducing and pressure regulating valves. These valves are operated by the steam or condensate in the piping system. Temperature regulating valves convert a change in temperature into a pressure signal by the expansion or contraction of a heat sensitive fluid contained in a sensor. Flow regulating valves maintain constant flow rates. Safety and Relief Valves can be included in this category, except that their purpose is protection and not process control. They are discussed in the chapter 5.

PRESSURE REDUCING AND REGULATING VALVES

Pressure reducing valves are used for automatic delivery of a constant selected reduced pressure at the downstream side of the valve. Designs vary from the completely self contained unit shown in Figure 9-1 to the pilot operated valves for accurate regulation. Constant downstream pressures can be maintained even with fluctuating upstream loads when using a pressure regulating valve.

Steam passes from the inlet port to the valve outlet, while the valve disc throttles the process. See Figure 9-1. An adjustment screw compresses the valve's main spring which acts on the valve's disc to open the valve to the flow of steam. The valve spring located under the disc insures that the disc closes completely when the up stream pressure decreases below a predetermined value. These valves have a limited use: maintaining only one down stream pressure at a time. Re-adjustment to a new pressure set point can be time consuming. For this reason, the pressure regulating valve is more suitable for controlling steam pressure than the pressure reducing valve.

Other designs of pressure regulating or control valves range: from the obsolete weight loaded valves to electronic pressure sensors. Other models include features such as pressure loading, where a constant fluid pressure is the loading element.

TEMPERATURE REGULATING VALVES

Temperature regulating valves are designed to control temperature in a treated system, these are constructed in much the same manner as the pressure regulating valves already discussed. The valves are operated by a change in temperature causing the expansion or contraction of a heat-sensitive fluid which delivers an increased or decreased pressure to the bellows which in turn act on the pilot valve. A simple type of temperature regulator suitable for steam is shown in Figure 9-2 complete with sensing bulb, capillary tube, and bellows.

The electronically actuated pressure regulating valve has a pressure sensitive element. The element is an electric resistor that expands length wise when pressure is lowered and contracts by the force of a high pressure. Electrically, the long resistor opposes the flow of current while the short resistor permits easy current flow. This current can be amplified a

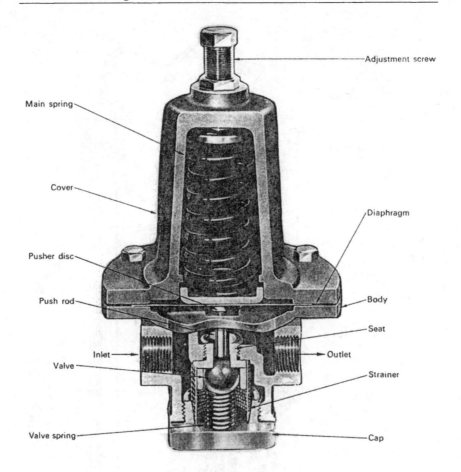

Figure 9-1. Pressure Reducing Valve for Steam.

number of ways, and transformed to a higher voltage to drive a motor on the valve. Figure 9-3 is a schematic of a valve control loop. Valve motor drives were discussed in Chapter 8.

FLOW REGULATING VALVES

These valves regulate a rate of flow of steam or water and maintain this rate constant, regardless of pressure fluctuations in the piping system. Various designs exist, one type being operated by means of a constant pressure drop across an orifice, this being graduated by use of a

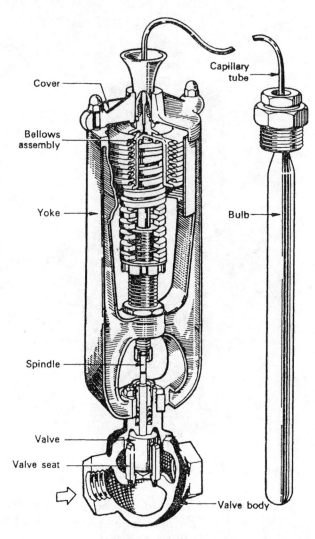

Cover

Bellows
assembly

Yoke

Spindle

Valve

Valve seat

Capillary
tube

Bulb

Valve body

Figure 9-2. Temperature Regulating with Sensor.

sleeve which proportionately covers the orifice slot, with the setting
generally being shown on a dial. Another method to maintain a constant
flow rate, despite pressure variations, is to regulate the port opening of
the sleeve control valve through an impeller, which is spring loaded.
Forces acting on the impeller are balanced so that upstream pressure
acting downwards directly onto the impeller is counteracted by downstream pressure across the orifice, plus the force exerted by the sprint.

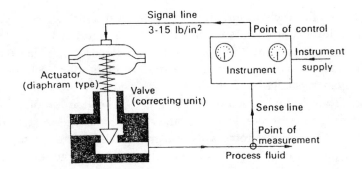

Figure 9-3. Simple Control Loop for Valve Control.

This total force then equals the pressure drop across the orifice and a constant flow rate is maintained.

LEVEL CONTROLLERS

The simplest type of level controller is the ball float valve, where a floating ball attached to a lever maintains an already determined liquid level by opening or closing a piston valve. This principle is shown in connection with steam traps in Figure 10-6, Chapter 10.

Refinement to the principles of level measurement, and control include direct operation of pilot valves, switches and similar devices, and the initiation of alarm signals. Variations in type include a double seated valve, pressure operated, where the float and lever actuate a pilot valve allowing line pressure to position the main valve, and a carefully machined valve where there is normally no leakage possible when the valve is closed.

The Tri-level Regulator device is a displacement type of controller. It is widely used because of its adaptability and flexibility for the indication and recording of, liquid levels. As well as functioning as a level control and use as an 'on-off' device the controller possesses a fail safe feature. It is based on the principle of displacement of an immersed body. The principal element is the body itself, called the displacer, which is suspended from a torsional measuring spring. The displacer always weighs more than the upward buoyant force developed by the liquid in which it is submerged and it is usually of cylindrical shape and constant cross sectional area, so that for each equal increment of submersion depth, an equal increment of buoyancy change will result. Thus it yields the desired linear characteristic of this type level controller. A diagram of a liquid level controller assembly is shown in Figure 9-4.

Another type of controller is the torsional spring, known as a torque tube. It is designed to twist a specific amount for each increment of buoyancy change, and to be insensitive to pressure changes in the vessel. The degree of rotation of the inner end of the torque tube is in linear proportion to the degree of buoyancy. The design is such that the degree of torque tube rotation is transmitted with accuracy, to the outside of the vessel in such a way that the interior of the torque tube is carried through the vessel at atmospheric pressure. The requirement of modern processes for longer range level measurement, greater accuracy, and wider versatility has lead to the development of the displacement type of controller. What is presented here, are descriptions of the more common types of liquid level controllers. A schematic of a control loop is shown in Figure 9-3.

SOLENOID VALVES

Solenoid valves are used for on-off service, particularly for emergency shut-off or opening and the design is principally applied to sliding stem globe valves, or similar types of valves. In this case, the valve stem is fitted with an armature, which is drawn into the magnetic field of the solenoid winding when current is passed through the coil. The force exerted by a spring holds the valve plug either in the closed or open position—the solenoid therefore is required to move the valve plug against this force. Operating pressures are from extreme vacuum to as high as 10,000 PSI and temperature from cryogenic service to over 1000F. The usual size ranges from 3/16 in bore, and up. In larger valves where the operational forces are two great for a solenoid to overcome, a small pilot solenoid valve is used to control the fluid which pressurizes the actuator mounted on the main valve.

AUTOMATIC PROCESS CONTROL VALVES

These bring controls into the realm of instruments and a simple control loop assembly. They consists of an instrument, which senses any deviation in the process conditions from a desired value previously set

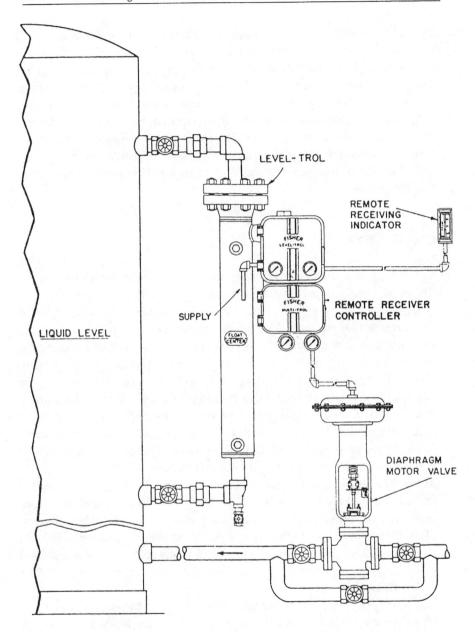

Figure 9-4. Fluid Level Controller. May be used in monitoring and controlling the liquid level in a boiler or other steam equipment.

in the instrument. A pressure signal is then sent to the valve actuator in order to increase or decrease the exposed portion of the bore of the valve at any point from maximum flow to zero. Any type of valve can be arranged to control steam or condensate in a pipe line, but the degree of extent of control required effectively narrows the choices of available valves. If a large amount of throttling is required, then this will immediately define a few types of valves. The types of valves used for automatic process control are: the globe value, angle and 'Y' valve, the ball valve, the plug valve, the butterfly valve and, to a more limited degree, the diaphragm valve.

GLOBE VALVE

Globe and angle valves satisfy the majority of applications and are the most used valves, particularly for fine and accurate control. The globe valve can be either single or double ported. The single ported valve is less likely to leak. The double ported valve, Figure 9-5, requires less power to operate since the force acting on the two discs are opposed to each other. This makes it easier to turn the screw in the valve stem. The double ported valve is more likely to leak since there is twice as much area to seal when the valve is closed. If a high degree of tightness is necessary, then a single ported valve may be required. This type of valve will require a larger more powerful drive if the valve is automatic. If the degree of leakage is not critical then a double seated design, where the forces on the valve discs are equalized, then a smaller less costly driver can be used. It should be noted, that the double ported valve is usually more costly than the single ported valve. When considering both performance and economics the double ported valve would be the most likely choice over the single valve if throttling is a requirement. Usually, in automatic systems a manually operated single seated valve is installed before a throttling valve to provide the security and safety when part of the plant is off the line.

For applications where a high degree of leak tightness and high pressure drop are encountered a balanced piston valve is a good selection. It closes tightly and has a good throttling characteristic. See Figure 9-6.

Another type of throttling valve used in steam systems is the plug valve. Figure 9-7 shows a series of inner valve plugs which perform a selection of services. They are: Equal percentage, Modified linear V-port,

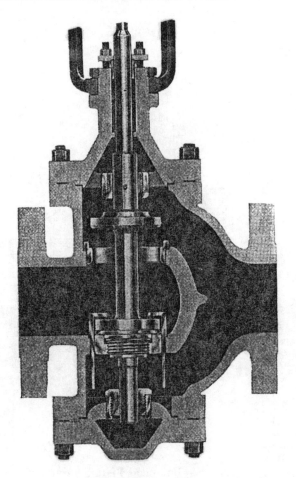

Figure 9-5. Balanced, Double Ported Valve.
(often used as a steam engine throttle)

Linear contoured, and Quick opening. These valves perform according to the characteristics shown on the valve characteristic curves in the figure. They are plotted on a semi-logarithmic scale format. The three types of characteristics shown are for the three inner valve plugs illustrated in the Figure 9-7.

The valve flow characteristic is defined as the ratio of the movement of the valve disc off its seat-vs-flow of fluid through the valve. The movement of the valve disc can be measured by the movement of the valve stem.

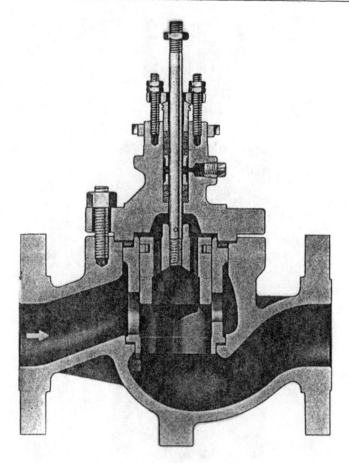

Figure 9-6. Balanced Piston Valve.

The linear characteristic is used where a true Linear characteristic is required and where the pressure drop in the system is fairly constant and concentrated at the valve seat. However, due to losses in the piping system upstream and downstream of the valve, the available pressure drop across the valve usually decreases as the flow increases, and in order to allow for cases where a large portion of the system pressure drop is not concentrated at the valve seat, the Equal percentage characteristic is preferred. This characteristic is also used if long lags exist in the complete control loop. There are some cases where it is desirable to furnish automatic control plug valves with smaller capacities than would normally be supplied with a given valve. This is done by modifying the trim to restrict the flow through the valve to 40 percent of the full capac-

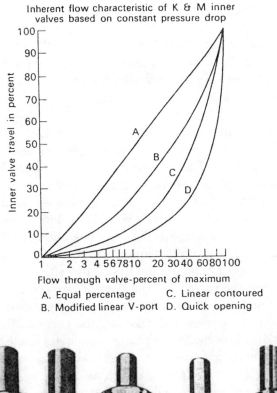

Inherent flow characteristic of K & M inner valves based on constant pressure drop

A. Equal percentage C. Linear contoured
B. Modified linear V-port D. Quick opening

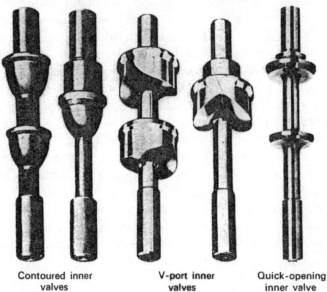

Contoured inner valves

V-port inner valves

Quick-opening inner valve

Figure 9-7. Selection of Plug Valve Stems with Different Flow Characteristics.

ity known as 40 percent trim. Some of the reasons for this are as follows:

(a) To provide structurally strong bodies while still retaining correctly sized valves. To make provision for future requirements by providing a body large enough, but with the valve plug sized for current conditions.

(b) To provide valve bodies large in size to reduce inlet and outlet velocity. To avoid the use of expensive line reducers and to correct over-sizing errors. There are inner valve forms to meet many different requirements; examples are the fitting of a ported cage to meet a special characteristic. The valve plug moves inside the ported cage and can give the valve a number of flow characteristics, depending upon the shape of the plug. In addition to the trim, there is another element which can be incorporated in single or double seated valves to minimize the destructive effects of high pressure drop across the valve seat. The element is a single or double cage with a number of radial holes to allow liquid to pass through and progressively release and dissipate the energy. This greatly increases the life of the inner valve and promotes stable control while reducing noise and vibration. Most manufacturers define control valve size in terms of Cv. The Fluid Control Institute (FCI) has quantified the term Cv and defined it as the flow of water in US gallons per minute at standard conditions which a valve will pass with 1 psi pressure drop. Once this factor is established for a specific valve, the fluid flow which that valve will pass can be determined from a standard fluids flow handbook, such as *Cameron Hydraulic Data* published by the Ingersoll Rand Corp., or other reliable source. The flow factor then is the basis for determining the flow through control valves. When using gas, steam, or vapor formulas, it is important to ensure that the pressure drop across the valve is low enough to avoid sonic velocities causing unwanted noise and vibration. If the downstream pressure of the valve is maintained at more than one half the upstream pressure, sonic velocities can usually be avoided.

For sizing purposes, valve pressure drop must always be taken as equal to or less than one-half the supply pressure. It is this value which is substituted in the equation to calculate the required Cv. When sizing for liquid flow there are other physical properties of a liquid which can effect the valve sizing, that are not in the standard handbooks. These are discussed in the following:

VISCOSITY

The liquid sizing formula does not take viscosity into account. It is known that below a Reynolds Number of 2000, liquid flow changes from turbulent to viscous and under this condition a correction factor must be applied. Where this is suspected, first calculate the Cv using the method for non viscous flow. Then determine the Reynolds Number as would be done for standard pipe of the size of the valve opening. Figure 9-8 shows relationship of viscosity vs. Reynolds number.

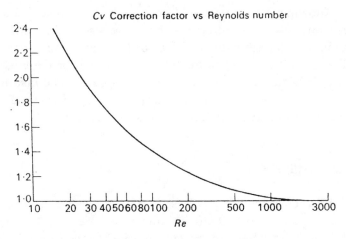

Figure 9-8. Viscosity Correction Factor.

Valve sizing is also affected by the flashing, of liquid to vapor when passing through the valve opening. This occurs when liquid passing through a valve experiences a pressure drop which is large enough to lower the resultant downstream pressure below the liquid's vaporization point.

The calculation of the effect of flashing on valve sizing has a number of different approaches, all of which lead to little change in the calculated Cv. However, probably the simplest and most commonly used method is the 'heat balance' method. This assumes that the percentage of liquid which flashes is equal to the difference in total heat per pound at the inlet and outlet conditions, divided by the latent heat of vaporization at the outlet condition. This is represented by the formula

$$X = S \ \frac{\left(T1 + T2\right)}{L}$$

Where   X = The fraction by weight of fluid vaporized.
         S = The specific heat of the fluid.
         T1 = The upstream heat of the fluid.
         T2 = Outlet temperature, F.
         L = The latent heat of vaporization of the fluid at
             the downstream condition.

Knowing the percentage of liquid, which has changed to a gas or vapor, and assuming all of this change occurs at the valve seat, it is possible to calculate individual *Cvs* for the liquid and vapor phase. By adding these *Cvs* together, the total *Cv* required for the mixture is obtained. This method assumes that the total heat of the mixture leaving the valve is the same as the total heat of the liquid entering it, thereby assuming that a true throttling process has occurred.

RANGE ABILITY

Whenever possible the *Cv* should be calculated for maximum and minimum service conditions in order to ensure that these both come within the range ability of the valve plug. The range ability of a valve is defined as the ratio of the maximum and minimum controllable flow permitted by *Cv* of the valves. This is specified by the manufacturer but is usually of the order of 50: 1 for a single seated globe valve.

FINAL SELECTION

Having established the *Cv* and type of valve required, it is now possible to make the final selection of valve size from manufacturer's tables. One final check is necessary, to ensure that the inlet velocity to the valve is not too high as this may cause excessive vibration resulting in noise and wear in the valve body. The maximum allowable inlet veloci- ties obviously varied with the fluid being handled, but safe guidelines are as follows:

Fluid	Body Size (inches)	Allowable Velocity	
		CS Bodies	Chrome Moly
Steam, Air, Gas	1-2	20000 ft/min	25000 ft/min
	21-4	15000 ft/min	20000 ft/min
	6-12	13500 ft/min	15000 ft/min
Water	1-2	17 ft/s	25 ft/s
	21-4	15 ft/s	20 ft/s
	6-12	13 ft/s	25 ft/s

The mean inlet velocity in valves can be calculated by use of the following formula:

$$V_m = \frac{0 \bullet 408Q}{d2\,ft/sec}$$

Where Q = flow in gallons per minute
$\quad\quad\quad\;\,$ d = inlet diameter in inches.
$\quad\quad\quad\;$ V_m = mean inlet velocity entering the valve

Steam

$$V_m = \frac{33 \bullet 06WV}{dQ\,ft/sec}$$

Where W = Steam flow in pounds per hour.
$\quad\quad\quad\;\,$ d = inlet diameter in inches.
$\quad\quad\quad\;\,$ V = specific volume in cubic feet per pound

$$V = \frac{0 \bullet 595T}{P} - \frac{\left(1 + 0\bullet0513p\frac{1}{2}\right) \times 6\bullet684 \times 10^\circ}{T4}$$

Where T = absolute temperature (°F=460).
$\quad\quad\quad\;\,$ P = absolute pressure (Gauge +14•7).

If a check shows that the inlet velocities based on the selected valve body size are in excess of those shown in the table then an increase in valve size should be considered.

GENERAL

Control valve bodies, trim, sealing methods, end connections, bonnet types, etc., are essentially as already described for manually operated valves with refinements referred to in this section. Reliability is paramount and engineers should give close attention to this important feature. However, safeguards should be taken to ensure: fail safe features in the event of power loss failures, manually operated bypass valves, remote and local alarms. The usual materials for valve bodies are: chrome-molybdenum, steel, brass and stainless steels, but they are also manufactured in practically all metals and plastics.

ACCESSORIES

Accessories include limit switches for operating alarm signals, panel lights, relays, indication of inner valve position; filters and instruments, safety devices such as a lockup of valves to hold the air in a pneumatic system in the event of failure of air supply and unloading valves to release air pressure to atmosphere and thus permit manual movement of the valve position in the event of system air failure.

CHECKLIST FOR
PREPARING A CONTROL VALVE SPECIFICATION

1. Service-on/off, throttling, diverting, mixing, regulation of pressure, temperatures or level.

2. Fluid to be handled—liquid, vapor or gas; corrosive properties, specific gravity at 15.6-C (60°F), viscosity at operating temperature, particle size, if any; specific heat, latent heat; if steam, inlet steam pressure and state (wetness fraction, superheat) if gas molecular weight, compressibility at operating condition.

3. Maximum, normal, and minimum flow rates.

4. Pressure drop through valve at closed and operating condition (max./min.).

5. Operating temperature.

6. Permissible seat leakage.

7. Failure mode.

8. Type of inner valve.

9. Calculation of required Cv (check for viscosity and flashing.)

10. Selection of valve size (check inlet velocity.)

11. Valve body selection—style, material.

12. Trim details, seal or packing, bonnet type.

13. End connections—flanged, weld end, screwed, etc.

14. Available actuator pressure.

15. Selection of actuator type to give required failure mode and handle unbalanced forces

16. Direct or reverse action.

17. Fail safe, air to open or air to close, etc.

18. Electric power supply.

19. Is a positioner required?

20. Special conditions.

21. Accessories, hand wheel, top or side mounted, filter regulator,

unloading valve, solenoid valve, lock-up valve, and limit switches are all accessories which may be found at a control valve station. The arrangement of a Control Valve Station is important as regards the space to be occupied by each control valve and its operating equipment, the provision of isolation valves and their orientation to enable easy removal of the control valve itself. Valuable design time can be saved by standardizing control valve arrangements. See Figure 9-9a, b, and c, together with related Table 9-3a, b, and c. These illustrate the standard dimensions for assembling control valve station

Table 9-1. Control Valve Station Dimensions.

Line size (inches)	A	B	C
1	910	1080	1360
1-1/2	1040	1315	1600
2	1030	1340	1640

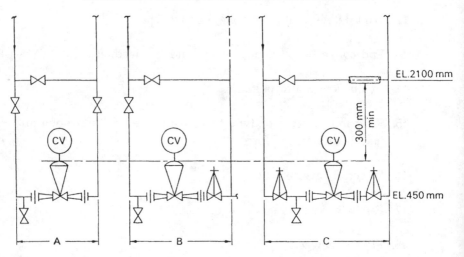

(a). Control valve arrangements: screwed and socket welding. (Courtesy of Foster-Wheeler LTd.)

Figure 9-9. Control Valve Arrangements.

Figure 9-9. Control Valve Arrangements (*Continued*).

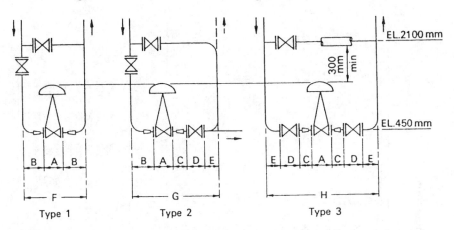

(b). Control valve arrangements: for raised face flanges. (Courtesy of Foster-Wheeler LTd.)

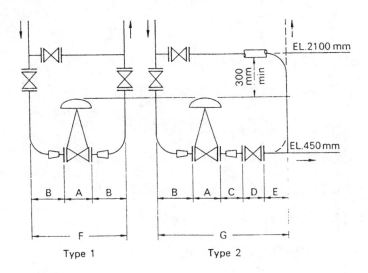

Table 9.2 Control Valve Assembly Dimensions.

Rating: 150 ANS LINE: 150ANS CV

Line size (inches)	CV size (inches)	A	B	C	D	E	F	G	H
1-1/1	1	187	176	181	168	119	539	831	1123
2	1	187	208	195	181	140	603	911	1219
2	1-1/2	225	214	202	181	140	653	962	1271
3	2	257	267	222	206	184	791	1136	1481
3	2-1/2	279	273	229	206	184	825	1171	1517
4	3	301	324	248	232	229	949	1334	1719
6	4	355	445	305	270	318	1245	1693	2141
8	6	454	546	343	295	406	1546	2044	2542
10	8	546	660	381	333	483	1866	2403	2940

Rating: 150 ANS LINE: 300 ANS CV

Line size (inches)	CV size (inches)	A	B	C	D	E	F	G	H
1-1/1	1	200	183	187	168	119	566	857	1148
2	1	203	214	202	181	140	628	937	1246
2	1-1/2	238	221	208	181	140	680	988	1296
3	2	270	273	229	184	184	816	1162	1508
3	2-1/2	295	279	235	184	184	853	1199	1545
4	3	320	333	257	229	229	986	1371	1756
6	4	371	454	314	270	318	1279	1727	2175
8	6	476	556	352	295	406	1588	2085	2582
10	8	571	670	391	333	483	1911	2448	2985

Rating: 300 ANS

Line size (inches)	CV size (inches)	A	B	C	D	E	F	G	H
1-1/1	1	200	183	194	194	125	566	896	1226
2	1	200	214	208	219	146	628	987	1346
2	1-1/2	238	221	214	219	146	680	1038	1396
3	2	270	273	238	286	194	816	1261	1706
3	2-1/2	295	279	244	286	194	853	1298	1743
4	3	320	333	267	308	238	986	1466	1946
6	4	371	454	324	406	327	1279	1882	2485
8	6	476	556	362	422	416	1588	2232	2376
10	8	571	670	406	460	498	1911	2605	3299

Rating: 600ANS

Line size (inches)	CV size (inches)	A	B	C	D	E	F	G	H
1-1/1	1	213	189	208	244	133	591	987	1383
2	1	213	221	224	295	156	655	1109	1563
2	1-1/2	254	229	232	295	156	712	1166	1620
3	2	289	283	257	359	203	855	1391	1927
3	2-1/2	314	289	264	359	203	892	1429	1966
4	3	340	343	298	435	260	1026	1676	2326
6	4	397	476	371	562	352	1349	2158	2967
8	6	511	581	416	663	445	1673	2616	3559
10	8	613	699	476	790	540	2011	3118	4225

300 ANS RTJ FLANGES

Line size (inches)	CV size (inches)	A	B	C	D	E	F	G
1-1/2	1	221	189	206	214	132	599	962
2	1	221	221	222	243	154	663	1065
2	1-1/2	259	227	229	243	154	713	1112
3	2	294	281	254	309	202	856	1340
3	2-1/2	319	287	260	309	202	893	1377
4	3	344	341	283	332	246	1026	1546
6	4	395	462	340	430	335	1219	1962
8	6	500	564	378	446	424	1628	2321
10	8	595	678	422	484	506	1951	2685

600 ANS RTJ FLANGES

Line size (inches)	CV size (inches)	A	B	C	D	E	F	G
1-1/2	1	221	189	208	252	133	599	1003
2	1	221	221	225	306	157	663	1130
2	1-1/2	262	229	233	306	157	720	1187
3	2	300	284	260	369	205	868	1418
3	2-1/2	325	291	267	369	205	907	1457
4	3	351	344	302	446	262	1039	1705
6	4	408	478	375	573	354	1364	2188
8	6	522	583	419	675	446	1688	2645
10	8	624	700	479	802	541	2024	3146

METERING

Orifice, Venturi and Pitot Tube

These are devices which are fitted into pipe lines to operate in conjunction with other instruments for measuring and controlling fluids and are therefore properly within the scope of this book. However, provision has to be made in pipe lines for inserting or fastening these sensors.

The Orifice Plant

An orifice is a thin plate containing an aperture slightly smaller than the pipe bore through which fluid passes. See Figure 9-10. The orifice diameter in steam distribution systems is usually sized larger than an orifice in a test device since the large diameter orifice will only produce a slight pressure drop and less loss in steam energy potential. Because of this, it is necessary to incorporate a very sensitive pressure sensor.

A typical electronic pressure sensor could consist of a resistor which would compress when a force is applied. In this case the force would be generated by the steam pressure. The shortened, compressed resistor would permit a larger flow of current signaling the controller or operator that the steam flow through the pipe has increased. In test apparatus the engineer may choose a small orifice in a large pipe where a simple manometer can be installed to measure a relatively large pressure drop.

With either the large or small orifice Bernoulli's equation is used to convert the pressure drop to a dimension of fluid flow (gas or liquid).

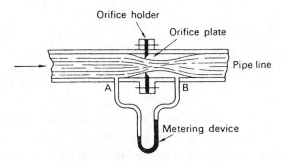

Figure 9-10. Orifice Meter, Consisting of an Orifice Plate and Differential Manometer.

Figure 9-10 shows a test rig with an orifice and U tube manometer. Since the orifice bore is large compared to the pipe bore the assembly could be for a distribution system pipeline. However, the velocity would have to be very high to produce a useful reading on a U tube manometer to calculate the required volume of liquid. The orifice is a very simple device, which is easily installed.

Venturi

The Venturi eliminates the principal disadvantages of the orifice in power lost and eddy currents, as the change in velocity is gradual. See Figure 9-11. The complete Venturi meter, including the two-way taper pipe, is generally supplied by an instrument manufacturer as it must be carefully proportioned and manufactured.

The Venturi in comparison with the orifice is costly, as the orifice is easy to manufacture and can be done on site in most machine shops.

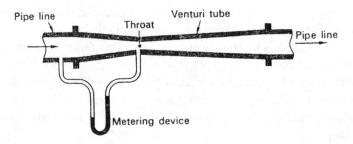

Figure 9-11. Venturi Meter.

Pitot Tube

The orifice meter and Venturi meter measure the average velocity of the entire fluid steam, while the Pitot tube measures velocity at one point only, from which an average velocity can be calculated. The Pitot tube is the one meter discussed that does not require a reduction in the size of the pipe diameter. It produces accurate readings; the pipe can be trammed to obtain readings across its entire diameter. A mean pressure drop can then be calculated from which the fluid velocity can be accurately calculated. The principle, as shown in Figure 9-12, can also be used for routine flow indication. With the Pitot tube inserted in the center of the pipe, a single reading can be corrected with a factor developed from the acceptance test mentioned above. The flow calculated by this method should be acceptable for most routine plant operations. As with the other meters the Pitot tube requires enough straight pipe upstream and down-stream of the meter to ensure laminar flow at the measuring point.

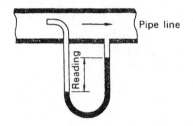

Figure 9-12. Pitot Tube, for Measuring Flow Velocities.

INSTRUMENTS AND CONTROLS

The purpose of instrumentation and process control devices is to main-tain the operation of a process plant at an efficient and safe level. It employs measuring devices or sensors which provide measurements or information regarding qualities, quantities of interest to plant operation, and regulating devices which are usually control valves by which the operation of the process is modulated or trimmed to the desired value or required set point. Manual control consists of an operator observing some measured value and operating a regulating device to achieve the desired conditions. Many plants are now automated with devices that can be depended upon to act consistently in the prescribed manner. They can be designed so that when they fail they will not cause serious inter-ruption of the plant process. Most devices can be self-contained units but in many plants a form of centralized supervision at a control panel is preferred. For this, some form of transmission is required and two forms of signal are usually employed. These are pneumatic signals and electric signals. These signals may merely indicate the state of the process or provide input to a controller which can send a signal to the valve or other device to change the flow or other function. Many plants have the entire control apparatus located in a control room. The control apparatus together with any other logging and computing apparatus is located in the control room. In most plants an engineer or technician is on duty in the control room and a final signal, manual or automatic, can be sent to the valve or other device in the plant.

The importance of the above to engineers is that the control system is largely dependent on the value of the signal that it receives and the quality of this measurement is highly dependent on the care with which the sensing device is connected to the steam or condensate piping.

Figure 9-13 illustrates the time components by which controls and instruments operate.

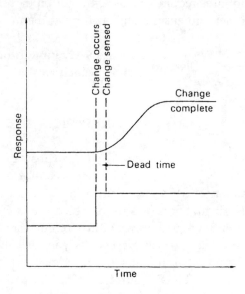

Figure 9-13. Exponential Curve Defining Valve Flow Characteristics.

Chapter 10

Steam Traps

A complete steam trap piping and valve arrangement diagram is shown in Figure 10-1. Although all the equipment shown in the this Figure may not be absolutely necessary, engineering judgment should be used to fit the components to the steam system.

Steam distribution systems link the boilers to the steam consuming equipment being used; transporting it to the location in the plant where its heat energy is needed.

The three components of steam distribution systems are the boiler headers, steam mains and branch lines. Each of these fulfill certain requirements of the system and together with steam separators and steam traps contribute to efficient steam use.

BOILER HEADERS

A boiler header is a specialized application of the steam distribution system. It receives the steam from one or more boilers and acts as a collecting chamber from which the steam mains receive the steam and to distribute it to the branch lines.

Effective boiler header steam trapping assures that carryover (boiler water and its dissolved and suspended solids) are removed before the steam is distributed to the steam piping network. Steam traps which serve this header must be capable of discharging large quantities of carryover from the boiler during start-up. It must be capable of resisting hydraulic shock or what is commonly known as water hammer. Water hammer had been discussed in detail in a previous chapter. The first choice would be an inverted bucket trap. If this selection were not possible then the alternative would be a float and thermostatic trap.

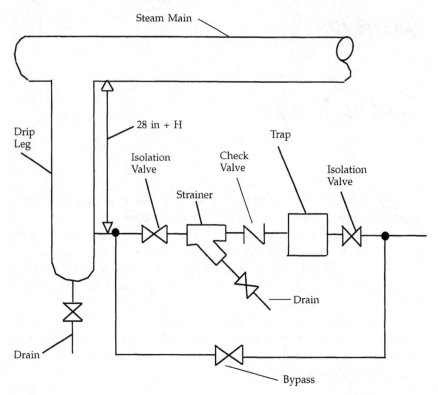

Figure 10-1. Ideal Steam Trap Piping and Fittings Arrangement.

TRAP SIZING FOR BOILER HEADERS

A 1.5:1 safety factor is recommended for boiler header applications. The factor is assigned to insure adequate protection since different boilers will have different amounts of carryover during start-up. Boiler water carryover is guaranteed by boiler manufacturers when the boiler is at operating conditions. Over sizing the steam trap does not necessarily mean that there will be an undesirable loss of steam during operation. The primary function of the trap is to separate steam from water, therefore excessive loss of steam during operation is prevented.

In large steam systems, a steam trap by-pass will usually be installed to assist the trap in discharging condensate during start-up. The by-pass valve can be manually or power operated; remotely or automatically controlled.

The required trap capacity can be obtained by: multiplying the safety factor by design steam load and by the anticipated carryover. For example:

Required trap capacity $\quad = 1.5 \times 50,000 \times 0.10$
$\qquad\qquad\qquad\qquad\quad\; = 7,500 \text{ lbs/hr}$

Response to the presence of condensate, resistance to hydraulic shock and tolerance to foreign matter and efficient operation under low load conditions are needed features found in the inverted bucket trap.

STEAM TRAP INSTALLATION

Steam traps are installed on a boiler header as shown in Figure 10-2. Since there are two traps connected to the boiler header the required capacity for each trap would be one half the capacity of what it would be if there were only one boiler and one steam distribution pipe.

Figure 10-2 shows the boiler header with two drip legs. These drip legs should be sized according to Table 10-1. For instance, if the steam header were eight inches in diameter then a six inch diameter drip leg would be adequate. If the start up procedure were manually supervised then the recommended drip pipe would be 15 inches in length. For an automatic start up the drip leg should be twenty-eight inches in length. The drip leg should be insulated the same way as the header.

STEAM MAINS

The most common uses of steam traps is in trapping steam mains. These lines need to be kept free of air and condensate in order to keep steam consuming machinery in safe and efficient operation and to prevent damage to the piping system from water hammer. Control valves and process sensors such as pressure gauges are the most likely types of equipment to be damaged from water hammer. Extreme water hammer can damage the piping as well. Water hammer has been discussed in detail in a previous chapter.

There are two methods used to warm steam mains during start up. They are: supervised and automatic.

Boiler Headers

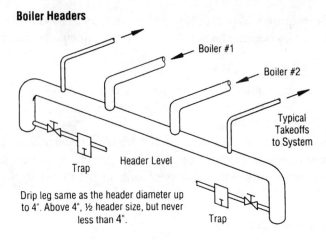

Figure 10-2.

Table 10-1. Recommended Steam Main and Branch Line Drip Leg Sizing.

M	D	H	
Steam Main Size (in)	Drip Leg Diameter (in)	*Drip Leg Length Min. (in)*	
		Supervised Warm-Up	Automatic Warm-Up
1/2	1/2	10	28
3/4	3/4	10	28
1	1	10	28
2	2	10	28
3	3	10	28
4	4	10	28
6	4	1 0	28
8	4	12	28
10	6	15	28
12	6	18	28
14	8	21	28
16	8	24	28
18	10	27	28
20	10	30	30
24	12	36	36

Table 10-2. Condensation in Insulated Pipes Carrying Saturated Steam in Quiet Air at 70°F (Insulation Assumed to be 75% Efficient).

Pressure, psig		15	30	60	125	180	250	450	600	900
Pipe Size (in)	sq ft per Lineal ft	\	\	\	\	Pounds of Condensate Per Hour Per Lineal Foot				
1	.344	.05	.06	.07	.10	.12	.14	.186	.221	289
1-1/4	.434	.06	.07	.09	.12	.14	.17	.231	.273	.359
1-1/2	.497	.07	.08	.10	.14	.16	.19	.261	.310	.406
2	.622	.08	.10	.13	17	.20	.23	.320	.379	.498
2-1/2	.753	.10	.12	.15	.20	.24	.28	.384	.454	.596
3	.916	.12	.14	.18	.24	.28	.33	.460	.546	.714
3'/2	1.047	.13	.16	.20	.27	.32	.38	.520	.617	.807
4	1.178	.15	.18	.22	.30	.36	.43	.578	.686	.897
5	1.456	.18	.22	.27	.37	.44	.51	.698	.826	1.078
6	1.735	.20	.25	.32	.44	.51	59	.809	.959	1.253
8	2.260	.27	.32	.41	.55	.66	.76	1.051	1.244	1.628
10	2.810	.32	.39	.51	.68	.80	.94	1.301	1.542	2.019
12	3.340	.38	.46	.58	.80	.92	1.11	1.539	1.821	2.393
14	3.670	.42	.51	.65	.87	1.03	1.21	1.688	1.999	2.624
16	4.200	.47	.57	.74	.99	1.19	1.38	1.927	2.281	2.997
18	4.710	.53	.64	.85	1.11	1.31	1.53	2.151	2.550	3.351
20	5.250	.58	.71	.91	1.23	1.45	1.70	2.387	2.830	3.725
24	6.280	.68	.84	1.09	1.45	1.71	2.03	2.833	3.364	4.434

Table 10-3. Pipe Weights Per Foot in Pounds.

Pipe Size (in)	Schedule 40	Schedule 80	Schedule 160	XX Strong
1	1.69	2.17	2.85	3.66
1-1/4	2.27	3.00	3.76	5.21
1-1/2	2.72	3.63	4.86	6.41
2	3.65	5.02	7.45	9.03
2-1/2	5.79	7.66	10.01	13.69
3	7.57	10.25	14.32	18.58
3-1/22	9.11	12.51	—	22.85
4	10.79	14.98	22.60	27.54
5	14.62	20.78	32.96	38.55
6	18.97	28.57	45.30	53.16
8	28.55	43.39	74.70	72.42
10	40.48	54.74	116.00	—
12	53.60	88.60	161.00	—
14	63.00	107.00	190.00	—
16	83.00	137.00	245.00	—
18	105.00	171.00	309.00	—
20	123.00	209.00	379.00	—
24	171.00	297.00	542.00	—

Figure 10-4. The Warming-Up Load from 70°F, Schedule 40 Pipe.

Steam Pressure, psig		2	15	30	60	125	180	250
Pipe Size (in)	wt of Pipe per ft (lbs)			Pounds of water Per Lineal Foot				
1	1.69	.030	.037	.043	.051	.063	.071	.079
1-1/4	2.27	.040	.050	.057	.068	.085	.095	.106
1-1/2	2.72	.048	.059	.069	.082	.101	.114	.127
2	3.65	.065	.080	.092	.110	.136	.153	.171
2-1/2	5.79	.104	.126	.146	.174	.215	.262	.271
3	7.57	.133	.165	.190	.227	.282	.316	.354
3-1/2	9.11	.162	.198	.229	.273	.339	381	.426
4	10.79	.190	.234	.271	.323	.400	.451	.505
5	14.62	.258	.352	.406	.439	.544	.612	.684
6	18.97	.335	.413	.476	.569	.705	.795	.882
8	28.55	.504	.620	.720	.860	1.060	1.190	1.340
10	40.48	.714	.880	1.020	1.210	1.500	1.690	1.890
12	53.60	.945	1.170	1.350	1.610	2.000	2.240	2.510
14	63.00	1.110	1.370	1.580	1.890	2.340	2.640	2.940
16	83.00	1.460	1.810	2.080	2.490	3.080	3.470	3.880
18	105.00	1.850	2.280	2.630	3.150	3.900	4.400	4.900
20	123.00	2.170	2.680	3.080	3.690	4.570	5.150	5.750
24	171.00	3.020	3.720	4.290	5.130	6.350	7.150	8.000

Supervised warm-up is widely used when mechanical equipment is steam driven. The suggested method is to bring the boilers up to operating pressure. With the header and/or steam main drip valves open to the atmosphere, carefully open the boiler steam stop valves; admitting steam to the piping network. The drip leg drains are left open until all the condensate is cleared from the piping. When visual inspection of the drains indicate that the collected condensate is removed from the piping, The steam traps can be put in operation. At this point, steam pressure should develop at an increased rate until the steam distribution system reaches its operating pressure and temperature. If the boiler is used to produce superheated steam, the temperature increase in the steam distribution system should be carefully monitored. If the temperature rise in the piping is too rapid, unwanted stress can be the result.

Automatic start-up permits the boiler and steam distribution piping to warm-up together. This is done by aligning the boiler and distribution system valves to allow the flow of steam to the distribution

Table 10-5.

	Col. 1 Gauge Pressure	Col. 2 Absolute Pressure (psia)	Col. 3 Steam Temp. (°F)	Col. 4 Heat of Sat. Liquid (Btu/lb)	Col. 5 Latent Heat (Btu/lb)	Col. 6 Total Heat of Steam (Btu/lb)	Col. 7 Specific Volume of Sat. Liquid (cu ft/lb)	Col. 8 Specific Volume of Sat. Steam (cu ft/lb)
Inches of Vacuum	29.743	0.08854	32.00	0.00	1075.8	1075.8	0.096022	3306.00
	29.515	0.2	53.14	21.21	1063.8	1085.0	0.016027	1526.00
	27.886	1.0	101.74	69.70	1036.3	1106.0	0.016136	333.60
	19.742	5.0	162.24	130.13	1001.0	1131.	0.016407	73.52
	9.562	10.0	193.21	161.17	982.1	1143.3	0.016590	38.42
	7.536	11.0	197.75	165.73	979.3	1145.0	0.016620	35.14
	5.490	12.0	201.96	169.96	976.6	1146.6	0.016647	32.40
	3.454	13.0	205.88	173.91	974.2	1148.1	0.016674	30.06
	1.418	14.0	209.56	177.61	971.9	1149.5	0.016699	28.04
PSIG	0.0	14.696	212.00	180.07	970.3	1150.4	0.016715	26.80
	1.3	16.0	216.32	184.42	967.6	1152.0	0.016746	24.75
	2.3	17.0	219.44	187.56	965.5	1153.1	0.016768	23.39
	5.3	20.0	227.96	196.16	960.1	1156.3	0.016830	20.09
	10.3	25.0	240.07	208.42	952.1	1160.6	0.016922	16.30
	15.3	30.0	250.33	218.82	945.3	1164.1	0.017004	13.75
	20.3	35.0	259.28	227.91	939.2	1167.1	0.017078	11.90
	25.3	40.0	267.25	236.03	933.7	1169.7	0.017146	10.50
	30.3	45.0	274.44	243.36	928.6	1172.0	0.017209	9.40
	40.3	55.0	287.07	256.30	919.6	1175.9	0.017325	7.79
	50.3	65.0	297.97	267.50	911.6	1179.1	0.017429	6.66
	60.3	75.0	307.60	277.43	904.5	1181.9	0.017524	5.82
	70.3	85.0	316.25	286.39	897.8	1184.2	0.017613	5.17
	80.3	95.0	324.12	294.56	891.7	1186.2	0.017696	4.65
	90.3	105.0	331.36	302.10	886.0	1188.1	0.017775	4.23
	100.0	114.7	337.90	308.80	880.0	1188.8	0.017850	3.88
	110.3	125.0	344.33	315.68	875.4	1191.1	0.017922	3.59
	120.3	135.0	350.21	321.85	870.6	1192.4	0.017991	3.33
	125.3	140.0	353.02	324.82	868.2	1193.0	0.018024	3.22
	130.3	145.0	355.76	327.70	865.8	1193.5	0.018057	3.11
	140.3	155.0	360.50	333.24	861.3	1194.6	0.018121	2.92
	150.3	165.0	365.99	338.53	857.1	1195.6	0.018183	2.75
	160.3	175.0	370.75	343.57	852.8	1196.5	0.018244	2.60
	180.3	195.0	379.67	353.10	844.9	1198.0	0.018360	2.34
	200.3	215.0	387.89	361.91	837.4	1199.3	0.018470	2.13
	225.3	240.0	397.37	372.12	828.5	1200.6	0.018602	1.92
	250.3	265.0	406.11	381.60	820.1	1201.7	0.018728	1.74
		300.0	417.33	393.84	809.0	1202.8	0.018896	1.54
		400.0	444.59	424.00	780.5	1204.5	0.019340	1.16
		450.0	456.28	437.20	767.4	1204.6	0.019547	1.03
		500.0	467.01	449.40	755.0	1204.4	0.019748	0.93
		600.0	486.21	471.60	731.6	1203.2	0.02013	0.77
		900.0	531.98	526.60	668.8	1195.4	0.02123	0.50
		1200.0	567.22	571.70	611.7	1183.4	0.02232	0.36
		1500.0	596.23	611.60	556.3	1167.9	0.02346	0.28
		1700.0	613.15	636.30	519.6	1155.9	0.02428	0.24
		2000.0	635.82	671.70	463.4	1135.1	0.02565	0.19
		2500.0	668.13	730.60	360.5	1091.1	0.02860	0.13
		2700.0	679.55	756.20	312.1	1068.3	0.03027	0.11
		3206.2	705.40	902.70	0.0	902.7	0.05053	0.05

system freely. As the boiler pressure increases so will the temperature and pressure in the distribution system. Drain valves on the distribution system drip legs are left closed and the steam traps alone are used to clear the steam lines of unwanted condensate.

TRAP SELECTION AND SIZING

As with boiler headers, two very effective traps for steam main drainage are the inverted bucket trap and the float and thermostatic trap. If the automatic warm-up method is to be used, the float and thermostatic trap offers an additional advantage of purging air from the piping during start-up and operation. However, it is not designed to resist water hammer. The inverted bucket trap does not have a float that can be damaged by water hammer. Unlike the float and thermostatic trap, the inverted bucket trap will not be damaged by freezing. Heat tracing is required on both traps to permit operation during periods of below freezing temperatures.

Trap sizing can be accomplished by using Table 10-2 to determine the amount of condensate formation based on the operating pressure and the pipe size assuming a 75 percent insulation effectiveness. The following equation can be used to determine condensate rate of formation:

$$C = \frac{A \times U \times \left(T_1 - T_2\right) \times E}{H}$$

Where: C = Condensate in lbs/hr-foot.
      A = External area of pipe in sq ft (Table 10-2, col. 2)
      U = Btu/sq.ft./°F temperature difference/hr from Table 10-2
    T1 = Steam temperature, °F
    T2 = Air temperature, °F
      E = One minus the efficiency of the insulation (example: 75% efficient insulation, 1-.75 = 0.25 or E = 0.25)
      H = Latent heat of steam; see Steam Tables, Figure 10-5, when approximate values of latent heat will suffice (as with safety factors of 1.0 to 3.0).

When doing precise calculations, use the following equation:

$$q = \frac{0.523 \, r_2 \left(t_p - t_a\right)}{\dfrac{r_2 \, \text{LOG} \, \dfrac{r_2}{r_1}}{R} + 0.6}$$

$$w_c = \frac{q}{h_{fg}}$$

Where:

q = Btu loss per linear foot

t_p = Pipe temperature

t_a = Ambient air temperature

r_2 = Radius of the outer surface of the insulation

r_1 = Radius of the outer surface of the pipe

K = conductivity of the insulation in: Btu/(sq.ft.)(hr)(degree °F)

w_c = Condensate flow in lbs/hr

h_{fg} = Latent heat of steam in the pipe (Btu/lb)

SAFETY FACTOR

For traps installed between the boiler and the end of the main distribution line apply a 2:1 safety factor. Apply a 3:1 safety factor for traps installed at the end of the main steam line. Traps should also be installed before any reducing valves or stop valves.

INSTALLATION

Steam traps should be installed in the main distribution lines as follows: Ahead of risers,
          End of mains,
          Ahead of expansion joints or bends,
          Ahead of regulators, and main stop valves.

Drip legs and steam traps should be installed at intervals of 300 to 500 feet. Good engineering judgment and knowledge of the plant and piping layout should be exercised in the selection of trap installation locations. Again, the inverted bucket trap would be the first choice for most steam line trap installations. Drip legs with sizes according to Table 10-1 should be used with traps according to Fig 10-1 and 10-3.

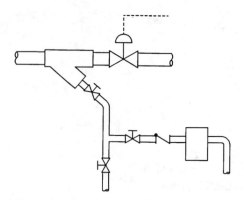

Figure 10-3. Trap Draining Strainer ahead of PRV.

BRANCH LINES

Branch linear take-offs from the steam mains supplying specific steam machines and equipment. The flexible nature of steam systems is that additional branches and steam equipment can be added or subtracted from the system. This can be done without redesigning major portions of the steam system. As stated in a preceding chapter, this cannot be done with a hydropic system.

TRAP SELECTION AND SIZING

The equations used to calculate condensate load in branch lines are the same as the equations used to calculate that load in steam mains.

Trap selection would be the same as are used on the main steam line.

INSTALLATION

The recommended piping from the steam main to a control valve is shown in Figure 10-4, for pipe lengths under 10 feet. Figure 10-5 shows a schematic for pipe lengths over 10 feet. Figure 10-6 shows a schematic of a piping network to use when a control valve is arranged below the steam main.

Install a full pipe size strainer upstream of control valves and pressure reducing valves. The strainers should be blown down regularly. This is particularly important during commissioning and start-up after a prolonged outage.

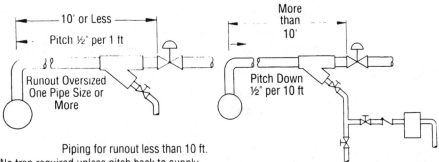

Piping for runout less than 10 ft.
No trap required unless pitch back to supply
header is less than ½" per ft.

Figure 10-4. Branch Lines.

Figure 10-5.

Piping for runout greater than 10'
Drip leg and trap required ahead of control valve.
Strainer ahead of control valve can serve as drip
leg if blowdown connection runs to an inverted
bucket trap. This will also minimize the strainer
cleaning problem. Trap should be equipped with
an internal check valve or a swing check installed
ahead of the trap.

Figure 10-6.

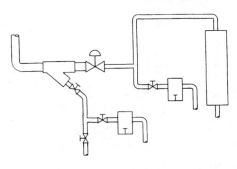

Regardless of the length of the
runout, a drip leg and trap are required ahead of
the control valve located below steam supply. If
coil is above control valve, a trap should also be
installed at downstream side of control valve.

STEAM SEPARATORS

Separators are designed to remove condensate from saturated steam to be used in steam machinery. They are also used for processes requiring very dry steam and on secondary steam lines where large amounts of condensate can form.

Separators are seldom used in systems where superheated steam is generated in the boiler or where external desuperheaters are used.

TRAP SELECTION

The inverted bucket trap is recommended for steam separators. They permit the release of large amounts of condensate, offer good thermal and mechanical shock resistance and perform well on light loads.

Separators often remove 10 to 20 percent condensate from steam. The accepted capacity factor is 3:1. Therefore, calculating the steam trap capacity for a steam rate of 10,000 lbs/hr is

$$
\begin{aligned}
\text{Trap capacity} \quad &= 0.10 \times 10,000 \times 3 \\
&= 3,000 \text{ lbs/hr}
\end{aligned}
$$

INSTALLATION

The basic trap installation is shown in Figure 10-7. The clean out down spout below the trap connection should be cleaned often especially during commissioning. Good judgment based upon commissioning experience, will dictate how often the dirt collector will require cleaning.

STEAM TRAPS FOR
TRACER LINES

Tracer lines are intended to protect outdoor piping from freezing in sub-freezing temperatures. Tracers are installed on the pipe to be protected before the piping insulation is installed. There are two types of tracers: electric resistance heating and steam heating. Both come as coils or as flat sheets that can be wrapped around a pipe or the latter fitted around the machine to be protected from low temperatures. The selection of the steam or electric tracing depends upon a number of factors. Some questions to be answered are: availability of the source of power, will drains be collected and recycled, will the electrical source require transformers or capacitors, will the temperature and

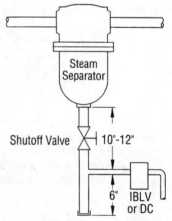

Drain downstream side of separator. Full size drip leg and dirt pocket are required to assure positive and fast flow of condensate to the trap.

Figure 10-7. Steam Separator.

pressure of the steam be adequate for the tracing. If steam is chosen, it must be carefully maintained and operated to avoid freezing the product or the product piping. Product systems can be very costly when tracer failures result in product damage and lost production time, when tracing equipment is neglected.

If steam tracers are chosen over electric then a steam trap must be installed to retain the steam-within the coil until the latent heat is fully utilized and then to discharge it as condensate along with the collected non condensable gases, including air.

In selecting and sizing steam traps it is important to consider their compatibility with the objectives of the system. Usually these are:

- Conserve energy by operating reliably over a long period of time,
- Withstand abrupt periodic discharge in order to purge condensate and air from tracer piping,
- Operate under light load conditions, and
- Resist damage from freezing if the entire steam system is shut down.

TRAP SELECTION FOR STEAM TRACER LINES

The condensate load to be processed on a steam tracer system can be determine from the heat loss from the product pipe by using the following equation:

$$Q = \frac{L \times U \times \Delta t \times E}{S \times H}$$

Where:

Q = Condensate load, lbs/hr

L = Length of product pipe between tracer line steam traps in feet

U = Heat transfer factor in Btu/sq ft/°F/hr

ΔT = Temperature differential in °F.

E = One minus the efficiency of the insulation. Say, the insulation is 75 percent then $1 - (0.75) = 0.25$

S = Linear feet of pipe per sq.ft. of surface. See Table 10-6.

H = Latent heat of steam in Btu/lb. See Table 10-5.

On most tracer lines the flow to the steam trap is surprisingly low; therefore, a small steam trap is normally adequate. Based on its ability to conserve energy, reliability over a wide range of loads and freeze resistance an inverted bucket trap is considered the most likely choice for trapping tracer lines.

INSTALLATION NOTES

Install tracer feed lines above the product line requiring the steam tracing.

Lines from the tracer to the steam trap should drain downward.

Also, insure low points in the product line are traced, since this is a likely point for the product to freeze.

See Figures 10-8, 10-9, and 10-10, for details of piping tracer lines on product piping requiring freeze protection.

STEAM TRAP SCHEMATICS

A selection of steam traps frequently found in steam distribution systems are show in Figures 10-12 through 10-18.

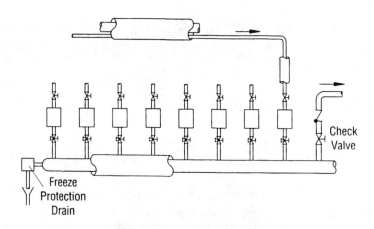

Figure 10-8. Typical Tracer Line.

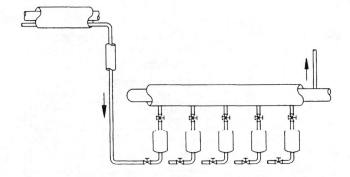

Figure 10-9. Typical Tracer Line.

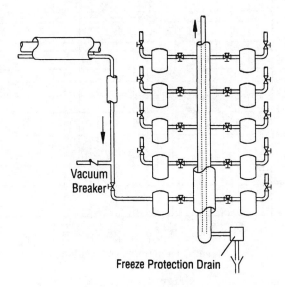

Vacuum
Breaker

Freeze Protection Drain

Figure 10-10. Typical Tracer Installation.

Btu Heat Loss Curves

Unit heat loss per sq ft of surface of uninsulated pipe of various diameters (also flat surface) in quiet air at 75°F for various saturated steam pressures or temperature differences.

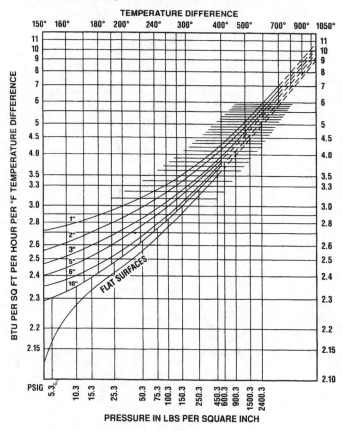

Figure 10-11.

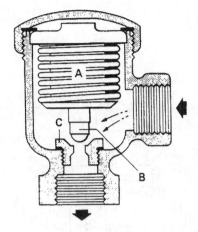

Figure 10-12. Balanced Pressure Thermostatic Trap in Open Position, Permitting Condensate Flow.

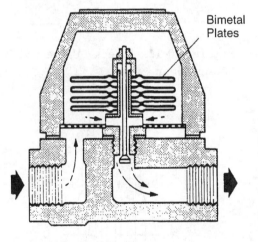

Bimetal Plates

Figure 10-13. Thermostatic Trap with Bimetal Plates. Shown in Open Position with the Bi-metallic Plates Contracted.

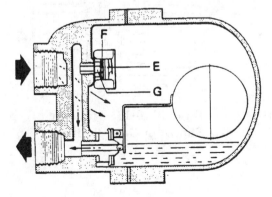

Figure 10-14. Float Trap with Thermostatic Air Vent. Shown with Air and Float Valves both Open.

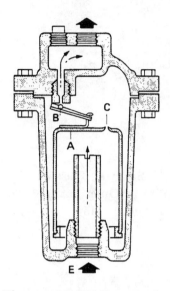

Figure 10-15. Inverted Bucket Trap.

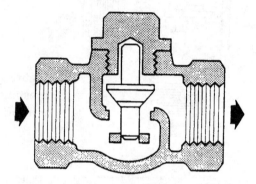

Figure 10-16. Check Valve with Guided Stem.

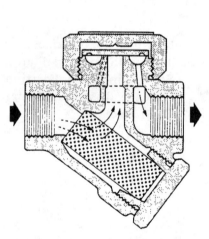

Figure 10-17. Thermodynamic Steam Trap with Built-in Strainer.

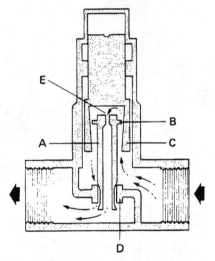

Figure 10-18. Impulse Trap Shown in Open Position.

Thermodynamic Fundamentals For Steam Transport

INTRODUCTION

Before commencing the study of thermodynamics for steam trap piping, first consider the quantity of steam and its cost estimated at $10 per million Btu. From Table A-1, it can be seen that a trap (failed open) with a 1/2-inch orifice wastes 35,000 lbs of steam/month at a cost of $50,100 per year. To understand traps and how they function, this appendix gives a brief review of the characteristics of how gases and liquids (steam and condensate) behave under different and ideal conditions. It gives the reader the fundamentals of energy and steam which is necessary in order to understand the steam-condensate phenomena and the steam trap.

MASS AND WEIGHT RELATIONSHIP

This text uses weight, usually in pounds, as the unit of measure rather than pounds mass. The weight-mass relationship for those users of the book who need to know, the mass of steam and/or condensate involved in the steam trapping process is:

From Newton's Law:

$$°F = ma/g_c$$

229

Table A-1. Cost of Various Sized Steam Trap Leaks at 100 psi.
(assuming steam costs of $5.00/1,000 lbs.)

Size of Orifice (in)	Lbs Steam Wasted Per Month	Total Cost Per Month	Total Cost Per Year
1/2	835,000	$4,175.00	$50,100.00
7/16	637,000	3,185.00	38,220.00
3/8	470,000	2,350.00	28,200.00
5/16	325,000	1,625.00	19,500.00
1/4	210,000	1,050.00	12,600.00
3/16	117,000	585.00	7,020.00
1/8	52,500	262.50	3,150.00

The steam loss values assume clean, dry steam flowing through a sharp-edged orifice to atmospheric pressure with no condensate present. Condensate would normally reduce these losses due to the flashing effect when a pressure drop is experienced. Courtesy: ARMSTRONG INTERNATIONAL, INC.

where

$°F$ = Pounds force due to gravity (weight),

m = Pounds mass,

g_c = Acceleration due to gravity at sea level, and

a = Acceleration due to gravity at the reference level, where a
= $32.174 - (3.32 \times 10^{-6})z$

PROPERTIES AND MEASUREMENTS

At times heat/power engineers need to understand the following thermodynamic properties of substances: pressure, specific volume, density, temperature, internal energy, enthalpy, entropy and specific heat. There are other properties, but the foregoing are those most used. The fundamental properties are pressure, volume and temperature; other properties are dependent on these.

Properties are invariable when a substance is at a specific state point. Conversely, the state point is fixed by two properties; for example, pressure and temperature, volume and temperature or pressure and volume. Thus one pound of gas at a certain pressure and temperature will have only one possible specific volume. Also, at any predetermined state point, internal energy, enthalpy and entropy each have only one value for the defined state point. They will have the same value if one pound of a gas is brought to the same pressure and temperature again.

MEASUREMENTS

There are two systems of measurement used in America. They are the United States Customary System (USCS) and the Systemi Internationale (SI). The fundamental quantities in the USCS are the pound, foot and second and the quantities used in the SI are the gram, meter and second. Two other units used are those measuring thermal energy: the British Thermal Unit (Btu) in the USCS and the Calorie used in the SI. These and other physical constants and conversion factors for both the USCS and SI are contained in Appendix B. The USCS is used in this text. The SI will be of interest to engineers involved in certain engineering specialties such as Bioengineering, research and overseas projects.

PROPERTIES OF SUBSTANCES

Pressure

Pressure is defined as force per unit area, measured as pounds per square inch (PSI) in the USCS. See Figure A-1 for a vector diagram illustrating how force acts in an enclosed container. It can also be measured as the height of a column of water or mercury pumped above a reference plane. This is referred to as the head of water (in feet of water) or mercury (in inches of mercury). Many other units are used, but they can all be related to a force per unit area.

In practice, pressures above and below atmospheric are determined by means of a pressure gauge (Figure A-2) or a manometer. The dial of a pressure gauge is marked to read the gauge pressure, usually in pounds per square inch. The gauge pressure is the difference between the pressure inside a vessel and the atmospheric pressure outside. Thus, to find the absolute pressure when this pressure is above atmospheric pressure, one must add the atmospheric pressure to the gauge reading; that is,

Absolute pressure = atmospheric pressure + gauge pressure.

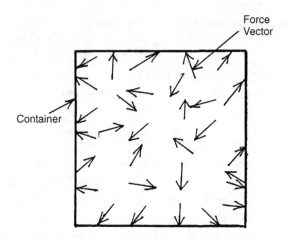

Figure A-1. Vector Analysis of Pressure.

Pressure is defined as a force (lbs) applied to a unit area. Consider the vectors of force in the diagram to be of equal magnitude, acting in all directions at once. Then the pressure will be the force per unit area (lbs/sq. in.). Pressure is measured as positive when the force per unit area exceeds that of the atmosphere at sea level. Negative pressure exists when the net force per unit area is less than atmospheric. Atmospheric pressure is approximately 14.7 psi.

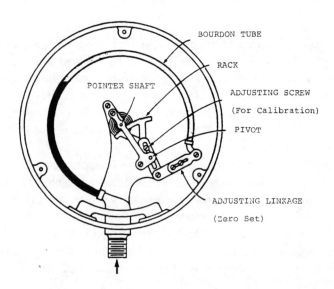

Figure A-2. Bourdon Tube Pressure Gauge.

Understand that in the equations in this book, the unit of pressure is generally pounds per square foot, equaling pounds per square inch multiplied by 144 sq.in./sq.ft.

If the absolute pressure is less than atmospheric pressure, the gauge reading is spoken of as the vacuum pressure. In this instance, the absolute pressure is found from

Absolute pressure = atmospheric pressure – gauge pressure

The gauge still measures the difference between the pressure inside the vessel and the pressure of the atmosphere.

Some other commonly used pressure measuring devices are shown in Figures A-3 and A-4.

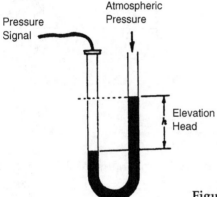

Figure A-3. U-tube Manometer.

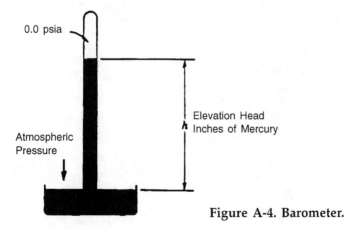

Figure A-4. Barometer.

Temperature

Temperature may be defined as a measure of the potential to transfer thermal energy. It is caused by the motion of particles of a substance rubbing and impacting on each other, generating heat through friction within the substance. This heat is measured as temperature. The greater the particle motion, the greater the heat generated and the higher the temperature.

There are four common temperature scales: Fahrenheit (USCS), Rankine (USCS), Kelvin (SI) and Celsius (SI). Some approximate critical points for water on these scales are shown in Table A-2.

Table A-2. Temperature of Water at Atmospheric Pressure.

POINT	F	C	K	R
Boiling	+212	+100	+373	+672
Freezing	+32	0	+273	+492
Absolute Zero	-460	-273	0	0

The conversion formulae relating the temperature of the four commonly used scales are:

Fahrenheit (°F) = 32 + 9/5C
Rankine (R) = °F + 460
Rankine (R) = 9/5K
Kelvin (K) = C + 173

Some temperature measuring devices are shown as follows in Figures A-5 through A-7.

Volume

The volume of a substance is defined as the quantity of space it occupies. Volumes are three dimensional and their fundamental unit of measurement is the cubic foot. There are many volumetric measures used by engineers including gallons, liters, cubic inches, cubic centimeters and others. These and other volume interrelationships are expressed

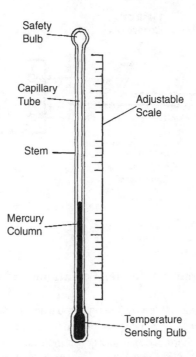

Figure A-5. Mercury in Glass Thermometer.

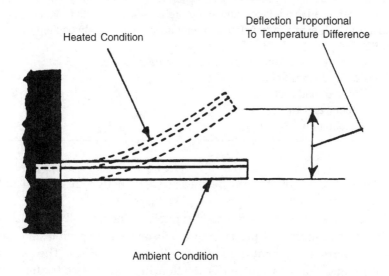

Figure A-6. Bimetallic Strip Temperature Indicator.

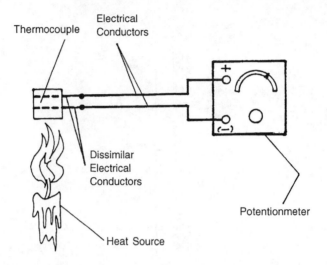

Figure A-7. Thermocouple and Potentiometer Thermoelectric Thermometer.

A thermocouple is assembled by joining two dissimilar metals. When the junction is heated, an electrical potential is generated which can be measured in millivolts proportional to the thermocouple junction temperature.

A potentiometer is a millivolt meter with a built in thermoelectric ambient temperature measuring device. The millivoltage generated by the ambient temperature serves as a reference for higher temperatures for most engineering work. When used in the laboratory, the more accurate ice bath reference junction is used to measure the freezing temperature of water at atmospheric pressure (31.69°F).

Some typical thermocouple wire combinations are:

Iron and Constantin,
Chromel and Alumel,
Copper and Nickel,
Rhodium and Platinum.

---

in Appendix B. Volume can be measured, although in general not as easily as temperature and pressure.

The specific volume can be measured mathematically by relating it to the temperature and pressure of a substance at a specific state point. This is particularly simple for fluids (liquids or gases). The pressure, volume and temperature (PVT) relationship is discussed later in this chapter.

Specific volume is equated as:

$$Cu.\ ft./lb. \times lbs. = cu.\ ft.$$

When working with liquids, the density of the substance is important. Density is defined as the weight per unit volume (lbs./cu.ft.). It is the reciprocal of specific volume and identified by the letter ρ (rho).

The secondary properties, internal energy, enthalpy and entropy are discussed under Energy, later in this appendix.

THERMODYNAMIC PROCESS

Steam traps perform functions known in science and engineering as processes; a review of what a process is and how steam traps and machinery connected to them facilitate these processes is important. If the reader finds it necessary to make a steam trap application to a heat exchanger, draining condensate from a branch or steam main, trapping steam in an evaporator or any one of a multitude of other jobs delegated to the steam trap, a working knowledge of thermodynamic processes can be most helpful.

A process is a series of changes in state points following fundamental laws of nature as governed primarily by pressure and temperature. A process can be represented by the locus of a point traveling from one steady state point to another. See Figure A-8. The figure is a graphic representation of Boyle's Law and demonstrates the constant temperature process. In Figure A-8, two common coordinates are used: pressure and volume. State point (1) defined by the intersection of P1 and V1 is the starting place. In proceeding to state point (2), defined by P2 and V2 the process follows certain laws of nature.

In nature, there are five known processes: isothermal ($T = C$), shown in Figure A-8; isobaric ($P = C$), Figure A-9; isometric ($V = C$), Figure A-10; isentropic ($S = C$), Figure A-11 and polytropic ($N = C$), Figure A-12. The isentropic and polytropic processes are addressed with respect to the throttling process later in this appendix. Their significance in steam trapping is there delineated.

The use of the Temperature-Entropy (T-S) plane in the above figures is significant in dealing with steam processes. For almost all steam machinery, the (T-S) plane is used more extensively than any other in

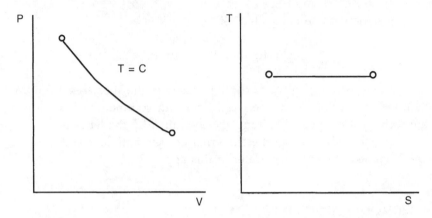

Figure A-8. Isobar Thermal Process on Pressure-volume and Temperature-entropy Planes.

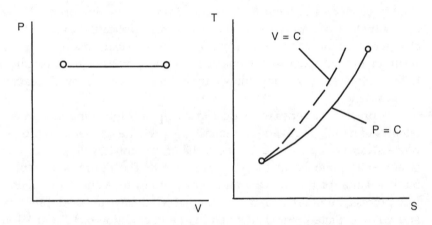

Figure A-9. Isobaric Process on Pressure-volume and Temperature-entropy Planes.

steam engineering practice. An illustration of the Mollier Chart is presented in Appendix D. A detailed Mollier Chart is shown in Appendix A. The Mollier Chart is useful in tracking the flow of steam to condensate through a steam trap to determine the energy saved or lost depending on whether the trap is functioning or has failed.

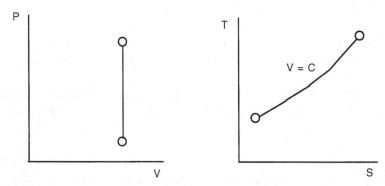

Figure A-10. Isometric Process on Pressure-volume and Temperature-entropy Planes.

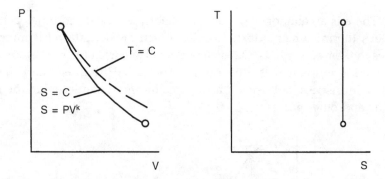

Figure A-11. Isotropic Process on Pressure-volume and Temperature-entropy Planes.

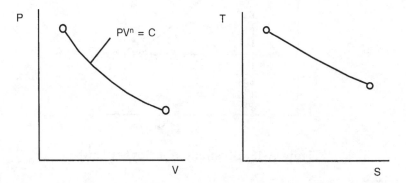

Figure A-12. Polytropic Process on Pressure-volume and Temperature-entropy Planes.

BOYLE'S LAW

The law first discovered with respect to the thermodynamics of gas behavior was Boyle's Law. Robert Boyle (1627-1691) observed that:

If a given quantity of gas was maintained at a constant Temperature, the Absolute Pressure would vary inversely with the Volume.

Mathematically:

$$P_1/P_2 = V_1/V_2 \text{ or } P_1V_1 = P_2V_2, \text{ where } T = C.$$

It can be said that when $T = C$ that $P_1V_1 = C$ also. See Figure A-13.

There is an abundance of data available on steam and other gases Modern thermodynamic textbooks list much of the data on common gasses in their appendices. Other data is available from the publishers of engineering journals. Also the ASME, ASCE, ASHRAE, etc. and their related industries publish data on the thermodynamic behavior of steam* and other gases.

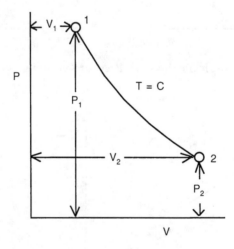

Figure A-13. Illustration of Boyle's Law.

---

*The American Society of Mechanical Engineers (ASME) Steam Tables are an excellent source of modern steam data.

CHARLES LAW

Jacques A. Charles (1746-1823) discovered another phenomena concerning gas behavior about 100 years after Boyle formulated his law.

Charles Law states:

> If the pressure of a given weight of gas is held constant, the volume will vary directly with the absolute temperature.

See Figure A-14a.

Mathematically:

$$V_1/V_2 = T_1/T_2 \text{ or } T/V = C, \text{ for } P = C. \text{ See Figure A-9.}$$

And:

> If the volume of a given weight of gas is held constant, the absolute pressure will vary directly with the absolute temperature.

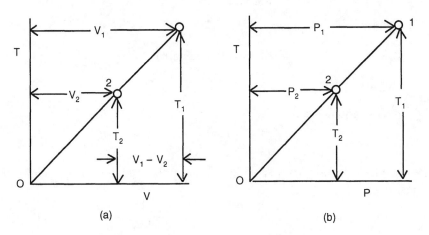

(a)   (b)

Figure A-14. Illustration of Charles' Law.
In Figure A-14-a, the line 1-2 represents a constant pressure process for a perfect gas as it appears on the TV plane. In A-14b, the line 1-2 represents a constant volume process as it appears on the T-P plane. The slope of the line in A-14a is $(T_1 - T_2)/(V_1 - V_2) = C$.

Mathematically:

$P_1/P_2 = T_1/T_2$, or $T/P = C$, when $V = C$. See Figure A-14b.

The equations for the isobaric and isometric processes formulated from the above are straight lines ($y = cx$) and apply to the ideal gas. As with Boyles Law, gases behaving closely to Charles Law are: Air, Argon, Helium, Oxygen, Nitrogen and Carbon dioxide.

Although Charles Law refers to absolute zero as the point of origin from which his measurements were made, the law applies to a limited temperature range because of changes of phase in the gas. Charles Law is most accurate in the mid-range between the liquid-gaseous state and the critical state point.*

The characteristic or characteristic constant of an ideal gas is formulated by combining Charles and Boyles Laws for a predetermined weight of gas. From the constant pressure process it is seen that:

$T_1/T_2 = V_1/V_2$, or $T_1 = V_1T_2/V_2$.

And from the constant volume process:

$P_1/P_2 = T_1/T_2$, or $T_1 = P_1T_2/P_2$.

Therefore:

$P_1/V_2/T_1 = P_2V_2/T_2 = R$ (a constant).

See Figure A-15.

This equates to:

$PV = RT$

And when dealing with specific volume or a predetermined weight of gas, then:

$PV = wRT$.

The preceding is the characteristic equation of a perfect gas. See Figure A-15.

---

*The critical state point is the Temperature point on the T-S plane where liquid changes to gas without requiring the latent heat of evaporation.

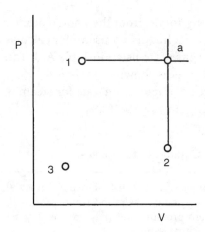

Figure A-15. Characteristic of a Perfect Gas.

THE UNIVERSAL GAS CONSTANT

Often, when selecting and sizing a steam trap for a heat exchanger it is necessary to understand something of the nature of the substance being heated as is the case with air heaters, space heaters and many chemicals that are heated in the processing industries. The following is offered for readers selecting steam traps for heat exchangers used to heat air or other substances.

The constant R is known as the Universal Gas Constant and remains unchanged for gases having the same molecular weight. This is expressed as:

$$MR = PwV/T$$

The USCS values in the foregoing equations are:

P   =   Pounds per square foot,
V   =   Cubic feet,
T   =   Degrees Rankine
M   =   Molecular weight and
w   =   Weight, lbs per hour

For the more nearly perfect gases, MR = 1544 is an accepted value of the universal gas constant. Thus, if an approximation is satisfactory (note the values of MR for CO_2 and C_2H_2), the specific gas constant R for any gas may be found from the relation:

$$R = 1544/M.$$

Since the molecular weight is easily found from the known atomic weights, this relationship (MR = PwV/T) is frequently useful for calculations. Some characteristic constants for gases are listed in Table A-3. The variation in gas constants for some real gases is shown in Table A-4.

As noted previously, a good source of engineering data for steam is the American Society of Mechanical Engineers (ASME).

Table A-3. Characteristic Constants for Gases.

(a) For the gases marked (a), the values of c_p are instantaneous values at $540°R = 80°F$ taken from spectroscopic data. Then c_v was found from $c_v = c_p - R/J$, and k from c_p /c_v. These values are for zero pressure, but they are suitably accurate for ordinary computations at ordinary pressures.

(b) For the gases marked (b), the values of c_p and k were taken from the *International Critical Tables*, Vol. V, for standard atmospheric pressure and 15°C (59°F). Then c_v was computed from $c_v = c_p k$.

The gas constant R for each gas was computed from $R = p/(\rho T)$, where $\rho = 1/v$ was taken from the International Critical Tables, Vol. III, for standard atmospheric pressure and 32°F. M = approximate molecular weight: c_p and c_v are in Btu per lb.-degree F.

Gas		M	c_p	c_v	k	R	MR
Air	(a)	29	0.24	0.1715	1.4	53.3	1545
Carbon monoxide (CO)	(a)	28	0.2487	0.1779	1.398	55.1	1543
Hydrogen (H_2)	(a)	2.016	3.421	2.4354	1.405	767	1546
Nitric oxide (NO)	(a)	30	0.2378	0.1717	1.384	51.4	1542
Nitrogen (N_2)	(a)	28	0.2484	0.1776	1.4	55.1	1543
Oxygen (O_2)	(a)	32	0.2193	0.1573	1.394	48.25	1544
Argon (A)	(b)	39.9	0.125	0.0749	1.668	38.7	1544
Helium (He)	(b)	4	1.25	0.754	1.659	386	1544
Carbon dioxide (CO_2)	(a)	44	0.202	0.157	1.29	34.9	1536
Hydrogen sulfide (H_2S)	(a)	34	0.328	0.270	1.21	44.8	1523
Nitrous oxide (N_2O)	(a)	44	0.211	0.166	1.27	34.9	1536
Acetylene (C_2H_2)	(a)	26	0.409	0.333	1.23	58.8	1529
Ethane (C_2H_4)	(a)	30	0.422	0.357	1.18	50.8	1524
Ethylene (C_H_4)	(a)	28	0.374	0.304	1.23	54.7	1532
Iso-butane (C_4H_x)	(a)	58.1	0.420	0.387	1.00	25.8	1499
Methane (CH_4)	(a)	16	0.533	0.409	1.30	96.2	1539
Propane (C_3H_8)	(a)	44.1	0.404	0.360	1.12	34.1	1504

Source: Applied Thermodynamics, Faires, 1947

Table A-4. Variation in the Universal Gas Constant for Typical Gases.

Gas	M	R	MR
Hydrogen (H_2)	(2.016)	(767)	1546
Carbon monoxide (CO)	(28)	(55.1)	1543
Nitrogen (N_2)	(28)	(55.1)	1543
Oxygen (O_2)	(32)	(48.25)	1544
Helium (He)	(4)	(386)	1544
Carbon dioxide (CO_2)	(44)	(34.9)	1536
Acetylene (C_2H_2)	(26)	(58.8)	1529

Source: Physics, Hausmann and Slack, 1948

SPECIFIC HEAT

As with the Universal Gas Constant, it is important to have an understanding of a substances specific heat when using steam (and its resultant condensate) to heat it. Steam Traps play an important role in governing the behavior of the steam in the heat exchanger. (See Heat Recovery and the Steam Trap, Chapter 2.) The following are some thoughts on specific heat which should be helpful to the engineer/technician as a student of specific heat fundamentals for use when referring to later chapters which deal with trapping steam from heat exchangers.

The Specific Heat is the number of Btus required to raise one pound of a substance one degree Fahrenheit. By definition, Specific Heat is:

$$c = Q/w(T_2 - T_1),$$

where:

c = Specific Heat of the substance,

Q = Quantity of thermal energy (Btus) released or absorbed,

w = Weight of the substance: for this purpose, $w = 1$ lb,

T_1 = Temperature of the substance at state point 1, the beginning of the process (°F), and

T_2 = Temperature of the substance at state point 2, the terminus of the process (°F).

It is evident from the above equations that Specific Heat is process rather than state point oriented.

Specific heat varies with the temperature. It is 1.0 Btu/lb-°F for water at 70°F and 0.938 at 400°F. Table A-5 lists the specific heat of some common substances for the Temperature ranges indicated. For most engineering work it suffices to use the mean specific heat between the starting point and the terminus. Laboratory accuracy requires calculus, or plotted data.

SPECIFIC HEAT OF GASES

The specific heat of gases as with other substances is the thermal energy necessary to raise one pound of the gas one degree Fahrenheit. However, with gases this is not single valued. There is a range of values of specific heat depending on the constraints placed upon it during the thermal energy transfer. For instance, a gas heated within a closed container, Figure A-16a has a specific heat different than that heated in a cylinder with a piston of predetermined weight upon it. See Figure A-16b. In the former case, there is no motion or movement of mass as heat Q is added. The container remains stationary as does all its parts and there is no work performed. It is a constant volume process. In this instance the specific heat is defined as:

$$C_v = Q/T_2 \ T_1$$

In Figure A-16b the piston moves within the cylinder so that a constant pressure expansion occurs as heat Q is added. In this case, work is performed, since a mass, the piston, is moved through a distance as the volume of the enclosure enlarges during the expansion of the gas.

Work $W = ws$ where:

w   =   weight in lbs and
s   =   distance through which the mass is moved in feet.

Therefore, the heat added has been used to perform two functions. It has increased the temperature of the gas as in the former case and it has also performed work. It can be said that:

$$Q = W + c_p(T_2 - T_1) \quad \text{or} \quad c_p = (Q - W)/(T_2 - T_1)$$

Table A-5. Ideal Gas Specific Heats in Btu/lb-F.

1. Zero-pressure specific heats for six common gases, where $k = c_p/c_v$

Temp., °F	c_p	c_v	k	c_p	c_v	k	c_p	c_v	k	Temp., °F
	Air			Carbon dioxide (CO_2)			Carbon monoxide (CO)			
40	0.240	0.171	1.401	0.195	0.150	1.300	0.248	0.177	1.400	40
100	0.240	0.172	1.400	0.205	0.160	1.283	0.249	0.178	1.399	100
200	0.241	0.173	1.397	0.217	0.172	1.262	0.249	0.179	1.397	200
300	0.243	0.174	1.394	0.229	0.184	1.246	0.251	0.180	1.394	300
400	0.245	0.176	1.389	0.239	0.193	1.233	0.253	0.182	1.389	400
500	0.248	0.179	1.383	0.247	0.202	1.223	0.256	0.185	1.384	500
600	0.250	0.182	1.377	0.255	0.210	1.215	0.259	0.188	1.377	600
700	0.254	0.185	1.371	0.262	0.217	1.208	0.262	0.191	1.371	700
800	0.257	0.188	1.365	0.269	0.224	1.202	0.266	0.195	1.364	800
900	0.259	0.191	1.358	0.275	0.230	1.197	0.269	0.198	1.357	900
1000	0.263	0.195	1.353	0.280	0.235	1.192	0.273	0.202	1.351	1000
1500	0.276	0.208	1.330	0.298	0.253	1.178	0.287	0.216	1.328	1500
2000	0.286	0.217	1.312	0.312	0.267	1.169	0.297	0.226	1.314	2000
	Hydrogen (H_2)			Nitrogen (N_2)			Oxygen (O_2)			
40	3.397	2.412	1.409	0.248	0.177	1.400	0.219	0.156	1.397	40
100	3.426	2.441	1.404	0.248	0.178	1.399	0.220	0.158	1.394	100
200	3.451	2.466	1.399	0.249	0.178	1.398	0.223	0.161	1.387	200
300	3.461	2.476	1.398	0.250	0.179	1.396	0.226	0.164	1.378	300
400	3.466	2.480	1.397	0.251	0.180	1.393	0.230	0.168	1.368	400
500	3.469	2.484	1.397	0.254	0.183	1.388	0.235	0.173	1.360	500
600	3.473	2.488	1.396	0.256	0.185	1.383	0.239	0.177	1.352	600
700	3.477	2.492	1.395	0.260	0.189	1.377	0.242	0.181	1.344	700
800	3.494	2.509	1.393	0.262	0.191	1.371	0.246	0.184	1.337	800
900	3.502	2.519	1.392	0.265	0.194	1.364	0.249	0.187	1.331	900
1000	3.513	2.528	1.390	0.269	0.198	1.359	0.252	0.190	1.326	1000
1500	3.618	2.633	1.374	0.283	0.212	1.334	0.263	0.201	1.309	1500
2000	3.758	2.773	1.355	0.293	0.222	1.319	0.270	0.208	1.298	2000

Source: Data adapted from *Tables of Thermal Properties of Gases*, NBS Circular 564, 1955.

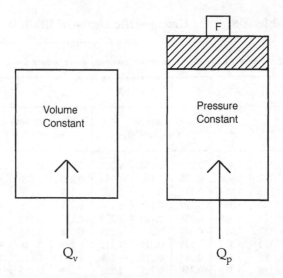

Figure A-16. Illustration of Variation in Specific Heats.
In case (a), to change the temperature of the substance, it is necessary to add
only enough heat to bring about the required change of internal energy, since no
energy is required to make the gas expand against some force. In case (b),
enough heat must be added not only to accomplish the required change of internal
energy, but also to cause the gas to do work in expanding, the work of
pushing the piston upward against the resistance of a constant force F.

$$Q_v = wc_v (T_2 - T_1)$$
$$Q_p = wc_p (T_2 - T_1)$$

Although the specific heat of a solid is substantially constant through a limited
range of temperatures, the specific heat of a gas may have any value, depending
upon the nature of the expansion and the useful work performed, if any.

The relationship between c_p and C_v can be derived from the Universal Gas Constant, R:

$$c_p = c_v (R/J),$$

where J is the Joule Constant, equal to 778 Ft-lbs/Btu, and can be
equated from the relationship between heat, work and internal energy,
discussed later in the chapter. Other expressions of the relationship are:

$$C_v = R/(k - 1)J \text{ and } c_p = kR/(k - 1)J.$$

Values for k and R can be found in Table A-3. The equation, $k=c_p/c_v$, is also of value since it shows the relationship between the constant pressure and constant volume specific heats.

Table A-5 lists Ideal Gas Specific Heats for a selection of gases. Table A-6 lists Specific Heat Data for Monatomic Gases and Table A-7 is a list of equations for calculating Variable Specific Heats. Of particular note in this table is the specific heat of water (steam) from 540 to 5400R.

Table A-6. Specific Heat for Monatomic Gases.

Over a wide range of temperatures at low pressures, the specific heats c_v and c_p of all monatomic gases are essentially independent of temperature and pressure. In addition, on a mole basis all monatomic gases have the same value for either c_v or c_p in a given set of units. One set of values is

$$c_v = 2.98 \text{ Btu(lbmol} \cdot {}^\circ\text{F)} \quad \text{and} \quad c_p = 4.97 \text{ Btu(lbmol} \cdot {}^\circ\text{F)}$$

Source: Data adapted from *Tables of Thermal Properties of Gases*, NBS Circular 564, 1955.

3. Constant-pressure specific-heat equations for various gases at zero pressure (USCS units)

$$c_p/R_u = a + bT + cT^2 + dT^3 + eT^4$$

where T is in degrees Rankine, equation valid from 540 to 1800°R

Gas	a	$b \times 10^3$	$c \times 10^6$	$d \times 10^9$	$e \times 10^{12}$
CO	3.710	−0.899	1.140	−0.348	0.0229
CO$_2$	2.401	4.853	−2.039	0.343	
H$_2$	3.057	1.487	−1.793	0.947	−0.1726
H$_2$O	4.070	−0.616	1.281	−0.508	0.0769
O$_2$	3.626	−1.043	2.178	−1.160	0.2054
N$_2$	3.675	−0.671	0.717	−0.108	−0.0215
Air (dry)	3.653	−0.741	1.016	−0.328	0.0262
NH$_3$	3.591	0.274	2.576	−1.437	0.2601
NO	4.046	−1.899	2.464	−1.048	0.1517
NO$_2$	3.459	1.147	2.064	−1.639	0.3448
SO$_2$	3.267	2.958	0.211	−0.906	0.2438
SO$_3$	2.578	8.087	−2.832	−0.136	0.1878
CH$_4$	3.826	−2.211	7.580	−3.898	0.6633
C$_2$H$_2$	1.410	10.587	−7.562	2.811	−0.3939
C$_2$H$_4$	1.426	6.324	2.466	−2.787	0.6429

Source: Adapted from the data in NASA SP-273, Government Printing Office, Washington, 1971.

Table A-7. Equations of Variable Specific Heats.

(a) The value marked (a) is derived from Spencer and Justice, "Empirical heat capacity equations for simple gases," Journal of the American Chemical Society, Vol. 56, p. 2311. This paper gives heat capacities for a number of other gases in the form $\alpha + \beta T + \delta T^2$ which are accurate up to about 2700°R.

(b) The values marked (b) are taken from Spencer and Flannagan, "Empirical heat capacity equations of gases," Journal of the American Chemical Society, Vol. 64, p. 2511; or (b') from Spencer, Journal of the American Chemical Society, Vol. 67, p. 1859.

(c) The value marked (c) is taken from Chipman and Fontana, "A new approximate equation for heat capacities at high temperatures," Journal of the American Chemical Society, Vol. 57, p. 48.

(d) The values marked (d) are taken from Sweigert and Bearddey, "Empirical specific heat equations based upon spectroscopic data." Bulletin No. 2, Georgia School of Technology.

Substance (Temp. Range)	Mol. Wt.	Btu per lb.-°R	Btu per mol.-°R
(a) Air (500–2700°R)	29	$c_p = 0.219 + 0.342T/10^4 - 0.293T^2/10^8$ $c_v = 0.1509 + 0.342T/10^4 - 0.293T^2/10^8$	$C_p = 6.36 + 9.92T/10^4 - 8.52T^2/10^8$ $C_v = 4.375 + 9.92T/10^4 - 8.52T^2/10^8$
(b) SO₂, Sul. Diox. (540–3400°R)	64.06	$c_p = 0.1875 + 0.0944T/10^4 - 1.336 \times 10^4/T^2$ $c_v = 0.1547 + 0.0944T/10^4 - 1.336 \times 10^4/T^2$	$C_p = 11.89 + 6.05T/10^4 - 85.6 \times 10^4/T^2$ $C_v = 9.91 + 6.05T/10^4 - 85.6 \times 10^4/T^2$
(b) NH₃, Ammonia (540–1800°R)	17.03	$c_p = 0.363 + 2.57T/10^4 - 1.319T^2/10^8$ $c_v = 0.247 + 2.57T/10^4 - 1.319T^2/10^8$	$C_p = 6.19 + 43.8T/10^4 - 22.47T^2/10^8$ $C_v = 4.20 + 43.8T/10^4 - 22.47T^2/10^8$
(c) H₂, Hydrogen (540–4000°R)	2.016	$c_p = 2.857 + 2.867T/10^4 + 9.92/T^{1/2}$ $c_v = 1.871 + 2.867T/10^4 + 9.92/T^{1/2}$	$C_p = 5.76 + 5.78T/10^4 + 20/T^{1/2}$ $C_v = 3.775 + 5.78T/10^4 + 20/T^{1/2}$

Table A-7. Equations of Variable Specific Heats (Continued).

(d) O₂, Oxygen (540–5000°R)	32	$c_p = 0.36 - 5.375/T^{1/2} + 47.8/T$ $c_v = 0.298 - 5.375/T^{1/2} + 47.8/T$	$C_p = 11.515 - 172/T^{1/2} + 1530/T$ $C_v = 9.53 - 172/T^{1/2} + 1530/T$
(d) N₂, Nitrogen (540–9000°R)	28.016	$c_p = 0.338 - 123.8/T + 4.14 \times 10^4/T^2$ $c_v = 0.267 - 123.8/T + 4.14 \times 10^4/T^2$	$C_p = 9.47 - 3470/T + 116 \times 10^4/T^2$ $C_v = 7.485 - 3470/T + 116 \times 10^4/T^2$
(d) CO, Carb. Mon. (540–9000°R)	28	$c_p = 0.338 - 117.5/T + 3.82 \times 10^4/T^2$ $c_v = 0.267 - 117.5/T + 3.82 \times 10^4/T^2$	$C_p = 9.46 - 3290/T + 107 \times 10^4/T^2$ $C_v = 7.475 - 3290/T + 107 \times 10^4/T^2$
(d) H₂O, Steam (540–5400°R)	18.016	$c_p = 1.102 - 33.1/T^{1/2} + 416/T$ $c_v = 0.992 - 33.1/T^{1/2} + 416/T$	$C_p = 19.86 - 597/T^{1/2} + 7500/T$ $C_v = 17.875 - 597/T^{1/2} + 7500/T$
(d) CO₂, Carb. Diox. (540–6300°R)	44	$c_p = 0.368 - 148.4/T + 3.2 \times 10^4/T^2$ $c_v = 0.323 - 148.4/T + 3.2 \times 10^4/T^2$	$C_p = 16.2 - 6530/T + 141 \times 10^4/T^2$ $C_v = 14.22 - 6530/T + 141 \times 10^4/T^2$
(b′) CH₄, Methane (540–2700°R)	16.03	$c_p = 0.211 + 6.25T/10^4 - 8.28T^2/10^8$ $c_v = 0.0873 + 6.25T/10^4 - 8.28T^2/10^8$	$C_p = 3.38 + 100.2T/10^4 - 132.7T^2/10^8$ $C_v = 1.396 + 100.2T/10^4 - 132.7T^2/10^8$

Table A-7. Equations of Variable Specific Heats (Continued).

Substance (Temp. Range)	Mol. Wt.	Btu per lb.-°R	Btu per mol.-°R
(b) C₂H₄, Ethylene (540–2700°R)	28.03	$c_p = 0.0965 + 5.78T/10^4 - 9.97T^2/10^8$ $c_v = 0.0262 + 5.78T/10^4 - 9.97T^2/10^8$	$C_p = 2.706 + 162T/10^4 - 279.6T^2/10^8$ $C_v = 0.721 + 162T/10^4 - 279.6T^2/10^8$
(b') C₂H₆, Ethane (540–2700°R)	30.05	$c_p = 0.0731 + 7.08T/10^4 - 11.3T^2/10^8$ $c_v = -0.0078 + 7.08T/10^4 - 11.3T^2/10^8$	$C_p = 2.195 + 212.7T/10^4 - 340T^2/10^8$ $C_v = 0.210 + 212.7T/10^4 - 340T^2/10^8$
(b') C₄H₁₀, n-Butane (540–2700°R)	58.08	$c_p = 0.075 + 6.94T/10^4 - 11.77T^2/10^8$ $c_v = 0.0419 + 6.94T/10^4 - 11.77T^2/10^8$	$C_p = 4.36 + 403T/10^4 - 683T^2/10^8$ $C_v = 2.37 + 403T/10^4 - 683T^2/10^8$
(b') C₃H₈, Propane (540–2700°R)	44.06	$c_p = 0.0512 + 7.27T/10^4 - 12.32T^2/10^8$ $c_v = -0.0073 + 7.27T/10^4 - 12.32T^2/10^8$	$C_p = 2.258 + 320T/10^4 - 543T^2/10^8$ $C_v = 0.273 + 320T/10^4 - 543T^2/10^8$
(b) C₂H₂, Acetylene (500–2300°R)	26.02	$c_p = 0.459 + 0.937T/10^4 - 2.89 \times 10^4/T^2$ $c_v = 0.383 + 0.937T/10^4 - 2.89 \times 10^4/T^2$	$C_p = 11.94 + 24.37T/10^4 - 75.2 \times 10^4/T^2$ $C_v = 9.96 + 24.37T/10^4 - 75.2 \times 10^4/T^2$

ENERGY

Definition of Energy

The energy of a substance can be defined as its capacity for doing work. Energy is inherent in all substances, both solids and fluids. A body may possess energy or have the capacity to do work because of a variety of states or conditions and it may be classified into groups such as mechanical, thermal, electric, chemical, atomic, etc. The main concern in this chapter is thermal energy and its relationship to mechanical energy (Thermodynamics). Therefore, a brief discussion of the forms of mechanical energy precedes the study of thermal energy.

Mechanical Energy

A review of mechanical energy is useful since it relates to the pressure-volume (PV/J) component of thermal energy. The total quantity of thermal energy is known as enthalpy (H) of which (PV) energy is a part. Mathematically $H = PV/J + U$ when (PV/J) is the mechanical equivalent energy and U is the inherent internal thermal energy. Engineers repeatedly deal with enthalpy (H) when selecting steam traps for heat exchangers. It is important to know which part of (H) is internal energy (U) used to transfer heat and which part is energy expended in the steam expanding in the heater (PV/J).

Mechanical energy can be divided into two forms, potential and kinetic. Potential energy is the energy of a body or substance due to its position with respect to another substance or reference datum. A body of weight, w at a height of Z elevation above a reference datum, possesses potential energy equal to:

$$PE = wZ$$

with respect to that datum. See Figure A-17 where the weight w possesses w lbs × Z feet or wZ_1 ft-lbs of energy relative to the plane A.

The potential energy of the body would be reduced if the weight were lowered to a different elevation above plane A, say to plane B in Figure A-18, where Z_2 is less than Z_1. The new potential energy would become wZ_2. It would have dissipated $w(Z_1-Z_2)$ ft-lbs of potential energy in the lowering from point 1 to plane B. The potential energy dissipated in the fall from point 1 to plane B would be transposed into kinetic energy. Kinetic energy equates to:

$$KE = w(V_s^2/2g)$$

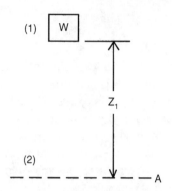

Figure A-17. Illustration of Potential Energy.

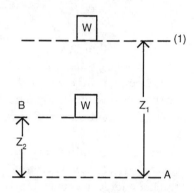

Figure A-18. Comparison of Potential and Kinetic Energy.

where

w = weight of body in lbs,
V_s = velocity in ft/sec and
g = acceleration due to gravity in ft/sec^2.

It is the energy of a substance in motion relative to another substance. In the case of the weight moving toward plane A, the kinetic energy equals the velocity squared times the weight ($v^2 \times w$) divided by two times the acceleration due to gravity at any point along the path of the object relative to plane A. If we express the kinetic energy equation in units of thermal energy:

$$KE = \frac{lbs \times ft^2/sec^2}{ft/sec^2 \times 778\ ft\text{–}lbs/Btu}$$

where 778 ft-lbs is the Joule constant. Its derivation is explained under the subject of thermal energy. A moving substance containing kinetic energy is capable of performing work. For example a jet stream impacting on blades or buckets of a wheel (Figure A-19) expends kinetic energy by nature of its change in velocity, and rotates the wheel.

$$KE = \frac{w\left(V_{s1}{}^2 - V_{s2}{}^2\right)}{2g}$$

In this case the energy of the jet stream equals the work W of the wheel in raising the weight:

KE = W = PK.

and

$$\frac{w\left(V_{s1}-V_{s2}\right)^2}{2g} = W = PE = w\,Z$$

where Z = the distance the weight (w) is raised in Figure A-19. This of course neglects friction and mechanical losses in the energy conversion.

THERMAL ENERGY

The most common engineering term used to express the total thermal energy of a substance is enthalpy. It is a summation of the internal energy (U) and the Pressure-Volume energy. The equation for enthalpy (H) is:

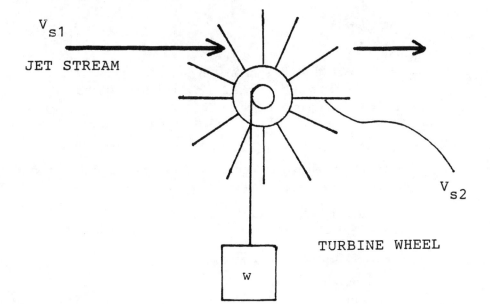

Figure A-19. Illustration of Kinetic Energy of a Jet Stream.

$$H = U + PV/J$$

where

    U = Internal energy in Btus,
    P = Pressure in lbs/sq.ft.,
    V = Volume in cu.ft. and
    J = The Joule Constant 778 ft-lbs/Btu.

Examining the two components of enthalpy is a systematic approach to an understanding of enthalpy.

INTERNAL ENERGY

Consider U, internal energy, as the temperature sensitive component. In order to have a U term, the substance must be at a temperature greater than its absolute zero. When all motion between particles ceases at zero R, the substance no longer has internal energy. The internal energy of a material comes from the molecules colliding and rubbing against one another, and the frictional (thermal) energy resulting is measured as temperature. The absolute value of a substance's internal energy is:

$$U = wcT$$

where

    w = Weight of the substance in pounds,
    c = Specific heat of the substance in Btu/lb/°F and
    T = Temperature in degrees R.

or

$$U = wc(T_2 - T_1)$$

If there is no change in temperature or no change in the action of the molecules of the substance then:

$$\Delta U = wc(\Delta T) \text{ and}$$
$$\Delta U = wc(0) = 0,$$

or if the object is at absolute zero, then

$$U = wcT \text{ and}$$
$$U = wc(0) = 0$$

and there is no motion between molecules of the substance.

PV ENERGY

Normally steam is not compressed in industrial processes. However, to maintain clarity this case is examined to help the reader understand the Pressure-Volume energy contained in steam. The more common process in industry is where steam expands, such as across a regulating valve to a steam heater or across a steam trap on the condensate outlet side of a heat exchanger. The expansion process is discussed along with reversibility and irreversibility later in the chapter after throttling processes. Then the PV term will be referred to again.

The PV/J term is the thermal equivalent of the mechanical energy the substance contains. For instance, if one pound of steam is contained under a piston of a compressor, Figure A-20, then it is at one level of energy at point A.

The energy level is $P_1 \times L_1 \times A$; where

P = Pressure in lb/sq.ft.,
L = Active length of piston in feet and
A = Cross-sectional area of cylinder in sq. ft.

The PV term then equals:

$$PV_A = P_1 \times L_1 \times A.$$

If the piston is driven to the right to position B while maintaining a constant temperature in the cylinder by cooling, then the PV energy becomes:

$$PV_B = P_2 L_1 A.$$

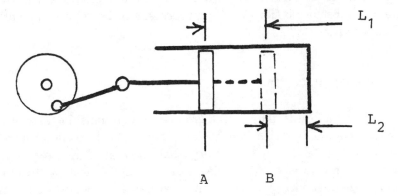

Figure A-20. Illustration of Irresistibility and Reversibility.

Since the temperature was maintained constant by design, the PV energy = H. The thermal energy removed from the cylinder has prevented a change in the internal energy by preventing a temperature increase. Although PVB contains more energy than PVA because of the compression, the energy in PVB cannot return the piston all the way to PVA because some energy was removed as thermal energy. In this case,

Q = H - PV/J, where:
Q = Transferred Heat in Btu.

HEAT, A TYPE OF ENERGY

According to *The Concise Columbia Encyclopedia*, heat is the "internal energy of a substance, associated with the positions and motions of its component molecules, atoms and ions. The average kinetic energy of the molecules or atoms, which is due to their motions, is measured by the temperature of a substance. The potential energy is associated with the state, or phase, of the substance." An undergraduate level physics text states that heat (conduction heat) is the "energy carried by molecules as random translational kinetic energy, and it is passed from molecule to molecule during collisions." But, the text cautions that we must make a distinction between internal energy (U) and heat (Q). The internal energy of a gas, is the collective energy carried by the gas molecules in their random motion. Heat, on the other hand, refers only to the translational kinetic energy that is being transported through the substance from a higher-temperature region to a lower temperature region. The temperature of a substance is, for practical purposes, a measure of its potential to transfer heat.

ENERGY IN TRANSITION

The term heat is defined as thermal energy in transition as a result of a temperature difference between two bodies. The distinction between internal energy and heat should be noted. Internal energy is a stored thermal energy; it refers to a single condition or state. Heat, on the other hand, is thermal energy in transition. Heat transfer requires a temperature difference between two points, which means two different conditions or states.

As an illustration, imagine a container with water at 90 degrees Fahrenheit. We say that this water has internal energy; by arranging some suitable process by which the temperature is reduced to a lower level, some of the energy stored in the water could be converted into work. Imagine that the air surrounding the wall of the container is also at 90 degrees Fahrenheit.

The water has internal energy, but there is no heat transfer and no flow of energy through the wall, because there is no temperature difference across the wall. On the other hand, if the outside air were at 60 degrees Fahrenheit, then the temperature difference of 30 degrees would indicate a flow of energy from the liquid through the wall to the outside air, and there would be a transfer of heat.

A unit quantity of heat is the British thermal unit (Btu), which is defined as the heat which must be transferred in order to raise the temperature of one pound of water one degree Fahrenheit. Many experiments have been made to measure the heat equivalent of work. Measurements indicate that one Btu is approximately equal to 778 foot-pounds.

Joule's experiments and the resultant Joule Constant shows the relationship between thermal and mechanical energy. A knowledge of this relationship is important to the understanding of the General Energy Equation. Basically all steam devices including the steam trap can be considered a black box in which energy flows in and out. For this reason consideration must be given to the Joule Constant and the General Energy Equation.

JOULE'S CONSTANT AND THE RELATIONSHIP
BETWEEN HEAT AND WORK

James P. Joule, a nineteenth century pioneer in work-energy experiments determined the relationship between thermal energy and mechanical energy. It is necessary in steam trapping to be reminded of the value of the mechanical equivalent of thermal energy. It is one Btu = 778 ft-lbs mechanical energy or mechanical work. This is so because, universally, steam traps are part of the metering components of heat exchangers which can be considered as "Black Boxes."

Work, like heat, involves a transfer of energy, or energy in transit.

James P. Joule, a nineteenth century pioneer in work/energy investigations, determined a relationship between heat and work with his paddlewheel experiment. See Figure A-21.

In the experiment, a known amount of mechanical energy (work) was converted completely into heat. The descending weight caused the paddle to turn, which then heated the water.

The thermometer shown was used to measure the temperature rise. It was assumed that the energy losses due to the pulley and bearings were negligible. The falling weight's loss of mechanical energy was determined from E = mgh. This value was then compared to the number of Btus (heat energy) that would have been required to raise the temperature of the known mass of water by the temperature change measured with the thermometer.

The conclusion of Joule's experiment is that 1 Btu = 778 ft-lbs. Although Joule's temperature measurement equipment was primitive compared to that available today, his measurement is very close to the currently accepted value of 777.9 (say 778) ft-lbs/Btu. This ratio is known as Joule's constant (J). It allows one to convert heat energy units to mechanical energy units, and thus describes the mechanical work equiva-

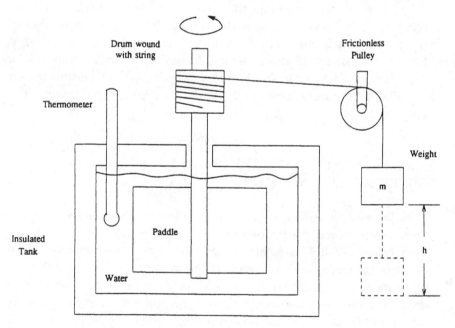

Figure A-21. Joule's Paddle Wheel Experiment.

lent of heat. Joule proved that the same amount of heat energy could be obtained from a given amount of work regardless of the method used to produce the work. He obtained the same constant (± 5 percent) when heating water by electrical current or by rubbing iron rings in a liquid.

Joule's experiments were further evidence that heat is a form of energy rather than a substance. If heat were a substance, it would not be possible to remove heat indefinitely from a system that does not change. Count Rumford (Benjamin Thompson) showed this in his cannon boring experiments. The continuous heating of the cannon during the boring process could not be accounted for by the flow of thermal energy into the cannon because it would have required some part of the surroundings to get colder to relinquish thermal energy. Similarly, using Joule's apparatus, one could obtain an infinite amount of heat out of the water by continuously performing mechanical work without changing the condition of the water.

ENERGY EQUATION FOR THE STEADY FLOW OF ANY FLUID

Figure A-22 illustrates the notation for steady flow through any apparatus. Fluid enters at section 1 and leaves at section 2. The weight rate of flow leaving equals that entering. Usually, the general energy equation is used for one pound of the substance. The results then can be applied to the total weight of the substance flowing.

Potential energy is measured with respect to an arbitrary horizontal datum. At entrance, as at exit, the energy of the fluid is the sum of the potential energy, kinetic energy and the internal energy. In between sections 1 and 2 some heat can be transferred across the walls of the apparatus, and some work can be done on or by the fluid by means of some external machine (such as a pump, compressor or turbine). As the fluid passes through the apparatus, there is a change in energy in the fluid. The fluid gains energy; that is, the energy at section 2 is greater than that at section 1. Something must have been added to the fluid between sections 1 and 2 to cause this energy gain. This energy equals the heat added to the fluid plus the work done upon the fluid. The work done on the fluid is the sum of flow-work at each section and other mechanical work added to or released by the fluid between the sections.

Let Q be the heat transferred to the fluid, per unit weight; P the pressure; v the specific volume; W the net mechanical work (or equiva-

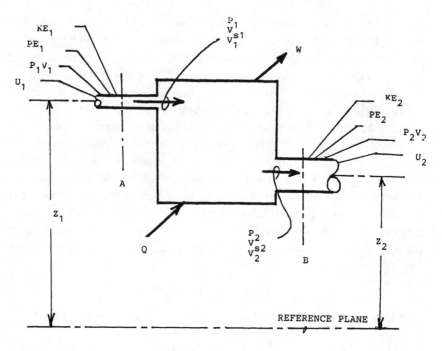

Figure A-22.

lent) done on the fluid by some external machine; U the internal energy; v_s the velocity of the fluid flowing; Z the elevation above datum; g the gravitational acceleration; and J the conversion factor or heat equivalent of work. Subscript 1 refers to the inlet condition; subscript 2 refers to the exit condition.

The term Q is heat, or thermal energy in transition, which passes through the walls of the apparatus. There is heat transfer Q only when there is a temperature difference across the walls of the apparatus. Internal frictional effects, which store internal energy in the fluid, contribute nothing to this term Q; such internal energy does not pass through the walls of the apparatus.

The term P_1v_1 represents the work transfer or flow-work at entrance and exit. The w term of the equation includes outside work done on or by the fluid with respect to the environment. For instance, W is positive for a turbine and negative for a pump.

The general energy equation is:

$$PE_1 + KE_1 + P_1v_1 + U_1 + w = PE_2 + KE_2 + P_2v_2 + U_2 + Q$$

where:

PE$_1$ Potential Energy = wZ,

KE$_1$ Kinetic Energy = $wv_s^2/2g$,

P$_1$v$_1$ = Flow work or flow energy,

(See Charles Laws and PV Energy)

U = Internal energy = wc(T),

w = Mechanical work and

Q = Transferred heat in Btu/lb.

Note that when using the general energy equation, terms must be consistent. They should all be ft-lbs or Btus. The Joule constant, ft-lbs/Btu, should be applied to obtain a consistency of terms.

THE LAWS OF THERMODYNAMICS

Since the steam trap involves a change in both thermal energy and the flow dynamics in a process, it is necessary to understand the relationship of thermal and flow energy, i.e., Thermodynamics.

There are two laws of thermodynamics. The first law states "Energy can neither be created or destroyed" and the second law states that "Energy cannot of its own accord flow from a lower temperature to a higher temperature."

THE FIRST LAW OF THERMODYNAMICS

Basically what is stated in this First Law is best illustrated by the black box concept in Figure A-23. It is a reiteration of the General Energy Equation studied earlier and states that all energy entering a black box or machine must come out during any steady state process. The General Energy Equation as previously stated is

$$Q_1 + KE_1 + PE_11 + (P_1v_1/J) + U_1 = KE_2 + PE_2 (P_2v_2/J) + U_2 + W$$

TRANSITIONAL ENERGY

It does not include transitional conditions. That is it does not in-

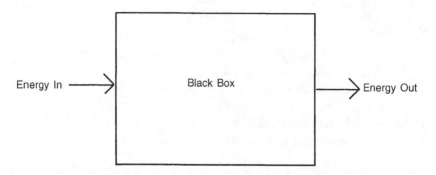

Figure A-23. Illustration of Simple Energy Balance.

clude energy transfer situations where the black box is at a changing energy level. Hydraulic analogy for instance, shows the tank in Figure A-24 would not be at steady state conditions. The flow entering the tank does not equal the flow leaving. While the level was rising in the tank, the potential energy would be increasing and the kinetic energy entering the tank would be decreasing. A boiler has to be at operating pressure and temperature to have steady state conditions. The warm up period or the cooling down period will be transitional, as would a steam turbine speeding up and slowing down where the contained energy levels were changing. Although steam traps experience transitional energy levels continually, the study is beyond the scope of this

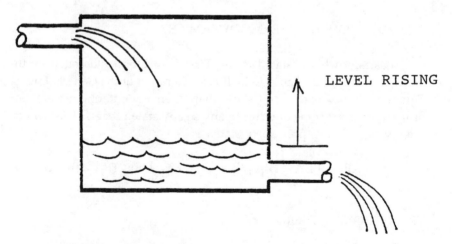

Figure A-24. Illustration of Potential Energy and Kinetic Energy.

text so let the above suffice. In practice transitional energy is studied at the laboratory level in practical experiments by trap manufacturers. The laboratory results emanating from a particular trap study are incorporated into the design and application of the trap.

The First Law states that "Energy can neither be created nor destroyed" but is converted from one form to another. The energy of a system may change when it absorbs heat from, discharges heat to or does work on the surroundings. If the energy content of the system or component does not change, output energy is equal to the input. Considering this principle it follows that First Law can be the simple accounting for the transfer or transformation of energy quantities. For example, in a coal-fired power plant, the chemical energy released in the combustion of coal is converted to thermal energy and in turn partly to mechanical work which, finally, is converted to electricity by a generator. The chemical energy in the coal not converted to electrical energy output is simply considered as system loss.

The First Law efficiently then is the ratio of the output energy in the desired forms to the energy applied to do the task, or:

$$e = \frac{\text{Energy in the desired outputs}}{\text{Energy applied to do the task}}$$

THE SECOND LAW OF THERMODYNAMICS

The Second Law states that "Thermal Energy cannot flow from a lower temperature to a higher temperature of its own accord. It takes energy analysis a step further; it accounts not only for the quantity of energy transferred or transformed but its quality as well. However, it is not necessary to pursue the concept of quantitative/qualitative analysis in detail, for the study of steam trapping. Let it be said simply that in natural processes energy flows from a state of higher potential to one of lower potential. Thus:

Fact 1. Electric charges flow from a high potential optimally to a potential of zero.

Fact 2. Water flows under gravitational potential ultimately to sea level.

Fact 3. Heat flows from a body at high temperature ultimately to ambient temperature.

It is only necessary to understand the similarity of Facts 1, 2 and 3 above to grasp the concept of thermal energy quality as it relates to steam traps.

STEAM AND THERMAL ENERGY

The engineer when dealing with steam traps is concerned primarily with steam, its thermal energy and to some extent air and carbon dioxide. To the extent, that steam traps are used to manage steam and condensate and their inherent energy when connected to heat exchangers and for draining condensate from steam mains, it is important to review the fundamentals of steam and thermal energy.

STEAM AND WATER AS A FLUID (GAS - LIQUID)

Most define a fluid as anything that flows. That is of course true, but lead is not a fluid, and yet under sufficient pressure lead may be made to flow through a small hole. The fact is that most things can be made to flow, so that to distinguish solids from fluids there must be some other characteristic. Try shape as a basis for definition. Fluids certainly have no shape, while solids have shape and they resist any attempt to deform them. It may be said that solids have rigidity of form, and fluids do not. But even this definition is not perfect, because some viscous fluids approach the behavior of very soft solids, and can have temporary form apart from any container. Tar is an example. However, in general, form is characteristic of solids and formlessness is characteristic of liquids.

The reason why fluids have little or no rigidity of form is that their molecules have great freedom of motion, while in solids the motion of molecules is restricted to a very limited region about a mean position of rest. The molecules are more dense in solids than in liquids.

LIQUIDS AND GASES

A liquid is a fluid which possesses a definite volume at a definite temperature, while a gas is a fluid which can occupy any volume by expanding or being compressed until it just fills it. This means that liquids resist a change in volume just like solids. Gases also resist compression, but they tend to expand indefinitely when the pressure is withdrawn.

The inner molecular structure which accounts for these differences is only partly understood, but at any rate the molecules of a liquid, though free to move among each other, cannot do so as well as gases can. They are held together by intimate bonds which keep their average distances from each other constant at constant temperature. In gases these bonds are extremely weak, and a molecule moves with almost perfect freedom in a straight line between collisions with other molecules, or with the walls of the containing vessel. The average distance between collisions is called the mean free path. It increases as the gas expands to occupy larger volumes when the pressure is reduced.

THEORY OF STEAM GENERATION

If heat is added to ice, the effect will be to raise its temperature until the thermometer registers 32°F. When this point is reached, a further addition of heat does not produce a noticeable increase in temperature until all the ice is changed into water, or in other words, the ice melts. It has been found experimentally that 144 Btu are required to change one lb. of ice into water at the constant temperature of 32°F. This quantity is called the latent heat of fusion of ice.

After the given quantity of ice, which for simplicity may be taken as 1 lb., has all been turned into water, it is found that if more heat is added, the temperature of the water will again increase, though not so rapidly as did that of the ice. While the addition of each Btu increases the temperature of ice 2°F, in the case of water an increase of only about one degree will be noticed for each Btu of heat added. This difference is due to the fact that the specific heat, or the resistance offered by ice to a change in temperature, is only one-half that offered by water. That is, the specific heat of ice is 0.5. If the water is heated in a vessel open to the atmosphere, its temperature will continue to rise until it reaches a temperature of about 212°F, the boiling point of water, when further addition of heat will not produce any temperature changes, but steam will issue from the vessel. It has been found that about 970 Btu are required to change 1 lb. of water at atmospheric pressure and at 212°F into steam at the same temperature. This quantity of heat which changes the physical state of water from a liquid to steam without changing the temperature is called the enthalpy of vaporization.

If the above operations are performed in a closed vessel, such as an ordinary steam boiler, water will boil at a higher temperature than 212°F, since the steam driven off cannot escape and is compressed, raising the pressure and, consequently, the temperature.

The fact that the boiling point of water depends on the pressure is well known. Thus, in a locality where the altitude is 6,000 ft. above sea level and the barometric pressure is 12.6 lb. per square inch absolute the boiling point of water is about 204°F as compared with 212°F at sea level where the barometric pressure is 14.7 lb. per square inch absolute.

Assuming that the pressure is increased to 60 lb. per square inch when measured by a gage, it is found that the boiling point of water is 307.02°F. At 100 lb. per square inch water will boil at 337.8°F and at 150 lb. the temperature will read 365.8°F before steam will be formed.

STEAM QUALITY

Dry Steam and Wet Steam

It must be explained that the Steam Tables show the properties of what is usually known as "dry saturated steam." This is steam which has been completely evaporated, so that it contains no droplets of liquid water.

In practice, steam often carries tiny droplets of water with it and cannot be described as dry saturated steam. Nevertheless, we find that it is usually important that the steam used for process or heating is as dry as possible. It will be shown later how this is sometimes achieved by the proper use of steam traps.

Steam quality is described by its "dryness fraction"—the proportion of completely dry steam present in the steam moisture mixture being considered. The steam becomes "wet" if water droplets in suspension are present in the steam space, carrying no latent heat. For example, the specific enthalpy of steam—at 100 psi with a dryness fraction of 0.95 can be calculated as follows:

Each lb. of wet steam will contain the full amount of sensible heat, but as only 0.95 lbs. of dry steam is present with 0.05 lbs. of water, there will only be 0.95 of the latent heat. Thus, the specific enthalpy of the steam will be:

$$h_g = h_f + (0.95 \times h_{fg})$$
$$= 309 + (0.95 \times 881.6)$$
$$= 1146.5 \text{ Btu/lb.}$$

This figure represents a reduction of 44.1 Btu/lb from the total heat of steam at 100 psi gauge shown in the Steam Tables. Clearly, the "wet steam" has a heat content substantially lower than that of dry saturated steam at the same pressure.

The small droplets of water in wet steam have weight but occupy negligible space. The volume of wet steam is, therefore, less than that of dry saturated steam.

Volume of wet steam =

Volume of dry saturated steam × Dryness fraction

It is the water droplets in suspension which make wet steam visible. Steam as such is a transparent gas but the droplets of water give it a white cloudy appearance due to the fact that they reflect light.

Superheated Steam

As long as water is present, the temperature of saturated steam will correspond to the figure indicated for that pressure in the Steam Tables. However, if heat transfer continues after all the water has been evaporated, the steam temperature will again rise. The steam is then called "superheated," and this "superheated steam" will be at a temperature above that of saturated steam at the corresponding pressure.

Saturated steam will condense very readily on any surface which is at a lower temperature, so that it gives up the latent heat which is the greatest proportion of its energy content. On the other hand, when superheated steam gives up some of its enthalpy, it does so by virtue of a fall in temperature. No condensation will occur until the saturation temperature has been reached, and it is found that the rate at which energy flow from superheated steam is often less than we can achieve with saturated steam, even though the superheated steam is at a higher temperature. Superheated steam, because of its other properties, is the natural first choice for power steam requirements, while saturated steam is ideal for process and heating applications.

Steam Tables

The relations existing between pressure, volume and temperature of saturated and superheated steam as well as the energy required to produce vaporization under various conditions have been expressed in the form of empirical equations from which steam tables have been computed. The fundamental properties determined of pressure and temperature are arrived at spectroscopically for a known volume.

Tables A-8 and A-9 give the most important properties of saturated and superheated steam, respectively, which include:

1) Pressure (P) of steam in pounds per square inch absolute, or inches of mercury for low pressures.

2) Temperature (T) of saturated steam in degrees Fahrenheit. This column shows the vaporization temperature, or the boiling point, at each given pressure.

3) Specific volumes. These include the specific volume of the saturated liquid (v_f) and the specific volume of saturated steam (v_g).

4) Enthalpy. The three columns for enthalpy are: The enthalpy of the liquid (h_f), enthalpy of evaporation (h_{fg}) and enthalpy of saturated steam (h_g). The enthalpy of the liquid (h_f) is the energy required to raise the temperature of 1 pound of water from freezing to the vaporization temperature, or boiling point, at the given pressure. The enthalpy of evaporation (h_{fg}) is the energy required to change 1 pound of water at its point of vaporization to dry saturated steam at the given pressure. The enthalpy of saturated steam (h_g) is the sum of h_f and h_{fg}, or the energy required to change 1 pound of water from the freezing point to dry saturated steam at the given pressure.

5) Entropy. The three columns for entropy include entropy of the liquid (S_f), entropy of vaporization (S_{fg}) and, entropy of saturated steam (S_g).

It should be noted that by applying the equation:

$$h = u + \frac{P_v}{J}$$

it is possible to determine the internal energy (u) for the respective pressure and temperature.

The tables that follow address condensate and steam as water (H_2O) as defined more generally in chemistry. Tables A-8 and A-9, although called Properties of Saturated Water, can be interpreted to be condensate and steam, Table A-9 applies only to the gaseous phase (steam) when heated above the saturation temperature for the given pressure. Table A-10 applies to condensate compressed to pressures above those corresponding to the values for the saturated liquid. A more complete set of tables are presented in Appendix D.

NOZZLES AND ORIFICES FOR STEAM TRAPPING

An understanding of the flow of steam and gases is basic to the study of steam traps. It is discussed here, how gases flow through constrictions (nozzles and orifices). All steam traps incorporate a form of constriction, usually an orifice; however, nozzles are used in some designs.

A nozzle, orifice or generally a constriction in the path of a gas flow in a closed conduit is utilized to change the pressure from a higher to a lower level. In a steam trap, however, the constriction is utilized to maintain a higher pressure on the upstream side of the trap, thereby guaranteeing the formation of condensate before the fluid is permitted to flow through the trap. With the exception of the orifice steam trap or drain orifice type trap, steam traps usually incorporate a valve in conjunction with the orifice to regulate the flow of condensate.

It is obvious that the sizing of a steam trap constriction is critical to trap performance. If the constriction is too small, the trap will not permit all the condensate to drain from the heater, pipe or other equipment to which it is attached. If the constriction is too large, rapid cycling and shortened trap life will result. Since flashing often occurs across the nozzle or orifice, the flow most often must be treated as a mixture of gas and liquid (steam and water) and usually is determined in the trap manufacturer's laboratory. However, using engineering data developed from a treatise similar to the following produces the necessary information to size a trap's orifice or nozzle. In the case of the orifice steam trap or drain orifice, most manufacturers will select orifices and orifice drains to suit the customer's needs. However, plant engineers may wish to size the traps themselves. In this case (orifice steam traps or drain orifices), it is suggested the reader consult Chapter 6. Consult Chapter 5 for orifice applications.

Table A-8. Sample Thermodynamic Properties of Steam—Temperature-oriented.

Temp. Fahr. t	Abs. Pressure		Specific Volume			Enthalpy			Entropy			Temp. Fahr. t
	Lb./Sq.In. p	In. Hg.	Sat. Liquid v_f	Evap. v_{fg}	Sat. Vapor v_g	Sat. Liquid h_f	Evap. h_{fg}	Sat. Vapor h_g	Sat. Liquid s_f	Evap. s_{fg}	Sat. Vapor s_g	
90°	0.6982	1.4215	0.01610	468.0	468.0	57.99	1042.9	1100.9	0.1115	1.8972	2.0087	90°
91	0.7204	1.4667	0.01611	454.4	454.4	58.99	1042.4	1101.4	0.1133	1.8927	2.0060	91
92	0.7432	1.5131	0.01611	441.2	441.3	59.99	1041.8	1101.8	0.1151	1.8883	2.0034	92
93	0.7666	1.5608	0.01611	428.5	428.5	60.98	1041.2	1102.2	0.1169	1.8838	2.0007	93
94	0.7906	1.6097	0.01612	416.2	416.2	61.98	1040.7	1102.6	0.1187	1.8794	1.9981	94
95°	0.8153	1.6600	0.01612	404.3	404.3	62.98	1040.1	1103.1	0.1205	1.8750	1.9955	95°
96	0.8407	1.7117	0.01612	392.8	392.8	63.98	1039.5	1103.5	0.1223	1.8706	1.9929	96
97	0.8668	1.7647	0.01612	381.7	381.7	64.97	1038.9	1103.9	0.1241	1.8662	1.9903	97
98	0.8935	1.8192	0.01613	370.9	370.9	65.97	1038.4	1104.4	0.1259	1.8618	1.9877	98
99	0.9210	1.8751	0.01613	360.4	360.5	66.97	1037.8	1104.8	0.1277	1.8575	1.9852	99
100°	0.9492	1.9325	0.01613	350.3	350.4	67.97	1037.2	1105.2	0.1295	1.8531	1.9826	100°
101	0.9781	1.9915	0.01614	340.6	340.6	68.96	1036.6	1105.6	0.1313	1.8488	1.9801	101
102	1.0078	2.0519	0.01614	331.1	331.1	69.96	1036.1	1106.1	0.1330	1.8445	1.9775	102
103	1.0382	2.1138	0.01614	321.9	321.9	70.96	1035.5	1106.5	0.1348	1.8402	1.9750	103
104	1.0695	2.1775	0.01615	313.1	313.1	71.96	1034.9	1106.9	0.1366	1.8359	1.9725	104
105°	1.1016	2.2429	0.01615	304.5	304.5	72.95	1034.3	1107.3	0.1383	1.8317	1.9700	105°
106	1.1345	2.3099	0.01615	296.1	296.2	73.95	1033.8	1107.8	0.1401	1.8274	1.9675	106
107	1.1683	2.3786	0.01616	288.1	288.1	74.95	1033.3	1108.2	0.1419	1.8232	1.9651	107
108	1.2029	2.4491	0.01616	280.3	280.3	75.95	1032.7	1108.6	0.1436	1.8190	1.9626	108
109	1.2384	2.5214	0.01616	272.7	272.7	76.94	1032.1	1109.0	0.1454	1.8147	1.9601	109
110°	1.2748	2.5955	0.01617	265.3	265.4	77.94	1031.6	1109.5	0.1471	1.8106	1.9577	110°
111	1.3121	2.6715	0.01617	258.2	258.3	78.94	1031.0	1109.9	0.1489	1.8064	1.9553	111
112	1.3504	2.7494	0.01617	251.3	251.4	79.94	1030.4	1110.3	0.1506	1.8023	1.9529	112
113	1.3896	2.8293	0.01618	244.6	244.7	80.94	1029.8	1110.7	0.1524	1.7981	1.9505	113
114	1.4298	2.9111	0.01618	238.2	238.2	81.93	1029.2	1111.1	0.1541	1.7940	1.9481	114

Temp												Temp
115°	1.4709	2.9948	0.01618	231.9	231.9	82.93	1028.7	1111.6	0.1559	1.7898	1.9457	115°
116	1.5130	3.0806	0.01619	225.8	225.8	83.93	1028.1	1112.0	0.1576	1.7857	1.9433	116
117	1.5563	3.1687	0.01619	219.9	219.9	84.93	1027.5	1112.4	0.1593	1.7816	1.9409	117
118	1.6006	3.2589	0.01620	214.2	214.2	85.92	1026.9	1112.8	0.1610	1.7776	1.9386	118
119	1.6459	3.3512	0.01620	208.6	208.7	86.92	1026.3	1113.2	0.1628	1.7735	1.9363	119
120°	1.6924	3.4458	0.01620	203.25	203.27	87.92	1025.8	1113.7	0.1645	1.7694	1.9339	120°
121	1.7400	3.5427	0.01621	198.02	198.03	88.92	1025.2	1114.1	0.1662	1.7654	1.9316	121
122	1.7888	3.6420	0.01621	192.93	192.95	89.92	1024.6	1114.5	0.1679	1.7614	1.9293	122
123	1.8387	3.7436	0.01622	188.01	188.02	90.91	1024.0	1114.9	0.1696	1.7574	1.9270	123
124	1.8897	3.8475	0.01622	183.23	183.25	91.91	1023.4	1115.3	0.1714	1.7533	1.9247	124
125°	1.9420	3.9539	0.01622	178.59	178.61	92.91	1022.9	1115.8	0.1731	1.7493	1.9224	125°
126	1.9955	4.0629	0.01623	174.09	174.10	93.91	1022.3	1116.2	0.1748	1.7454	1.9202	126
127	2.0503	4.1745	0.01623	169.71	169.72	94.91	1021.7	1116.6	0.1765	1.7414	1.9179	127
128	2.1064	4.2887	0.01624	165.46	165.47	95.91	1021.1	1117.0	0.1782	1.7374	1.9156	128
129	2.1638	4.4055	0.01624	161.33	161.35	96.90	1020.5	1117.4	0.1799	1.7335	1.9134	129
130°	2.2225	4.5251	0.01625	157.32	157.34	97.90	1020.0	1117.9	0.1816	1.7296	1.9112	130°
131	2.2826	4.6474	0.01625	153.43	153.44	98.90	1019.4	1118.3	0.1833	1.7257	1.9090	131
132	2.3440	4.7725	0.01626	149.65	149.66	99.90	1018.8	1118.7	0.1849	1.7218	1.9067	132
133	2.4069	4.9005	0.01626	145.97	145.99	100.90	1018.2	1119.1	0.1866	1.7179	1.9045	133
134	2.4712	5.0314	0.01626	142.40	142.42	101.90	1017.6	1119.5	0.1883	1.7141	1.9023	134
135°	2.5370	5.1653	0.01627	138.93	138.95	102.90	1017.0	1119.9	0.1900	1.7102	1.9002	135°
136	2.6042	5.3022	0.01627	135.56	135.58	103.90	1016.4	1120.3	0.1917	1.7063	1.8980	136
137	2.6729	5.4421	0.01628	132.29	132.30	104.89	1015.9	1120.8	0.1934	1.7024	1.8958	137
138	2.7432	5.5852	0.01628	129.10	129.12	105.89	1015.3	1121.2	0.1950	1.6987	1.8937	138
139	2.8151	5.7316	0.01629	126.00	126.02	106.89	1014.7	1121.6	0.1967	1.6948	1.8915	139
140°	2.8880	5.8812	0.01629	122.99	123.01	107.89	1014.1	1122.0	0.1984	1.6910	1.8894	140°
141	2.9637	6.0341	0.01630	120.06	120.08	108.89	1013.5	1122.4	0.2000	1.6873	1.8873	141
142	3.0404	6.1903	0.01630	117.22	117.23	109.89	1012.9	1122.8	0.2016	1.6835	1.8851	142
143	3.1188	6.3500	0.01631	114.45	114.46	110.89	1012.3	1123.2	0.2033	1.6797	1.8830	143
144	3.1990	6.5132	0.01631	111.75	111.77	111.89	1011.7	1123.6	0.2049	1.6760	1.8809	144

Table A-9. Thermodynamic Properties of Steam—Pressure-oriented.

Abs. Press. Lb. Sq. In. p	Temp. Fahr. t	Specific Volume		Enthalpy			Entropy			Internal Energy			Abs. Press. Lb. Sq. In. p
		Sat. Liquid v_f	Sat. Vapor v_g	Sat. Liquid h_f	Evap. h_{fg}	Sat. Vapor h_g	Sat. Liquid s_f	Evap. s_{fg}	Sat. Vapor s_g	Sat. Liquid u_f	Evap. u_{fg}	Sat. Vapor u_g	
3.0	141.48	0.01630	118.71	109.37	1013.2	1122.6	0.2008	1.6855	1.8863	109.36	947.3	1056.7	3.0
3.5	147.57	0.01633	102.72	115.46	1009.6	1125.1	0.2109	1.6626	1.8735	115.45	943.1	1058.6	3.5
4.0	152.97	0.01636	90.63	120.86	1006.4	1127.3	0.2198	1.6427	1.8625	120.85	939.3	1060.2	4.0
4.5	157.83	0.01638	81.16	125.71	1003.6	1129.3	0.2276	1.6252	1.8528	125.70	936.0	1061.7	4.5
5.0	162.24	0.01640	73.52	130.13	1001.0	1131.1	0.2347	1.6094	1.8441	130.12	933.0	1063.1	5.0
5.5	166.30	0.01643	67.24	134.19	998.5	1132.7	0.2411	1.5951	1.8363	134.17	930.1	1064.3	5.5
6.0	170.06	0.01645	61.98	137.96	996.2	1134.2	0.2472	1.5820	1.8292	137.94	927.5	1065.4	6.0
6.5	173.56	0.01647	57.50	141.47	994.1	1135.6	0.2528	1.5699	1.8227	141.45	925.0	1066.4	6.5
7.0	176.85	0.01649	53.64	144.76	992.1	1136.9	0.2581	1.5586	1.8167	144.74	922.7	1067.4	7.0
7.5	179.94	0.01651	50.29	147.86	990.2	1138.1	0.2629	1.5481	1.8110	147.84	920.5	1068.3	7.5
8.0	182.86	0.01653	47.34	150.79	988.5	1139.3	0.2674	1.5383	1.8057	150.77	918.4	1069.2	8.0
8.5	185.64	0.01654	44.73	153.57	986.8	1140.4	0.2718	1.5290	1.8008	153.54	916.5	1070.0	8.5
9.0	188.28	0.01656	42.40	156.22	985.2	1141.4	0.2759	1.5203	1.7962	156.19	914.6	1070.8	9.0
9.5	190.80	0.01658	40.31	158.75	983.6	1142.3	0.2798	1.5120	1.7918	158.72	912.8	1071.5	9.5
10	193.21	0.01659	38.42	161.17	982.1	1143.3	0.2835	1.5041	1.7876	161.14	911.1	1072.2	10
11	197.75	0.01662	35.14	165.73	979.3	1145.0	0.2903	1.4897	1.7800	165.70	907.8	1073.5	11
12	201.96	0.01665	32.40	169.96	976.6	1146.6	0.2967	1.4763	1.7730	169.92	904.8	1074.7	12
13	205.88	0.01667	30.06	173.91	974.2	1148.1	0.3027	1.4638	1.7665	173.87	901.9	1075.8	13
14	209.56	0.01670	28.04	177.61	971.9	1149.5	0.3083	1.4522	1.7605	177.57	899.3	1076.9	14
14.696	212.00	0.01672	26.80	180.07	970.3	1150.4	0.3120	1.4446	1.7566	180.02	897.5	1077.5	14.696
15	213.03	0.01672	26.29	181.11	969.7	1150.8	0.3135	1.4415	1.7549	181.06	896.7	1077.8	15
16	216.32	0.01674	24.75	184.42	967.6	1152.0	0.3184	1.4313	1.7497	184.37	894.3	1078.7	16
17	219.44	0.01677	23.39	187.56	965.5	1153.1	0.3231	1.4218	1.7449	187.51	892.0	1079.5	17
18	222.41	0.01679	22.17	190.56	963.6	1154.2	0.3275	1.4128	1.7403	190.50	889.9	1080.4	18
19	225.24	0.01681	21.08	193.42	961.9	1155.3	0.3317	1.4043	1.7360	193.36	887.8	1081.2	19

20	227.96	0.01683	20.089	196.16	960.1	1156.3	0.3356	1.3962	1.7319	196.10	885.8	1081.9
21	230.57	0.01685	19.192	198.79	958.4	1157.2	0.3395	1.3885	1.7280	198.73	883.9	1082.6
22	233.07	0.01687	18.375	201.33	956.8	1158.1	0.3431	1.3811	1.7242	201.26	882.0	1083.3
23	235.49	0.01689	17.627	203.78	955.2	1159.0	0.3466	1.3740	1.7206	203.71	880.2	1083.9
24	237.82	0.01691	16.938	206.14	953.7	1159.8	0.3500	1.3672	1.7172	206.07	878.5	1084.6
25	240.07	0.01692	16.303	208.42	952.1	1160.6	0.3533	1.3606	1.7139	208.34	876.8	1085.1
26	242.25	0.01694	15.715	210.62	950.7	1161.3	0.3564	1.3544	1.7108	210.54	875.2	1085.7
27	244.36	0.01696	15.170	212.75	949.3	1162.0	0.3594	1.3484	1.7078	212.67	873.6	1086.3
28	246.41	0.01698	14.663	214.83	947.9	1162.7	0.3623	1.3425	1.7048	214.74	872.1	1086.8
29	248.40	0.01699	14.189	216.86	946.5	1163.4	0.3652	1.3368	1.7020	216.77	870.5	1087.3
30	250.33	0.01701	13.746	218.82	945.3	1164.1	0.3680	1.3313	1.6993	218.73	869.1	1087.8
31	252.22	0.01702	13.330	220.73	944.0	1164.7	0.3707	1.3260	1.6967	220.63	867.7	1088.3
32	254.05	0.01704	12.940	222.59	942.8	1165.4	0.3733	1.3209	1.6941	222.49	866.3	1088.7
33	255.84	0.01705	12.572	224.41	941.6	1166.0	0.3758	1.3159	1.6917	224.31	864.9	1089.2
34	257.58	0.01707	12.226	226.18	940.3	1166.5	0.3783	1.3110	1.6893	226.07	863.5	1089.6
35	259.28	0.01708	11.898	227.91	939.2	1167.1	0.3807	1.3063	1.6870	227.80	862.3	1090.1
36	260.95	0.01709	11.588	229.60	938.0	1167.6	0.3831	1.3017	1.6848	229.49	861.0	1090.5
37	262.57	0.01711	11.294	231.26	936.9	1168.2	0.3854	1.2972	1.6826	231.14	859.8	1090.9
38	264.16	0.01712	11.015	232.89	935.8	1168.7	0.3876	1.2929	1.6805	232.77	858.5	1091.3
39	265.72	0.01714	10.750	234.48	934.7	1169.2	0.3898	1.2886	1.6784	234.36	857.2	1091.6
40	267.25	0.01715	10.498	236.03	933.7	1169.7	0.3919	1.2844	1.6763	235.90	856.1	1092.0
41	268.74	0.01716	10.258	237.55	932.6	1170.2	0.3940	1.2803	1.6743	237.42	855.0	1092.4
42	270.21	0.01717	10.029	239.04	931.6	1170.7	0.3960	1.2764	1.6724	238.91	853.8	1092.7
43	271.64	0.01719	9.810	240.51	930.6	1171.1	0.3980	1.2726	1.6706	240.37	852.7	1093.1
44	273.05	0.01720	9.601	241.95	929.6	1171.6	0.4000	1.2687	1.6687	241.81	851.6	1093.4
45	274.44	0.01721	9.401	243.36	928.6	1172.0	0.4019	1.2650	1.6669	243.22	850.5	1093.7
46	275.80	0.01722	9.209	244.75	927.7	1172.4	0.4038	1.2613	1.6652	244.60	849.5	1094.1
47	277.13	0.01723	9.025	246.12	926.7	1172.9	0.4057	1.2577	1.6634	245.97	848.4	1094.4
48	278.45	0.01725	8.848	247.47	925.8	1173.3	0.4075	1.2542	1.6617	247.32	847.4	1094.7
49	279.74	0.01726	8.678	248.79	924.9	1173.7	0.4093	1.2508	1.6601	248.63	846.4	1095.0

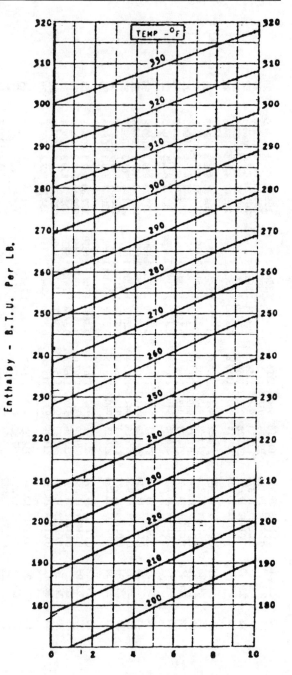

Table A-10. Enthalpy of Compressed Liquids.

Pressure—1000 lbs. per sq. in. gage

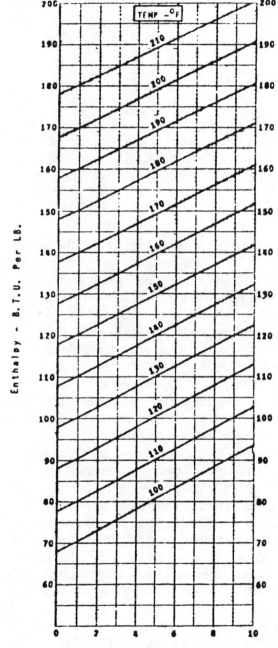

Pressure—1000 lbs. per sq. in. gage

The time spent in the study of this section can be helpful to the engineer in refreshing the fundamentals of steam and gas flow, prerequisite to commencing the sizing of orifices or nozzles for steam trap application.

THROTTLING PROCESS

The throttling process, a constant enthalpy process, is one in which no work is performed. Also, there is no heat added or subtracted from the medium undergoing the process; Q remains constant. Therefore:

$$KE_1 + H_1 + Q = KE_2 + H_2 + W$$

It follows that if $H_1 = H_2$ is stated as a parameter for the process and in theory there is no heat loss or gain, then $KE_1 = KE_2$. The purpose of throttling a gas is to reduce it from a higher pressure to a lower pressure to produce a substance in a more usable form. In theory this can be done at 100 percent efficiency. However, the throttling process is extremely inefficient. In most cases so much so that in reality when attempting to throttle the flow of a substance often times the process approaches isentropic. See Figure A-25.

What usually happens is that with the expansion of the gas from a higher pressure to a lower pressure there is a cooling effect causing a change in Q. However, useful work is not performed since the throttle in itself cannot convert heat energy to mechanical energy.

Therefore, a more profitable means of reducing the pressure of a gas is to expand it through a turbine or other type of engine. By doing so, the lower pressure is attained while producing mechanical work. Consider Figures A-26 and A-27, a reducing valve and a turbine, respectively. In the reducing valve, Fig. A-26, there is a large quantity of rejected heat while reducing the pressure from P_1 to P_2. In the turbine, Fig. A-27, where the machine can be thoroughly insulated, the rejected heat is negligible. Also the gas can be expanded isentropically from P_1 to P_2 and produce useful mechanical work in the rotating shaft of the turbine at the same time.

Where small quantities of gases are involved, typically steam at flow rates up to 500 lbs/hr, it would be foolish to expect to economically use a turbine to effect a pressure reduction. In this case, the economical

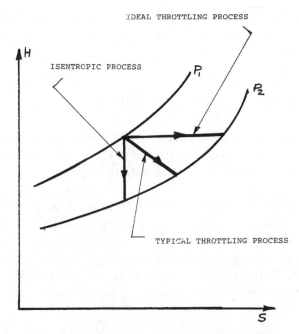

Figure A-25. Pressure Reducing Processes.

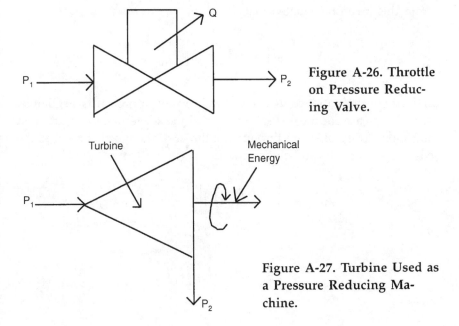

Figure A-26. Throttle on Pressure Reducing Valve.

Figure A-27. Turbine Used as a Pressure Reducing Machine.

thing would be to use a reducing valve. However, when flows are in millipounds, it is prudent to consider a turbine as a device for effecting a pressure reduction.

FLOW THROUGH NOZZLES AND ORIFICES

A nozzle, similar to a throttle, is a device used to lower the pressure of a gas. They are insulated to prevent heat loss, and because of their configuration, are useful in converting the heat energy of a gas to kinetic energy. Ideally, expansion through a nozzle or orifice is isentropic. In Fig. A-28, the gas expands ideally from point 1 to 2, but in reality entropy is generated as with all actual processes and expansion is along the dashed line from point 1 to 2'. Figures A-29 and A-30 are schematic cross-sections of a convergent-divergent nozzle and a knife edge orifice, respectively. In the nozzle itself, there is no work done and no heat transferred. Therefore, from the general energy equation, it follows:

$$\frac{v_{s1}^2}{2gJ} + H_1 = \frac{v_{s2}^2}{2gJ} + H_2$$

From this it can be equated that

$$v_{s2} = \left[2gJ \left(H_1 - H_2 \right) + v_{s1}^2 \right]^{1/2} \text{ft per sec.}$$

and where the entrance velocity or velocity of approach is negligible. This is the case for most steam traps, and the v_{s1} term can be dropped and the velocity of steam through the throat of the nozzle or across the edge of the orifice can be equated as:

$$v_{s2} = \left[2gJ \left(H_1 - H_2 \right) \right]^{1/2} \text{ft per sec.}$$

or

$$v_{s2} = 223.7 \left(H_1 - H_2 \right)^{1/2}$$

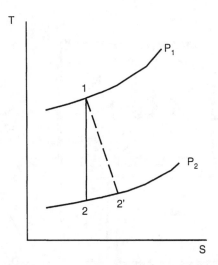

Figure A-28. Expansion of a Gas through a Nozzle. 1-2 Represents an Isentropic Expansion. 1-2' Represents an Irreversible Adiabatic Expansion.

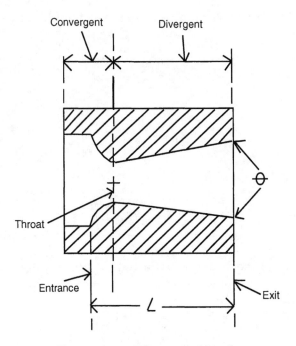

Figure A-29. Divergent Nozzle.

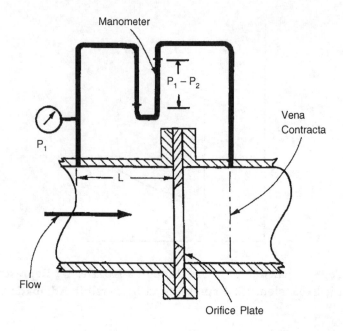

Figure A-30. Orifice Plate. The Distance ∠ Should be at Least One Pipe.

and since

$$\left(H_1 - H_2\right) = wc_p\left(T_1 - T_2\right)$$

it follows that

$$v_{s2} = 223.7\left[wc_p\left(T_1 - T_2\right)\right]^{1/2}$$

The velocity of the steam across a trap orifice or nozzle can also be equated in terms of pressure as:

$$v_{s2} = \left[2gRT_1\left(\frac{P_1 - P_2}{P_1}\right)\right]^{1/2}$$

Having found the velocity across the throat of the nozzle or across the orifice, it is simple to find the flow rate from:

$$w = Av_{s2}$$

where A is the cross-sectional area and v_{s2} the velocity. To equate this in terms of weight rate of flow:

$$w = \frac{223.7(A)}{vg} \left[H_1 - H_2 \right]^{1/2}$$

or any of the other previously derived velocity. Simply insert v_{s2} into the equation:

$$w = \frac{Av_{s2}}{vg}$$

where v_g is the specific volume of the steam from the steam tables.

One other factor to consider is the efficiency of the process e_n. It can have values between 0.85 and 0.98 depending upon the process. It must be determined by actual test of the steam trap orifice or nozzle.

To elaborate on the above equation and including the process efficiency (e_n):

$$w = \left(\frac{Av_{s2}}{vg} \right) e_n$$

Needless to say calculating steam trap orifice flow can be an extremely time consuming task even with today's modem computer technology. Therefore, it is easier to use manufacturers' data, if it is available...

Page-Konzo Steam Flow Charts

BACKGROUND OF STEAM FLOW FORMULAS

*T*he flow relationships that are in common use originated with Darcy and Weisbach, both of whom worked with flow of water. The Darcy-Weisbach equation for drop in pressure head for water flow has been carried forward to this time in its original form which included a friction factor *f* as follows:

$$\Delta h = \frac{fLv^2}{D_2g}$$

Δh	=	head loss in feet of fluid flowing,
f =		friction factor,
L =		length of pipe in feet,
D =		internal diameter of pipe in feet.

In 1857, Darcy proposed that the friction factor should be considered as a function of the diameter of the pipe. In 1876, Unwin modified the Darcy equation and returned the friction factor.

Unwin's equation is the same as that published by Babcock some 14 years later, and which has been commonly referred to as the Babcock equation. Babcock gave no derivation nor explanation of the equation; Page and Konzo are of the opinion that proper reference should be the "Unwin-Babcock" equation.

It is of interest to note that the equation was developed by Unwin for water flow and was adapted by Babcock for steam flow. However, this practice of taking an equation determined for one fluid and using it to some other fluid was common among early experimenters.

Unwin-Babcock equation for friction was:

$$f' = k \, (1 + 3.^6/d)$$

Where f was the friction factor, k a constant and d the pipe diameter. In 1804, Unwin re-evaluated the constant and presented a value of 0.0027, based largely on his analysis of experiments made by Riedler and Gutermuth with air flow. This value for k of 0.0027 is still in common use, over 60 years after the time it was first proposed.

One of the significant experiments of this early era was made by Carpenter and Sickles on steam flow. They were able to confirm and the k value of 0.0027 only by discarding extreme values. No good reason was given for discarding part of the data. Green and Stratford and McRae also reported experimental results with steam flow and were not successful in confirming the k value of 0.0027. However, since their reports were not widely circulated, any influence they might have had in questioning the Unwin constant was negligible in effort.

By the turn of the century, both the Unwin-Babcock equation and the Unwin constant had become well established and to some extent "frozen." For example, an interesting summary made by Gebhardt in 1909 shows not only that the Unwin-Babcock equation was widely accepted, but that the only other recognized form for friction factor was one in which f was assumed to be a constant and not a function of diameter.

The complete Unwin-Babcock equation was:

$$P = \frac{C_1 W_3 L}{pd^3} \left(1 + \frac{3.^6}{d} \right)$$

Where P – pressure drop in pounds per square inch, C, = 0.04839k, p = density of steam, m pounds per cubic foot, L = length of pipe in feet, W = steam flow rate in pounds per minute, and d = internal pipe diameter in feet. This equation is still in common use.

During the early part of the present century, numerous writers extended the application of this equation over a wide range of conditions without regard for the limitations of the empirical constants. For example, Carpenter in 1912 extended the equations to pipe sizes ranging from 1 inch to 60 inches in diameter, and to steam pressures ranging from 0.4 inch of mercury to 500 psi. Such extensions of the equation are highly questionable.

Unwin in 1907 indicated that the friction factor was affected by temperature, and thus the viscosity, of the fluid. For some strange reason, however, in the case of steam flow the restricted Unwin-Babcock concept persisted to such an extent that little work was done to correlate the existing experimental evidence with the newer concepts of flow.

In 1912, Blasius made a recalculation of the data obtained by two other experimenters with water flow and showed that the factor f in the Darcy-Weisbach equation was function of vDp/m, where m is the absolute viscosity of the fluid. This dimensionless expression is referred to as the Reynolds number. Another contribution was that by Stanton and Pannel, who adopted Reynolds' theories of 1883, and established experimentally that the friction factor for the flow of such diverse fluids as oil, water, and air was a function of Reynolds number.

In 1915, Lander presented experimental results with steam flow and water flow on the diagram first suggested by Stanton and Pannel. This research can be considered as the first valid evidence that the friction factor for steam flow was a function of something other than diameter alone.

Since 1914, the developments in fluid flow theory have been made at an accelerated rate. The contributions of Buckingham and Rayleigh in the field of dimensional analysis proved that the friction factor was a function of Reynolds number. In 1939, Keenan obtained experimental results with water and steam, and showed that the friction factor was not only a function of Reynolds number, but was also the same for incompressible and compressible flow. A slight modification in the form of presentation of data was given by Rouse in 1943, but essentially an agreement had been reached that friction factor was a function not a diameter alone, but of Reynolds number and the relative roughness of the pipe.

As a final step in the development of fluid flow up to this time, Moody in 1944 presented a convenient chart showing variations in friction factors for wide ranges of Reynolds number and for various values of the relative roughness of the pipe surface.

Page and Konzo concluded that the best approach to the study of steam flow in pipes consists in utilizing the recent concepts of fluid flow, and in particular the chart prepared by *Moody* (*Trans. ASME, 1944*). This chart gave the relation between friction factor and Reynolds number for various degrees of roughness of the surface of the pipe. They then proceeded to determine whether the original steam flow data of several investigators could be correlated with the Moody chart.

They showed that the old data of Carpenter and Sickles, originally used to justify the Unwin-Babcock equation, correlated better when analyzed by the Reynolds number concept. They also showed that the Unwin-Babcock equation result in discrepancies of practical significance. For example, with a 4-inch diameter pipe and a steam flow velocity of 60 fps the results are in substantial agreement with those from the Moody chart. However, with the same velocity of 60 fps and a 1-inch diameter pipe, the Unwin-Babcock values are about 70% greater than those from the Moody chart.

Further, Page and Konzo showed that the extrapolation of the Unwin-Babcock equation to any and all pressure and temperature conditions is not warranted. In some cases, the use of the equation may result in pipe sizes that are too small, and in other cases may result in oversized pipes.

They also concluded that the original data of Carpenter and Sickles, Eberle, Lander are in good agreement, as judged by the Reynolds number concept, and that the fault lies not in the data, but in the analysis made by others before fluid flow theory had been developed.

COMPUTATION BASED ON MODERN DATA

By the use of similarity concepts incorporated in the Moody chart for fluid flow, a new method can be derived for the sizing of steam pipes. Twenty-two working charts that summarize the method in graphical form are presented on subsequent pages. A detailed example of the necessary calculations that were made to establish a single point on the charts follows:

If the following variables are known:

(1)   Absolute viscosity, μ
(2)   Density, p
(3)   Velocity of steam, V
(4)   Inside diameter of pipe, D
(5)   Absolute roughness of pipe surface, e

Then the pressure drop per 100-ft length of pipe may be calculated by the Moody chart and equation (1).

The variables listed are functions of other parameters. For example, the first two variables of absolute viscosity and density depend upon the steam pressure and the steam condition (i.e., wet, dry, or superheated). The fourth variable of inside diameter depends upon the nominal pipe size and the schedule number of the pipe. For example, a two-inch Schedule 40 commercial-steel pipe has an inside diameter of 2.067 inches, while a two-inch Schedule 80 commercial-steel pipe has an inside diameter of 1.939 inches. The fifth variable of absolute surface roughness depends upon the material used in the pipe. Assume that the pressure drop is desired in 37 ft. of three-inch Schedule 40 commercial-steel pipe when 12 lb. per min. of saturated steam is flowing at an initial pressure of 15.7 psia. The first step of the solution is to determine steam properties. This can be done in the ten steps which follow:

(1) Density, $p = 1/25.27 = 0.03973$ lb. per cu ft (from steam tables by interpolation).

(2) Absolute viscosity, $\mu = 9.0 \times 10^{-6}$ lb per ft sec.

(3) Inside diameter, $D = 0.2557$ ft. (from piping tables).

(4) Velocity, $V = \dfrac{W}{pA}$, (from continuity equation).

Where W = pounds of steam flowing per minute,
A = cross sectional area of inside of pipe.

For the next step involving the Moody chart, the Reynolds number and relative roughness must be computed.

(5) $N_R = \dfrac{pVD}{H} = \dfrac{0.03973 \times 5{,}882 \times 0.2557}{9.0 \times 10^{-6} \times 60} = 1{,}107 \times 10^{6}$

(6) Absolute surface roughness, $e = 0.00015$ ft (from Moody).

(7) Relative roughness,

On the Moody chart, the intersection of the e/D line of 0.0005867 and the N_R line of 1.107×10^{6} is located.

(8) Friction factor, $f = 0.0206$ (from Moody chart). Appropriate values are substituted in equation (1) to obtain.

(9)   Head loss, $\Delta h = \dfrac{fLv^2}{D_2 g}$

$$= \frac{0.0206 \times 37 \times (5882)^2}{0.2557 \times 2 \times 32.2 \times 3600}$$

$$= 446\,\text{ft.}$$

(10) Pressure drop, $\Delta p = \dfrac{p\Delta h}{144}$

$$= \frac{0.03973 \times 446}{144}$$

$$= 0.123\,\text{psi}$$

THE PAGE-KONZO WORKING CHARTS

Tedious calculations of the preceding type can be practically eliminated by the use of 22 working charts as shown on subsequent pages, which were designed by Page and Konzo. The computations were made for Schedule 40 and 80 commercial-steel pipe. Chart scales show the following variables:

(1)   Weight rate of flow in pounds per minute (abscissa).

(2)   Pressure drop in psi per 100-ft length of pipe (left ordinate).

(3)   Pressure drop in ounces per square inch per 100-ft length of pipe, (right ordinate) for the convenience of those who utilize this unit of pressure drop.

(4)   Constant velocity lines (sloping downwards to the right).

(5)   Constant diameter lines (sloping upwards to the right).

Examples of Use of Charts
Case 1

When the pressure drop is the dependent variable. The sample problem already described can be readily solved by means of the working charts.

It was required to determine the pressure drop in 37 ft of 3 inch Schedule 40 commercial-steel pipe when 12 pounds per minute of saturated steam was flowing at an initial pressure of 15.7 psia. The first chart will be used since it applies to Schedule 40 pipe and to a pressure range of 14.7 to 16.6 psia.

Locate the flow rate of 12 pounds per minute on the right ordinate. Proceed to the left on the chart until the 3-inch diameter line is intersected. Then proceed down to obtain a pressure drop of 0.333 psi per 100 ft of length. The pressure drop for 37 ft of pipe is 0.37 × 0.333, or 0.123 psi. This result is the same as that of the detailed method previously given.

Case 2

When the weight flow rate is the dependent variable. What is the flow rate in pounds per minute of saturated steam at 18.0 psia through a Schedule 40 commercial-steel pipe of 1-inch diameter and 190 ft long, for a pressure drop of 1.1 psi?

The pressure drop per 100 ft length is 1.1 × 100/190, or 0.579 psi. Since the initial pressure is 3.3 psi," utilize the second chart. Locate the value of 0.579 psi at the bottom, and proceed vertically until the intersection with the 1-inch diameter line is reached. Then proceed horizontally to the right to obtain a weight flow rate of 0.98 pound per minute.

Case 3.

When the pipe diameter is the dependent variable. When given an initial pressure of 19 psia, a pressure drop of 0.5 psi per 100-ft length, and saturated steam to be transported at a rate of 25 pounds per minute, what are the pipe size and steam velocity?

Select the second chart for pressures between 16.7 to 19.7 psia. Locate 0.5 psi on the bottom and proceed vertically. Next locate 25 lb per min. on the ordinate and proceed horizontally. The intersection of these two lines lies between two commercial pipe sizes.

A four-inch pipe could be used, but the weight flow is 32.9 lb per min., which is considerably larger than necessary. On the other hand, a 3-1/2-inch pipe gives a weight flow of 23.1 pounds per minute. The 3-1/2-inch pipe will be preferred, since a flow rate of 25 pounds per minute would result in a pressure drop of 0.575 psi, which is only slightly larger that the desired value of 0.5 psi. The steam velocity is 8,000 fpm.

VERTICAL PIPES

If the pipe in a given piping system rises to a higher level in the direction of flow, an additional pressure drop is involved equal in magnitude to the product of the density and the change in elevation. Similarly, if the pipe drops to a lower level in the direction of flow, a pressure increase is involved equal to the density times the change in elevation. In Table B-1 is shown the relationship between the pressure drop (per 10 ft length) and the absolute pressure of saturated steam. For low pressures or small changes in elevation, the correction is negligible, but the high pressures and large changes in elevation the corrections may be significantly large.

Any correction for changes in elevation should be algebraically added to the pressure drop determined for frictional effects alone.

ADVANTAGES OF FORM OF WORKING CHARTS

Any given form of working charts for steam flow will have inherent advantages as well as limitations. The forms shown in the working charts were finally decided upon after a study of the limitations and

Table B-1. Pressure Drop in Saturated Steam for 10-Foot Change in Pipe Elevation.

Steam Pressure, psia	Pressure Drop, psi Per 10 Ft. Change
10	.0018
15	.0027
20	.0035
25	.0045
30	.0050
40	.0065
60	.0097
80	.0130
100	.0151
150	.0230
200	.0300
300	.0450
400	.0600
500	.0750

advantages of these as well as other forms.

The charts as given have two limitations, which have not been considered as insurmountable.

(1)   For complete coverage over a wide range of steam pressures, a large number of charts are required. The first chart for a range of pressures from 14.7 psia to 16.6 psia was actually based on values for 15.7 psia. Similarly, the second chart for a range of pressures from 16.7 psia to 19.7 psia was based on values for 18.0 psia. For higher steam pressures, successively large ranges of steam pressures were used to give the same order of accuracy as those embodied in the first two. In this connection, the maximum deviation in all the values shown was limited to seven per cent, which was regarded as reasonable considering the accuracy of the original data involved.

(2)   In order to cover the cases of pipes other than Schedule 40 (or, in the case of the last four charts on Schedule 80 pipe) commercial steel or wrought iron, Either separate charts or conversion factors would be necessary.

In spite of the obvious limitations of the form used, the advantages were considered to outweigh the disadvantages:

(1)   All the data for a particular steam condition are presented on one chart. For most practical applications, the engineer is able to solve a problem without reference to other charts.

(2)   The solution to any problem can be started at any place on the chart, providing any two of the four variables are known. As shown by the three sample problems, any variable on the chart can be the dependent variable.

(3)   The complete determination of the problem is made. Information is obtained not only about pressure drop, but also weight flow, diameter, and velocity. It is possible to observe how a change in any one of the four variables will effect the other three.

(4)   The flow charts are similar in form to those in common use for water and air flow, so that familiarity with such existing charts can be extended to these new charts for steam flow.

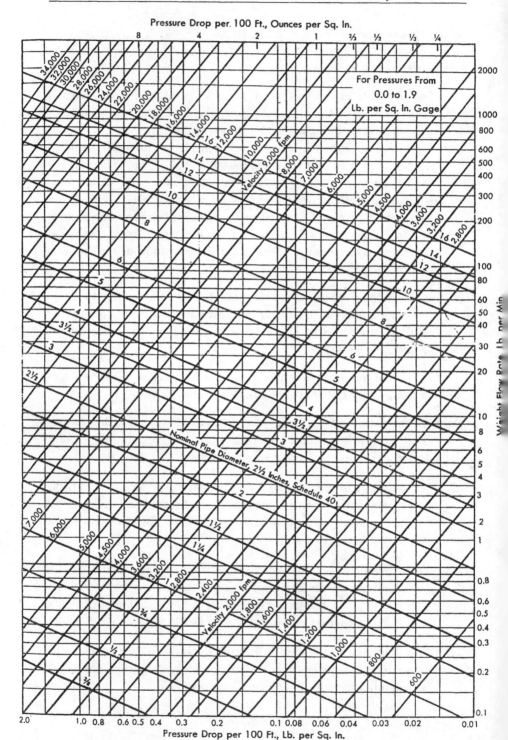

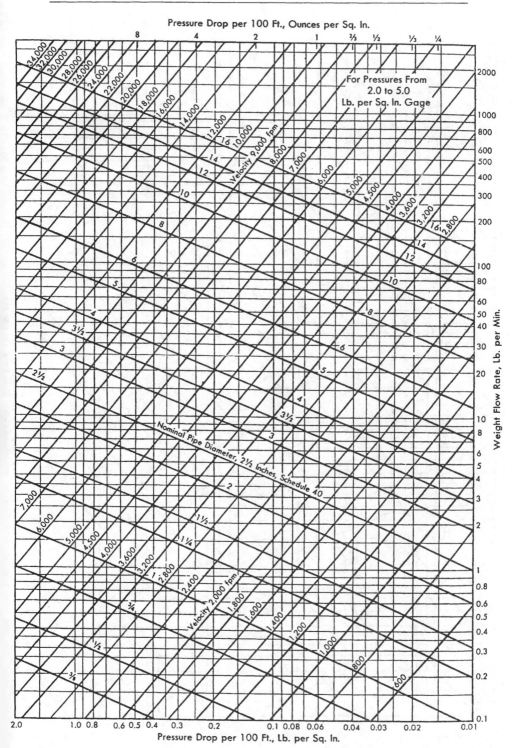

Pressure Drop per 100 Ft., Ounces per Sq. In.

For Pressures From 2.0 to 5.0 Lb. per Sq. In. Gage

Weight Flow Rate, Lb. per Min.

Pressure Drop per 100 Ft., Lb. per Sq. In.

Pressure Drop per 100 Ft., Ounces per Sq. In.

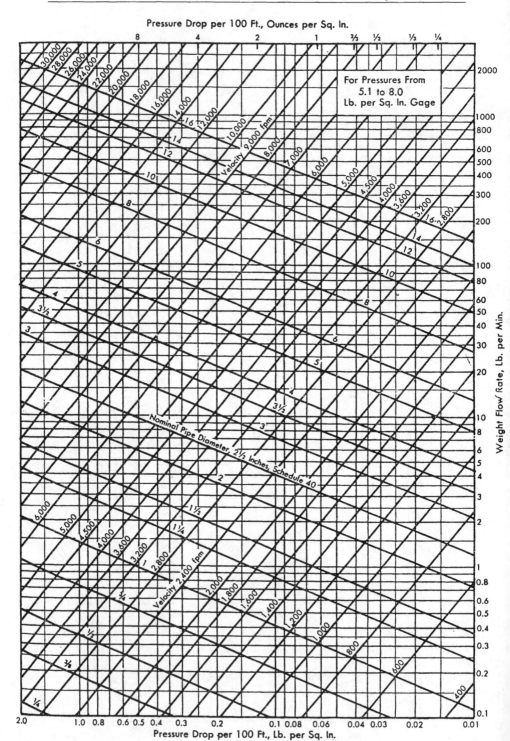

For Pressures From
5.1 to 8.0
Lb. per Sq. In. Gage

Weight Flow Rate, Lb. per Min.

Pressure Drop per 100 Ft., Lb. per Sq. In.

Pressure Drop per 100 Ft., Ounces per Sq. In.

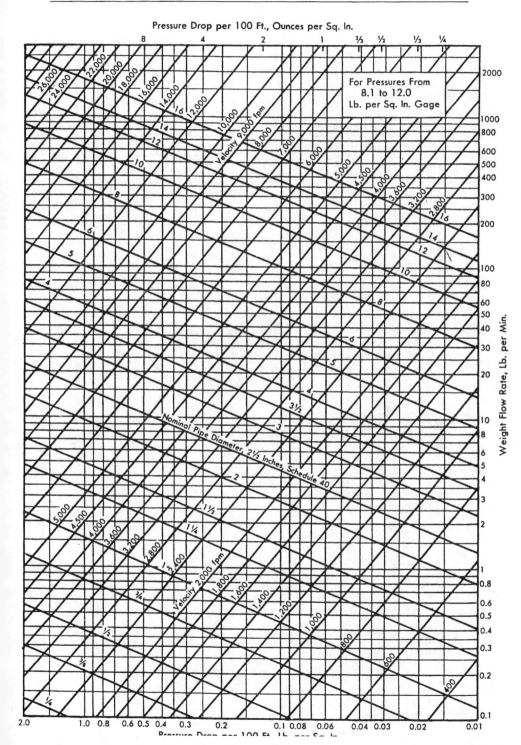

Pressure Drop per 100 Ft., Lb. per Sq. In.

Weight Flow Rate, Lb. per Min.

For Pressures From
8.1 to 12.0
Lb. per Sq. In. Gage

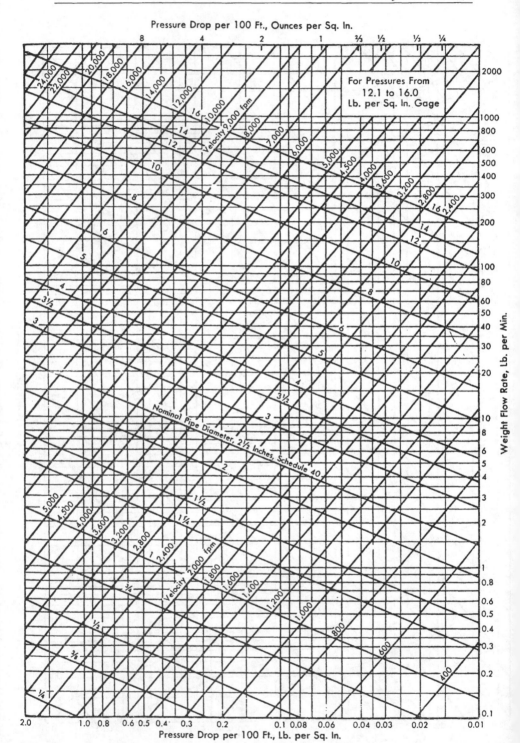

Pressure Drop per 100 Ft., Ounces per Sq. In.

For Pressures From
12.1 to 16.0
Lb. per Sq. In. Gage

Weight Flow Rate, Lb. per Min.

Pressure Drop per 100 Ft., Lb. per Sq. In.

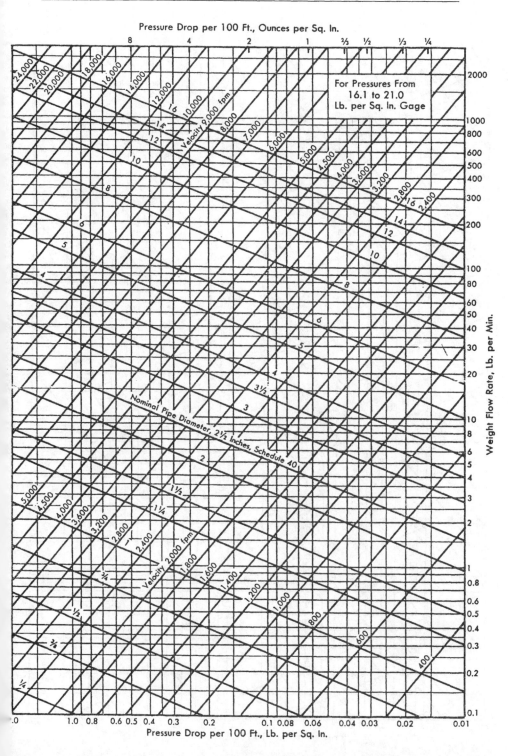

Pressure Drop per 100 Ft., Ounces per Sq. In.

For Pressures From
16.1 to 21.0
Lb. per Sq. In. Gage

Weight Flow Rate, Lb. per Min.

Pressure Drop per 100 Ft., Lb. per Sq. In.

Pressure Drop per 100 Ft., Ounces per Sq. In.

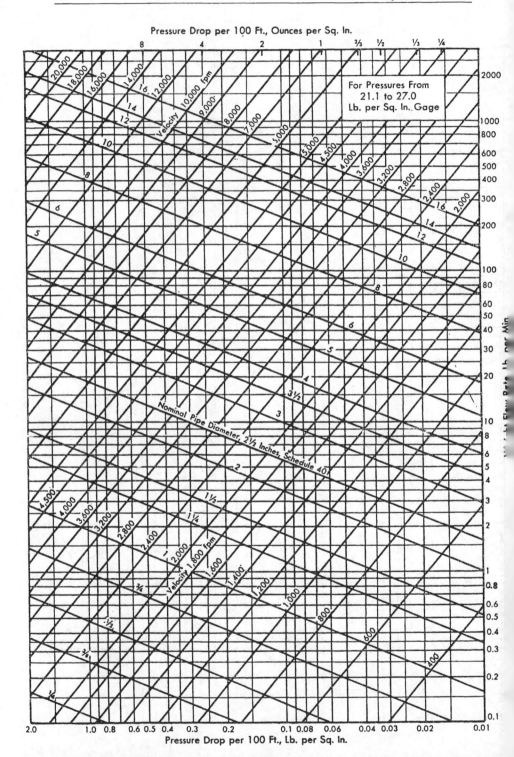

Pressure Drop per 100 Ft., Lb. per Sq. In.

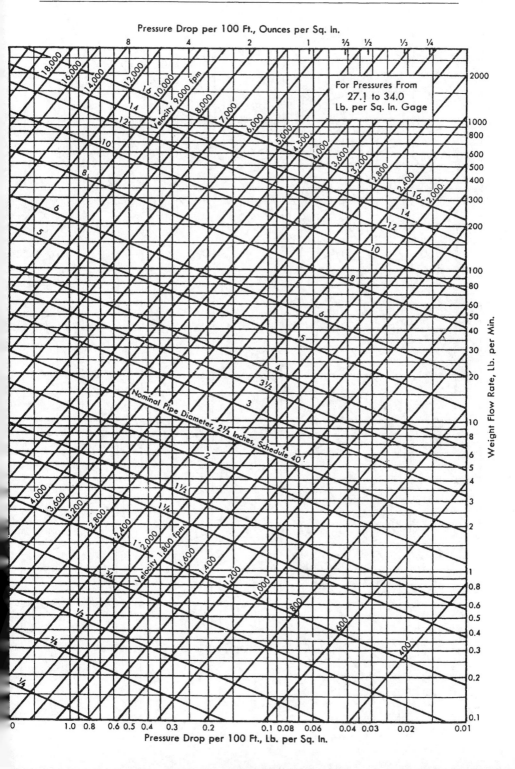

Pressure Drop per 100 Ft., Ounces per Sq. In.

For Pressures From
27.1 to 34.0
Lb. per Sq. In. Gage

Weight Flow Rate, Lb. per Min.

Pressure Drop per 100 Ft., Lb. per Sq. In.

Pressure Drop per 100 Ft., Ounces per Sq. In.

Pressure Drop per 100 Ft., Lb. per Sq. In.

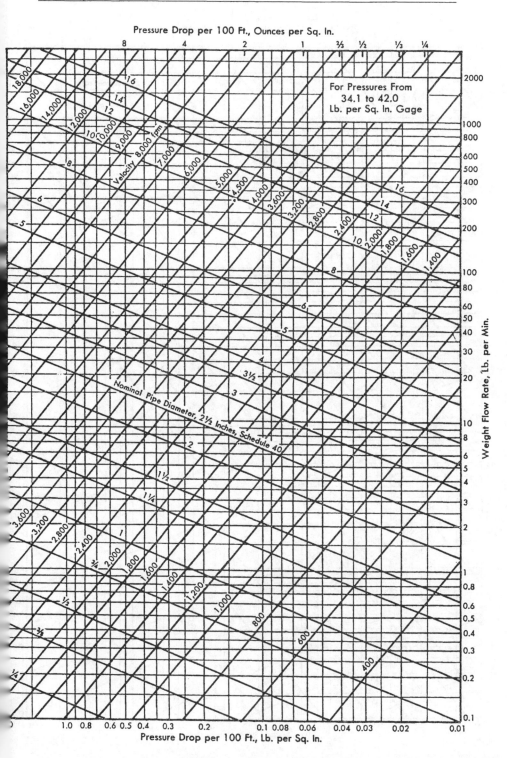

Pressure Drop per 100 Ft., Ounces per Sq. In.

For Pressures From
34.1 to 42.0
Lb. per Sq. In. Gage

Weight Flow Rate, Lb. per Min.

Pressure Drop per 100 Ft., Lb. per Sq. In.

Pressure Drop per 100 Ft., Ounces per Sq. In.

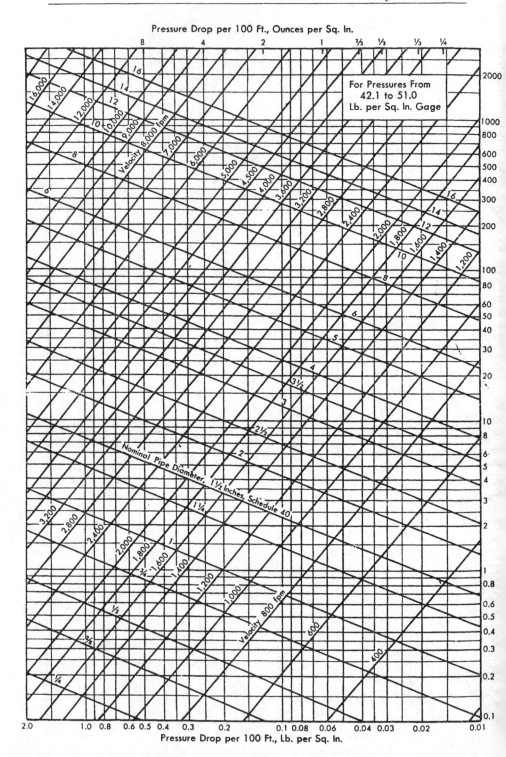

For Pressures From
42.1 to 51.0
Lb. per Sq. In. Gage

Pressure Drop per 100 Ft., Lb. per Sq. In.

Pressure Drop per 100 Ft., Ounces per Sq. In.

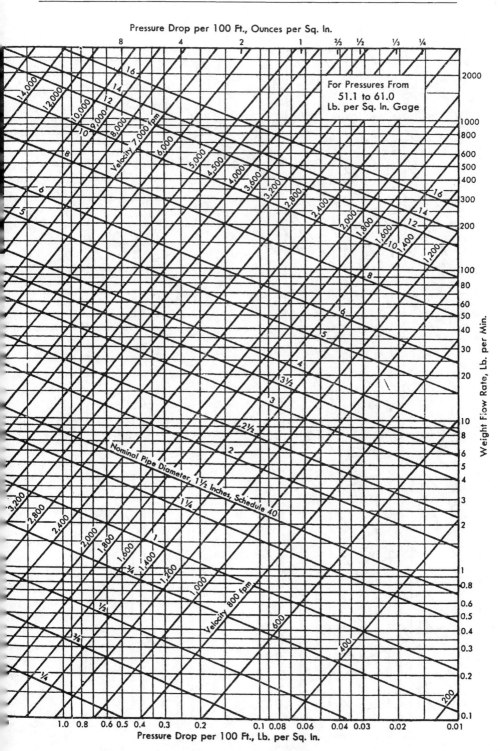

For Pressures From
51.1 to 61.0
Lb. per Sq. In. Gage

Weight Flow Rate, Lb. per Min.

Pressure Drop per 100 Ft., Lb. per Sq. In.

Pressure Drop per 100 Ft., Ounces per Sq. In.

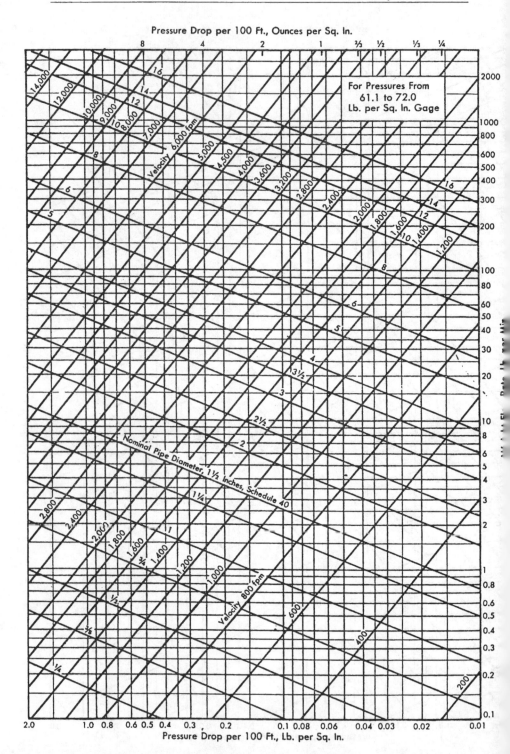

Pressure Drop per 100 Ft., Lb. per Sq. In.

Pressure Drop per 100 Ft., Ounces per Sq. In.

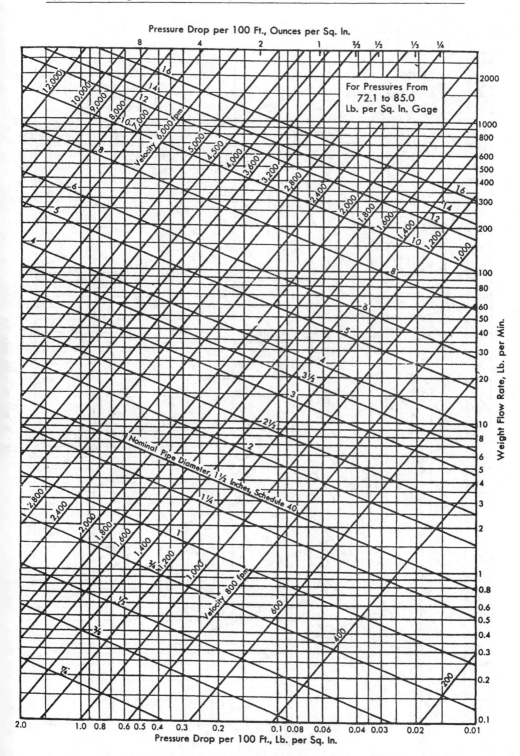

For Pressures From
72.1 to 85.0
Lb. per Sq. In. Gage

Weight Flow Rate, Lb. per Min.

Pressure Drop per 100 Ft., Lb. per Sq. In.

Pressure Drop per 100 Ft., Ounces per Sq. In.

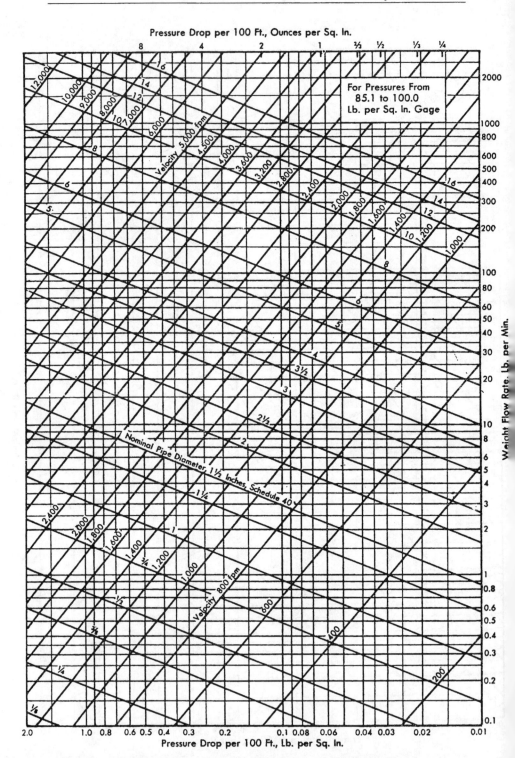

For Pressures From
85.1 to 100.0
Lb. per Sq. In. Gage

Weight Flow Rate, Lb. per Min.

Pressure Drop per 100 Ft., Lb. per Sq. In.

Pressure Drop per 100 Ft., Ounces per Sq. In.

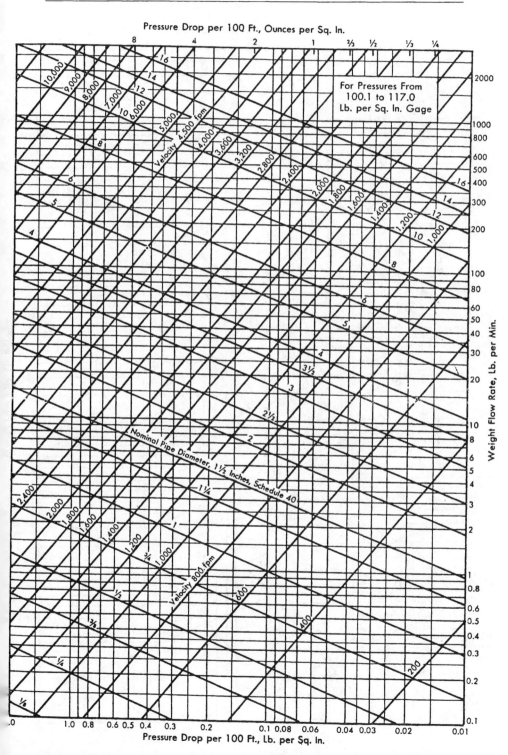

For Pressures From
100.1 to 117.0
Lb. per Sq. In. Gage

Weight Flow Rate, Lb. per Min.

Pressure Drop per 100 Ft., Lb. per Sq. In.

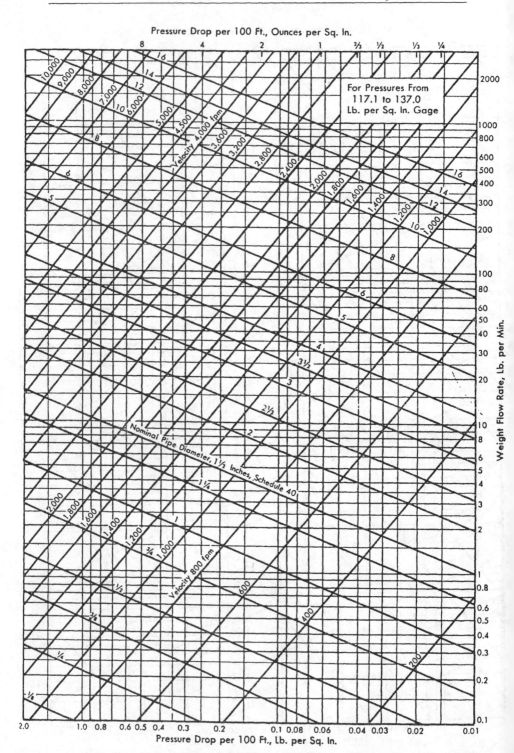

Pressure Drop per 100 Ft., Ounces per Sq. In.

For Pressures From
117.1 to 137.0
Lb. per Sq. In. Gage

Weight Flow Rate, Lb. per Min.

Pressure Drop per 100 Ft., Lb. per Sq. In.

Pressure Drop per 100 Ft., Ounces per Sq. In.

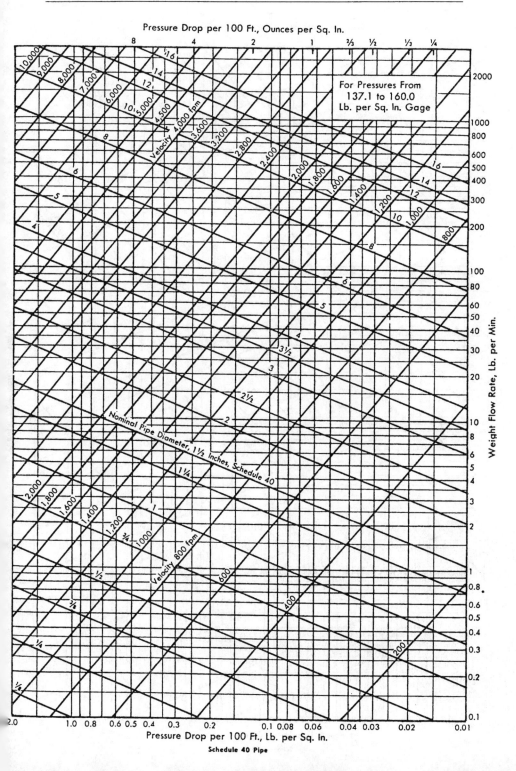

For Pressures From
137.1 to 160.0
Lb. per Sq. In. Gage

Weight Flow Rate, Lb. per Min.

Pressure Drop per 100 Ft., Lb. per Sq. In.

Schedule 40 Pipe

Pressure Drop per 100 Ft., Ounces per Sq. In.

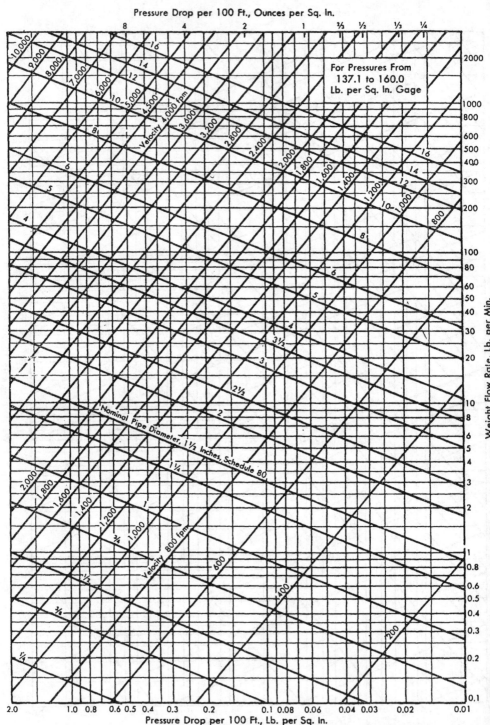

For Pressures From
137.1 to 160.0
Lb. per Sq. In. Gage

Weight Flow Rate, Lb. per Min.

Pressure Drop per 100 Ft., Lb. per Sq. In.

Schedule 80 Pipe

Pressure Drop per 100 Ft., Ounces per Sq. In.

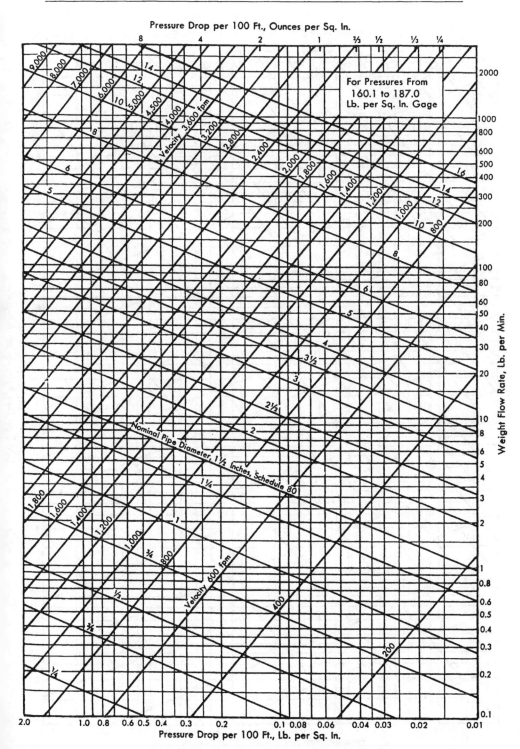

For Pressures From
160.1 to 187.0
Lb. per Sq. In. Gage

Weight Flow Rate, Lb. per Min.

Pressure Drop per 100 Ft., Lb. per Sq. In.

Pressure Drop per 100 Ft., Ounces per Sq. In.

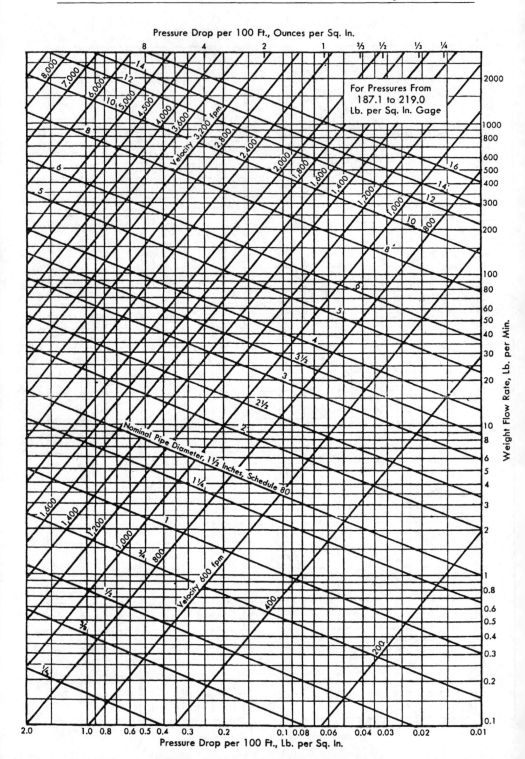

For Pressures From
187.1 to 219.0
Lb. per Sq. In. Gage

Weight Flow Rate, Lb. per Min.

Pressure Drop per 100 Ft., Lb. per Sq. In.

Pressure Drop per 100 Ft., Ounces per Sq. In.

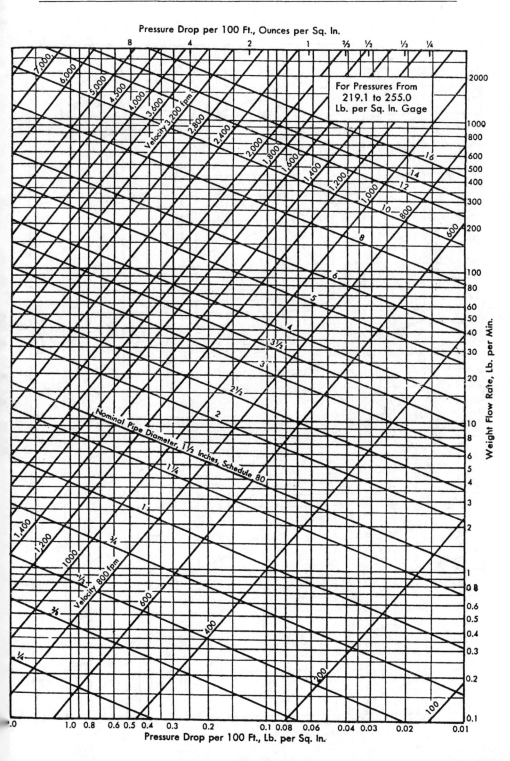

For Pressures From
219.1 to 255.0
Lb. per Sq. In. Gage

Weight Flow Rate, Lb. per Min.

Pressure Drop per 100 Ft., Lb. per Sq. In.

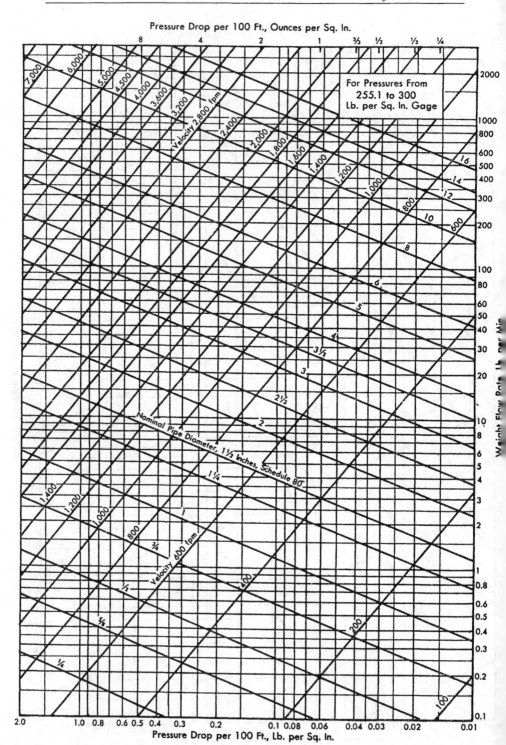

Pressure Drop per 100 Ft., Ounces per Sq. In.

For Pressures From 255.1 to 300 Lb. per Sq. In. Gage

Pressure Drop per 100 Ft., Lb. per Sq. In.

Appendix C

Engineering Conversion Tables

These tables include the United States Customary System (USCS) and conversions from the Systemé Internationalé (SI), more commonly referred to as the Metric System. There the conversion factors for the SI terms are expressed. The interrelationship of the more common terms (grams/sq cm, cycles/second, degrees temperature, etc.) and the newer nomenclature such as (Newton, Hertz, Celsius and Pascal) are expressed in easy form.

Generally, the USCS is still the dominant system in American industry. However, research engineering and certain chemical industries such as pharmaceuticals are using the SI more, to the exclusion of the USCS. Science education stresses use of SI and it is suggested that those with a need to know, commit the more commonly used SI terms to memory.

Table C-1. Engineering Conversion Factors

MULTIPLY	BY	TO OBTAIN
Acres	43,560	Square feet
Acres	4047	Square meters
Acres	1.562×10^{-3}	Square miles
Acres	4840	Square yards
Acre-feet	43,560	Cubic feet
Acre-feet	325,851	Gallons
Acre-feet	1233.49	Cubic meters

---

*From *The Steam Trap Handbook,* by James F. McCauley, P.E., JFM Technical Writing/Professional Engineering, Suite 193-B, Stonybrook, 803 Cooper Street, Deptford, NJ 08096; (609) 384-8174.

Table C-1. Engineering Conversion Factors (*Cont'd*)

Multiply	By	To Obtain
Atmospheres	76.0	Cms. of mercury
Atmospheres	29.92	Inches of mercury
Atmospheres	33.90	Feet of water
Atmospheres	10,333	Kgs./sq. meter
Atmospheres	14.70	Lbs./sq. inch
Atmospheres	1.058	Tons/sq. ft.
Barrels-oil	42	Gallons-oil
Barrels-cement	376	Pounds-cement
Bags or sacks-cement	94	Pounds-cement
Board-feet	144 sq. in. × 1 in.	Cubic inches
British Thermal Units	0.2520	Kilogram-calories
British Thermal Units	777.5	Foot-lbs.
British Thermal Units	3.927×10^{-4}	Horse-power-hrs.
British Thermal Units	107.5	Kilogram-meters
British Thermal Units	$2.928. \times 10^{-4}$	Kilowatt-hrs.
Btu/min.	12.96	Foot-lbs./sec.
Btu/min.	0.02356	Horse-power
Btu/min.	0.01757	Kilowatts
Btu/min.	17.57	Watts
Centares (Centiares)	1	Square meters
Centigrams	0.01	Grams
Centiliters	0.01	Liters
Centimeters	0.3937	Inches
Centimeters	0.01	Meters
Centimeters	10	Millimeters

Table C-1. Engineering Conversion Factors (*Cont'd*)

Multiply	By	To Obtain
Centimtrs. of mercury	0.01316	Atmospheres
Centimtrs. of mercury	0.4461	Feet of water
Centimtrs. of mercury	136.0	Kgs./sq. meter
Centimtrs. of mercury	27.85	Lbs./sq. ft.
Centimtrs of mercury	0.1934	Lbs./sq. inch
Centimeters/second	1.969	Feet/min.
Centimeters/second	0.03281	Feet/sec.
Centimeters/second	0.036	Kilometers/hr.
Centimeters/second	0.6	Meters/min.
Centimeters/second	0.02237	Miles/hr.
Centimeters/second	3.728×10^{-4}	Miles/min.
Cms./sec./sec.	0.03281	Feet/sec./sec.
Cubic centimeters	3.531×10^{-5}	Cubic feet
Cubic centimeters	6.102×10^{-2}	Cubic inches
Cubic centimeters	10^{-6}	Cubic meters
Cubic centimeters	1.308×10^{-6}	Cubic yards
Cubic centimeters	2.642×10^{-4}	Gallons
Cubic centimeters	10^{-3}	Liters
Cubic centimeters	2.113×10^{-3}	Pints (liq.)
Cubic centimeters	1.057×10^{-3}	Quarts (liq.)
Cubic feet	2.832×10^{4}	Cubic cms.
Cubic feet	1728	Cubic inches
Cubic feet	0.02832	Cubic meters
Cubic feet	0.03704	Cubic yards
Cubic feet	7.48052	Gallons
Cubic feet	28.32	Liters
Cubic feet	59.84	Pints (liq.)
Cubic feet	29.92	Quarts (liq.)
Cubic feet/minute	4,2.0	Cubic cms./sec.
Cubic feet/minute	0.12~7	Gallons/sec.

Table C-1. Engineering Conversion Factors (*Cont'd*)

Multiply	By	To Obtain
Cubic feet/minute	0.4720	Liters/sec.
Cubic feet/minute	62.43	Pounds of water/min.
Cubic feet/second	0.646317	Million gals./day
Cubic feet/second	4118.831	Gallons/min.
Cubic inches	16.39	Cubic centimeters
Cubic inches	5.787×10^{-4}	Cubic feet
Cubic inches	1.639×10^{-5}	Cubic meters
Cubic inches	2.143×10^{-5}	Cubic yards
Cubic inches	4.329×10^{-3}	Gallons
Cubic inches	1.639×10^{-2}	Liters
Cubic inches	0.03463	Pints (liq.)
Cubic inches	0.01732	Quarts (liq.)
Cubic meters	10^6	Cubic centimeters
Cubic meters	35.31	Cubic feet
Cubic meters	41.023	Cubic inches
Cubic meters	1.308	Cubic yards
Cubic meters	264.2	Gallons
Cubic meters	103	Liters
Cubic meters	2113	Pints (liq.)
Cubic meters	1057	Quarts (liq.)
Cubic yards	7.646×10^3	Cubic centimeters
Cubic yards	27	Cubic feet
Cubic yards	46.656	Cubic inches
Cubic yards	0.7646	Cubic meters
Cubic yards	202.0	Gallons
Cubic yards	764.6	Liters
Cubic yards	1616	Pints (liq.)
Cubic yards	807.9	Quarts (liq.)
Cubic yards/min.	0.45	Cubic feet/sec.
Cubic yards/min.	3.367	Gallons/sec.
Cubic yards/min.	12.74	Liters/sec.

Table C-1. Engineering Conversion Factors (*Cont'd*)

Multiply	By	To Obtain
Decigrams	0.1	Grams
Deciliters	0.1	Liters
Decimeters	0.1	Meters
Degrees (angle)	60	Minutes
Degrees (angle)	0.01745	Radians
Degrees (angle)	3600	Seconds
Degrees/sec.	0.01745	Radians/sec.
Degrees/sec.	0.1667	Revolutions/min.
Degrees/sec.	0.002778	Revolutions/sec.
Dekagrams	10	Grams
Dekaliters	10	Liters
Dekameters	10	Meters
Drams	27.34375	Grains
Drams	0.0625	Ounces
Drams	1.771845	Grams
Fathoms	6	Feet
Feet	30.48	Centimeters
Feet	12	Inches
Feet	0.3048	Miters
Feet	1/3	Yards
Feet of water	0.02950	Atmospheres
Feet of water	0.8826	Inches of mercury
Feet of water	304.8	Kgs./sq. meter
Feet of water	62.43	Lbs./sq ft.

Table C-1. Engineering Conversion Factors (*Cont'd*)

Multiply	By	To Obtain
Feet of water	0.4335	Lbs./sq. inch
Feet/min.	0.5080	Centimeters/sec.
Feet/min.	0.01667	Feet/sec
Feet/min.	0.01829	Kilometers/hr.
Feet/min.	0.3048	Meters/min.
Feet/min.	0.01136	Miles/hr.
Feet/sec.	30.48	Centimeters/sec.
Feet/sec.	1.097	Kilometers/hr
Feet/sec.	0.5921	Knots
Feet/sec.	18.29	Meters/min.
Feet/sec.	0.6818	Miles/hr.
Feet/sec.	0.01136	Miles/min.
Feet/sec./sec.	30.48	Cms./sec./sec
Feet/sec./sec.	0.3048	Meters/sec./sec.
Foot-pounds	1.286×10^{-3}	British Thermal Units
Foot-pounds	5.050×10^{-7}	Horse-power-hrs.
Foot-pounds	3.241×10^{-4}	Kilogram-calories
Foot-pounds	0.1383	Kilogram-meters
Foot-pounds	3.766×10^{-7}	Kilowatt-hrs.
Foot-pounds/min.	1.286×10^{-3}	Btus/min.
Foot-pounds/min.	0.01667	Foot-pounds/sec
Foot-pounds/min.	3.030×10^{-5}	Horse-power
Foot-pounds/min.	3.241×10^{-4}	Kg.-calories/min.
Foot-pounds/min.	2.260×10^{-5}	Kilowatts
Foot-pounds/sec.	7.717×10^{-2}	L. T. Units/min.
Foot-pounds/sec.	1.818×10^{-3}	Horse-power
Foot-pounds/sec.	1.945×10^{-2}	Kg.-calories/min.
Foot-pounds/sec.	1.356×10^{-3}	Kilowatts

Table C-1. Engineering Conversion Factors (*Cont'd*)

Multiply	By	To Obtain
Gallons	3785	Cubic centimeters
Gallons	0.1337	Cubic feet
Gallons	231	Cubic inches
Gallons	3.785×10^{-3}	Cubic meters
Gallons	4.951×10^{-3}	Cubic yards
Gallons	3.785	Liters
Gallons	8	Pints (liq.)
Gallons	4	Quarts (liq.)
Gallons, Imperial	1.20095	U.S. gallons
Gallons, U.S.	0.83267	Imperial gallons
Gallons water	8.3453	Pounds of water
Gallons/min.	2.226×10^{-3}	Cubic feet/sec.
Gallons min.	0.06308	Liters, sec.
Gallons/min.	8.0208	Cu. Ft./hr.
Gallons/min.	8.2208	Overflow rate (ft./hr.)
	Area(sq. ft.)	
Gallons water/mint	6.0086	Tons water/24 furs.
Grains troy	1	Grains (avoir.)
Grains troy	0.06480	Grams
Grains troy	0.04167	Pennyweights (troy)
Grains troy	2.0833×10^{-3}	Ounces (troy)
Grains/U.S. gal.	17.118	Parts/million
Grains/U.S. gal.	142.86	Lbs./million gal.
Grains/Imp. gal.	14.286	parts/million
Grams	980.7	Dynes
Grams	15.43	Grains

Table C-1. Engineering Conversion Factors (*Cont'd*)

Multiply	By	To Obtain
Grams	10^{-3}	Kilograms
Grams	10^3	Milligrams
Grams	0.03527	Ounces
Grams	0.03215	Ounces (troy)
Grams	2.205×10^{-3}	Pounds
Grams/cm.	5.600×10^{-3}	Pounds/inch
Grams/cu. cm.	62.43	Pounds/cubic foot
Grams/cu. cm.	0.03613	Pounds/cubic inch
Grams/liter	58.417	Grains/gal.
Grams/liter	8.345	Pounds/1000 gals.
Grams/liter	0.0652427	Pounds/cubic foot
Grams/liter	1000	parts/million
Hectares	2.471	Acres
Hectares	1.076×10^5	Square feet
Hectograms	100	Grams
Hectoliters	100	Liters
Hectometers	100	Meters
Hectowatts	100	Watts
Horse-power	42.44	Btus/min.
Horse-power	33,000	Foot-lbs./min.
Horse-power	550	Foot-lbs./sec.
Horse-power	1.014	Horse-power (metric)
Horse-power	10.70	Kg.-calories/min.
Horse-power	0.7457	Kilowatts
Horse-power	745.7	Watts

Table C-1. Engineering Conversion Factors (*Cont'd*)

Multiply	By	To Obtain
Horse-power (boiler)	33,479	Btu/hr.
Horse-power (boiler)	9.803	Kilowatts
Horse-power-hours	2547	British Thermal Units
Horse-power-hours	1.98×10^6	Foot-lbs.
Horse-power-hours	641.7	Kilogram-calories
Horse-power-hours	2.737×10^5	Kilogram-meters
Horse-power-hours	0.7457	Kilowatt-hours
Inches	2.540	Centimeters
Inches of mercury	0.03342	Atmospheres
Inches of mercury	1.133	Feet of water
Inches of mercury	345.3	Kgs./sq. meter
Inches of mercury	70.73	Lbs./sq. ft.
Inches of mercury	0.4912	Lbs./sq. inch
Inches of water	0.002458	Atmospheres
Inches of water	0.07355	Inches of mercury
Inches of water	25.40	Kgs./sq. meter
Inches of water	0.5781	Ounces/sq. inch
Inches of water	5.202	Lbs./sq. foot
Inches of water	0.03613	Lbs./sq. inch
Kilograms	980,665	Dynes
Kilograms	2.205	Lbs.
Kilograms	1.102×10^{-3}	Tons (short)
Kilograms	10^3	Grams
Kilograms-calories	3.968	British Thermal Units
Kilograms-calories	3086	Foot-pounds
Kilograms-calories	1.558×10^{-3}	Horse-power-hours
Kilograms-calories	1.162×10^{-3}	Kilowatt-hours
Kilogram-calories/min.	51.43	Foot-pounds/sec.

Table C-1. Engineering Conversion Factors (*Cont'd*)

Multiply	By	To Obtain
Kilogram-calories/min.	0.09351	Horse-power
Kilogram-calories/min.	0.06972	Kilowatts
Kgs./meter	0.6720	Lbs./foot
Kgs./sq. meter	9.678×10^{-5}	Atmospheres
Kgs./sq. meter	3.281×10^{-3}	Feet of water
Kgs./sq. meter	2.896×10^{-3}	Inches of mercury
Kgs./sq. meter	0.2048	Lbs./sq. foot
Kgs./sq. meter	1.422×10^{-3}	Lbs./sq. inch
Kgs./sq. millimeter	10^6	Kgs./sq. meter
Kiloliters	10^3	Liters
Kilometers	10^5	Centimeters
Kilometers	3281	Feet
Kilometers	10^3	Meters
Kilometers	0.6214	Miles
Kilometers	1094	Yards
Kilometers/hr.	27.78	Centimeters/sec.
Kilometers/hr.	54.68	Feet/min.
Kilometers/hr.	0.9113	Feet/sec.
Kilometers/hr.	0.5396	Knots
Kilometers/hr.	16.67	Meters/min.
Kilometers/hr.	0.6234	Miles/hr.
Kms./hr./sec.	27.78	Cms./sec./sec.
Kms./hr./sec.	0.9113	Ft./sec./sec.
Kms./hr./sec.	0.2778	Meters/sec./sec.
Kilowatts	56.92	Btus/min.
Kilowatts	4.425×10^4	Foot-lbs./min.
Kilowatts	737.6	Foot-lbs./sec.

Table C-1. Engineering Conversion Factors (*Cont'd*)

Multiply	By	To Obtain
Kilowatts	1.341	Horse-power
Kilowatts	14.34	Kg.-calories/min.
Kilowatts	10^3	Watts
Kilowatt-hours	3415	British Thermal Units
Kilowatt-hours	2.655×10^6	Foot-lbs.
Kilowatt-hours	1.341	Horse-power-hrs.
Kilowatt-hours	860.5	Kilogram-calories
Kilowatt-hours	3.671×10^5	Kilogram-meters
Liters	10^3	Cubic centimeters
Liters	0.03531	Cubic feet
Liters	61.02	Cubic inches
Liters	10^{-3}	Cubic meters
Liters	1.308×10^{-3}	Cubic yards
Liters	0.2642	Gallons
Liters	2.113	Pints (liq.)
Liters	1.057	Quarts (liq.)
Liters/min.	5.886×10^{-4}	Cubic ft./sec.
Liters/min.	4.403×10^{-3}	Gals/sec.

Lumber Width (in.) $\times \dfrac{\text{Thickness (in.)}}{12}$	Length (ft)	Board Feet
Meters	100	Centimeters
Meters	3.281	Feet
Meters	39.37	Inches
Meters	10^{-3}	Kilometers
Meters	10^3	Millimeters
Meters	1.094	Yards
Meters/min.	1. 667	Centimeters/sec.

Table C-1. Engineering Conversion Factors (*Cont'd*)

Multiply	By	To Obtain
Meters/min.	3.281	Feet/min.
Meters/min.	0.05468	Feet/sec.
Meters/min.	0.06	Kilometers/hr.
Meters/min.	0.03728	Miles/hr.
Meters/sec.	196.8	Feet/min.
Meters/sec.	3.281	Feet/sec.
Meters/sec.	3.6	Kilometers/hr.
Meters/sec.	0.06	Kilometers/min.
Meters/sec.	2.237	Miles/hr.
Meters/sec.	0.03728	Miles/min.
Microns	10^{-6}	Meters
Miles	1.609×10^5	Centimeters
Miles	5280	Feet
Miles	1.609	Kilometers
Miles	1760	Yards
Miles/hr.	44.70	Centimeters/sec.
Miles/hr.	88	Feet/min.
Miles/hr.	1.467	Feet/sec.
Miles/hr.	1.609	Kilometers/hr.
Miles/hr.	0.8684	Knots
Miles/hr.	26.82	Meters/min.
Miles/min.	2682	Centimeters/sec.
Miles/min.	88	Feet/sec.
Miles/min.	1.609	Kilometers/min.
Miles/min.	60	Miles/hr.
Milliers	10^3	Kilograms
Milligrams	10^{-3}	Grams

Table C-1. Engineering Conversion Factors (*Cont'd*)

Multiply	By	To Obtain
Milliliters	10^{-3}	Liters
Millimeters	0.1	Centimeters
Millimeters	0.03937	Inches
Milligrams/liter	1	Parts/million
Million gals./day	1.54723	Cubic ft./sec.
Miner's inches	1.5	Cubic ft./min.
Minutes (angle)	2.909×10^{-4}	Radians
Ounces	16	Drams
Ounces	437.5	Grains
Ounces	0.0625	Pounds
Ounces	28.349527	Grams
Ounces	0.9115	Ounces (troy)
Ounces	2.790×10^{-5}	Tons (long)
Ounces	2.835×10^{-5}	Tons (metric)
Ounces (troy)	480	Grains
Ounces (troy)	20	Pennyweights (troy)
Ounces (troy)	0.08333	Pounds (troy)
Ounces (troy)	31.103481	Grams
Ounces (troy)	1.09714	Ounces, avoir
Ounces (fluid)	1.805	Cubic inches
Ounces (fluid)	0.02957	Liters
Ounces/sq. inch	0.0625	Lbs./sq. inch
Overflow rate (ft./hr.) / 1	$0.12468 \times$ area (sq. ft.)	Gals./min.

Table C-1. Engineering Conversion Factors (*Cont'd*)

Multiply	By	To Obtain
Overflow rate (ft./hr.)	8.0208	Sq. Ft./gal./min.
Parts/million	0.05834	Grains/U.S. gal.
Parts/million	0.07016	Grains/Imp. gal.
Parts/million	8.345	Lbs./million gal.
Pennyweights	24	Grains
Pennyweights	1.55517	Grams
Pennyweights	0.05	Ounces (troy)
Pennyweights	4.1667×10^{-3}	Pounds (troy
Pounds	16	Ounces
Pounds	256	Drams
Pounds	7000	Grains
Pounds	0.0005	Tons (short)
Pounds	453.5924	Grams
Pounds	1.21528	Pounds (troy)
Pounds	14.5833	Ounces (troy)
Pounds (troy)	5760	Grains
Pounds (troy)	240	Pennyweights (troy)
Pounds (troy)	12	Ounces (troy)
Pounds (troy)	373.24177	Grams
Pounds (troy)	0.822857	Pounds (avoir.)
Pounds (troy)	13.1657	Ounces (avoir.)
Pounds (troy)	3.6735×10^{-4}	Tons (long)
pounds (troy)	4.1143×10^{-4}	Tons (short)
Pounds (troy)	3.7324×10^{-4}	Tons (metric)
Pounds of water	0.01602	Cubic feet
Pounds of water	27.68	Cubic inches
Pounds of water	0.1198	Gallons
Pounds of water/min.	2.670×10^{-4}	Cubic ft./see

Table C-1. Engineering Conversion Factors (*Cont'd*)

Multiply	By	To Obtain
Pounds/cubic foot	0.01602	Grams/cubic cm.
Pounds/cubic foot	16.02	Kgs./cubic meter
Pounds/cubic foot	5.787×10^{-4}	Lbs./cubic inch
Pounds/cubic inch	27.68	Grams/cubic cm.
Pounds/cubic inch	2.768×10^{4}	Kgs./cubic meter
Pounds/cubic inch	1728	Lbs./cubic foot
Pounds/foot	1.488	Kgs./meter
Pounds/inch	178.6	Grams/cm.
Pounds/sq. foot	0.01602	Feet of water
Pounds/sq. foot	4.883	Kgs./sq. meter
Pounds/sq. foot	6.945×10^{-3}	Pounds/sq. inch
Pounds/sq. inch	0.06804	Atmospheres
Pounds/sq. inch	2.307	Feet of water
Pounds/sq. inch	2.036	Inches of mercury
Pounds/sq. inch	703.1	Kgs./sq. meter
Quadrants (angle)	90	Degrees
Quadrants (angle)	5400	Minutes
Quadrants (angle)	1.571	Radians
Quarts (dry)	67.20	Cubic inches
Quarts (liq.)	57.75	Cubic inches
Quintal, Argentine	101.28	Pounds
Quintal, Brazil	129.54	Pounds
Quintal, Castile, Peru	101.43	Pounds
Quintal, Chile	101.41	Pounds
Quintal, Mexico	101.47	Pounds
Quintal, Metric	220.46	Pounds

Table C-1. Engineering Conversion Factors (*Cont'd*)

Multiply	By	To Obtain
Quires	25	Sheets
Radians	57.30	Degrees
Radians	3438	Minutes
Radians	0.637	Quadrants
Radians/sec.	57.30	Degrees/sec.
Radians/sec.	0.1592	Revolutions/sec.
Radians/sec.	9.549	Revolutions/min.
Radians/sec./sec.	573.0	Revs./min./min.
Radians/sec./sec.	0.1592	Revs./sec./sec.
Reams	500	Sheets
Revolutions	360	Degrees
Revolutions	4	Quadrants
Revolutions	6.283	Radians
Revolutions/min.	6	Degrees/sec.
Revolutions/min.	0.1047	Radians/sec.
Revolutions/min.	0.01667	Revolutions/sec.
Revolutions/min./min.	1.745×10^{-3}	Rads./sec./sec.
Revolutions/min./min.	2.778×10^{-4}	Revs./sec./sec.
Revolutions/sec.	360	Degrees/sec.
Revolutions/sec.	6.283	Radians/sec.
Revolutions/sec.	60	Revolutions/min.
Revolutions/sec./sec.	6.283	Radians/sec./sec
Revolutions/sec./sec.	3600	Revs./min./min.
Seconds (angle)	4.848×10^{-6}	Radians

Table C-1. Engineering Conversion Factors (*Cont'd*)

Multiply	By	To Obtain
Square centimeters	1.076×10^{-3}	Square feet
Square centimeters	0.1550	Square inches
Square centimeters	10^{-4}	Square meters
Square centimeters	100	Square millimeters
Square feet	2.296×10^{-5}	Acres
Square feet	929.0	Square centimeters
Square feet	144	Square inches
Square feet	0.09290	Square meters
Square feet	3.587×10^{-8}	Square miles
Square feet	1/9	Square yards
$\dfrac{1}{\text{Sq. ft./gal./min.}}$	8.0208	Overflow rate (ft./hr.)
Square inches	6.452	Square centimeters
Square inches	6.944×10^{-3}	Square feet
Square inches	645.2	Square millimeters
Square kilometers	247.1	Acres
Square kilometers	10.76×10^{6}	Square feet
Square kilometers	10^{6}	Square meters
Square kilometers	0.3861	Square miles
Square kilometers	1.196×10^{6}	Square yards
Square meters	2.471×10^{-4}	Acres
Square meters	10.76	Square feet
Square meters	3.861×10^{-7}	Square miles
Square meters	1.196	Square yards
Square miles	640	Acres
Square miles	27.88×10^{6}	Square feet
Square miles	2.590	Square kilometers
Square miles	3.098×10^{6}	Square yards

Table C-1. Engineering Conversion Factors (*Cont'd*)

Multiply	By	To Obtain
Square millimeters	0.01	Square centimeters
Square millimeters	1.550×10^{-3}	Square inches
Square yards	2.066×10^{-4}	Acres
Square yards	9	Square feet
Square yards	0.8361	Square meters
Square yards	3.228×10^{-7}	Square miles
Temp. (°C.) + 273	1	Abs. temp. (°C.)
Temp. (°C.) + 17.78	1.8	Temp. (°F.)
Temp. (°F.) + 460	1	Abs. temp. (°F.)
Temp. (°F.) -32	5/9	Temp. (°C)
Tons (long)	1016	Kilograms
Tons (long)	2240	Pounds
Tons (long)	1.12000	Tons (short)
Tons (metric)	10^3	Kilograms
Tons (metric)	2205	Pounds
Tons (short)	2000	Pounds
Tons (short)	32000	Ounces
Tons (short)	907.18486	Kilograms
Tons (short)	2430.56	Pounds (troy)
Tons (short)	0.89287	Tons (long)
Tons (short)	29166.66	Ounces (troy)
Tons (short)	0.90718	Tons (metric)

$\dfrac{1}{\text{Tons dry solids/24 hrs.}}$	Area (sq. ft.)	Sq. ft./ton/24 hrs.
Tons of water/24 hrs.	83. 333	Pounds water/hour
Tons of water/24 hrs	0.16643	Gallons/min.
Tons of water/24 hrs	1.3349	Cu. ft./hr.

Table C-1. Engineering Conversion Factors (*Cont'd*)

Multiply	By	To Obtain
Watts	0.05692	Btus/min
Watts	44.26	Foot-pounds/min.
Watts	0.7376	Foot-pounds/sec.
Watts	1.341×10^{-3}	Horse-power
Watts	0.01434	Kg.-calories/min.
Watts	10^{-3}	Kilowatts
Watt-hours	3.415	British Thermal Units
Watt-hours	2655	Foot-pounds
Watt-hours	1.341×10^{-3}	Horse- power- hours
Watt-hours	0.8605	Kilogram-calories
Watt-hours	367.1	Kilogram-meters
Watt-hours	10^{-3}	Kilowatt-hours
Yards	91.44	Centimeters
Yards	3	Feet
Yards	36	Inches
Yards	0.9144	Meters

Table C-2. Physical Constants and Conversion Factors

Physical Constants

Standard gravity $\quad= 32.174$ ft/s^2

Universal gas constant $\quad R_u \quad = 1545$ ft/lb$_f$/(lbmol. °R)

$\quad = 1.986$ Btu/(lbmol. °R)

$\quad = 0.730$ atm.ft^3/(lbmol. °R)

$\quad = 10.73$ psia.ft^3/(lbmol. °R)

Conversion Factors

Length	1 cm	= 0.397 in = 10^4μm = 10^8A
	1 in	= 2.540 cm
	1 ft	= 30.48 cm

(Continued)

Table C-2. Physical Constants and Conversion Factors (*Continued*)

Mass	1 lb_m	= 453.59 g_m = 7000 gr
	1 kg_m	= 2.205 lb_m
Force	1 lb_f	= 32.174 lb_m ft/s^2
	1 lb_f	= 444,800 dyne = 4.448 N
	1 torr	= 1 mmHg at 0°C = $10^3 \mu$m Hg
		= 1.933 × 10^{-2} psi
Volume	1 L	= 0.0353 ft^3 = 0.2642 gal = 61.025 in^3
	1 ft^3	= 28.316 L = 7.4805 gal = 0.02832 m^3
	1 in^3	= 16.387 cm^3
Density	1 lb_m/ft^3	= 0.01602 g_m/cm^3
Energy	1 Btu	= 778.16 ft. lb_f = 1055 J
	1 Btu/lb	= 2.32 kJ/kg
Power	1 W	= 1 J/s = 3.413 Btu/h
	1 hp	= 746 W = 550 ft.lb_f/s = 2545 Btu/h
Velocity	1 mi/h	= 0.447 m/s
Specific Heat	1 Btu/(lb_mF)	= 4.187 kJ/(kg.K)
Temperature	T(°R)	= T(°F) + 459.67 = 1.8[T(°C) + 273.1] = 1.8(K)

Table C-3. USCS (Engineering) Derived Units

Physical Quantity	Unit	Symbol	Definition
Force	pound-force	lb_f	32.174 lb_mft/s^2
Pressure	atmosphere	atm	68,087 lb_m/ft. s^2 (= 14.696 lb_f/in^2)
Energy	foot-pound-force	ft. lb_f	32.174 lb_mft^2/s^2
Power	foot-pound-force/second	ft. lb_f/s	32.174 lb_mft^2/s^3 (= 1.82 × 10^{-3} hp)

Table C-4. Physical Constants and Conversion Factors

Physical Constants

Avogrado's number	N_a	$= 6.023 \times 10^{26}$ atoms/kgmol
Universal gas constant	R_u	$= 0.08205$ L.atm/(gmol.K)
		$= 8.314$ kJ/(kgmol.K)
		$= 0.08314$ bar.m^3/(kgmol.K)
		$= 8.314$ kPa.m^3/(kgmol.K)
Planck's constant	h	$= 6.626 \times 10^{-34}$ J.s/molecule
Boltzman's constant	k	$= 1.380 \times 10^{-23}$ J/(K.molecule)
Speed of light	c	$= 2.988 \times 10^{10}$ cm/s
Standard gravity	g	$= 9.80665$ m/s^2

Conversion factors

Length	1 cm	$= 0.3937$ in $= 10^4 \mu$m $= 10^8$ A
	1 km	$= 0.6215$ mi $= 3281$ ft
Mass	1 kg	$= 2.205$ lb$_m$
Force	1 N	$= 1$ kg.m/s^2 $= 0.2248$ lb$_f$
Pressure	1 bar	$= 10^5$ N/m^2 $= 0.9869$ atm
		$= 100$ kPa
	1 torr	$= 1$ mmHg at 0°C $= 1.333$ mbar
		$= 1.933 \times 10^{-2}$ psi
	1 mbar	$= 0.402$ in H_2O
Volume	1 L	$= 0.0353$ ft^3 $= 0.2642$ gal $= 61.025$ in^3
		$= 10^{-3}$ m^3
Density	1 g/cm^3	$= 1$ kg/L $= 62.4$ lb$_m$/ft^3 $= 10^3$ kg/m^3
Energy	1 J	$= 1$ N.m $= 1$ v.c
		$= 0.7375$ ft.lb$_f$ $= 10$ bar.cm^3 $= 0.624 \times 10^{19}$eV
	1 kJ/kg	$= 0.431$ Btu/lb
Power	1 W	$= 1$ J/s $= 3.413$ Btu/h
	1 kW	$= 1.3405$ hp $= 737.3$ ft.lb$_f$/s
Velocity	1 m/s	$= 2.237$ mi/h $= 3.60$ km/h $= 3.281$ ft/s
Specific heat	1 kJ/(kg.K)	$= 0.2389$ Btu/(lb$_m$.F)
Temperature	T(K)	$= 5/9[T(°F) + 459.67] = T(°C) + 273.15$
		$= T(°R)/18$

Table C-5. SI Derived Units and Common Multipliers

SOME SI DERIVED UNITS

Physical Quantity	Unit	Symbol	Definition
Force	newton	N	1 kg.m/s^2
Pressure	pascal	Pa	$1 \text{ kg/m.s}^2 \ (= 1 \text{ N/m}^2)$
	bar	bar	$10^5 \text{ kg/m.s}^2 \ (= 10^5 \text{ N/m}^2)$
Energy	joule	J	$1 \text{ kg.m}^2/\text{s}^2 \ (= 1 \text{ N.m})$
Power	watt	W	$1 \text{ kg.m}^2/\text{s}^2 \ (= 1 \text{ J/s})$
Electric Quantity	coulomb	C	1 A.s
Electric potential difference	volt	V	$1 \text{ kg.m}^2/(\text{A.s}^3)(= 1 \text{ A.Q})$
Electric resistance	ohm	Q	$1 \text{ kg.m}^2/(\text{A}^2.\text{s}^3)(= 1 \text{ V/A})$
Electric capacitance	farad	F	$1 \text{ A}^2.\text{s}^{-4}/(\text{kg.m}^2) \ (= 1 \text{ C/V})$

NAMES AND SYMBOLS FOR COMMON MULTIPLIERS OF SI UNITS

Multiplier	Prefix	Symbol
10^9	giga	G
10^6	mega	M
10^3	kilo	k
10^{-1}	deci	d
10^{-2}	centi	c
10^{-3}	milli	m
10^{-6}	micro	μ
10^{-9}	nano	n

Table C-6. Engineering Conversion Factors for Pressure

Multiply	By	To Obtain
atmosphere	14.696	Lbs/sq. in. abs.
	1.01325×10^3	Newtons/sq. meter
	2116	Lbs/sq. ft.
N/m^2	1.0	Pascals
atmosphere	760	Millimeters of mercury
Bar	10^5	Newtons/sg meter
Microbar	1.0	Dyne/sq. centimeter
	2.089	Lbs force/sq. ft.
	0.1	Newtons/sq. meter
Millimeters of mercury	1333.22	Microbars
	133.322	Newtons/sq. meter
Micrometer of mercury	10^{-6}	Meters of mercury
	10^{-3}	Millimeters of mercury
	0.13322	Newtons/sq. meter
Micrometer of mercury	0.133	Pascals
Torr	1.0	Millimeters of mercury
Inches of mercury	70.73	pounds force/sq. ft.
Inches of Water	6.203	pounds force/sq. ft.
Pounds/sq. in. Abs	6894.76	Newtons/sq. meter
	6.894	Kilo Pascals
	0.070307	Kilograms force/centimeter

Appendix D

Steam Tables and Mollier Chart

from the A.S.M.E. Steam Tables

Courtesy of The American Society of Mechanical Engineers, New York, NY.

Table D-1. Properties of Saturated Steam and Saturated Water (Temperature). (Page 342-353.)

Table D-2. Properties of Saturated Steam and Saturated Water (Pressure, inches H$_g$). (Page 354-367.)

Table D-3. Specific Heat at Constant Pressure of Steam and of Water. (Page 368-369.)

Steam Charts. (Page 370-395.)

Table D-1. Properties of Saturated Steam and Saturated Water (Temperature)

Temp. F	Press. psia	Volume, ft³/lbm			Enthalpy, Btu/lbm			Entropy, Btu/lbm × R			Temp. F
		Water v_f	Evap. v_{fg}	Steam v_g	Water h_f	Evap. h_{fg}	Steam h_g	Water s_f	Evap. s_{fg}	Steam s_g	
705.47	3208.2	0.05078	0.00000	0.05078	906.0	0.0	906.0	1.0612	0.0000	1.0612	705.47
705.0	3198.3	0.04427	0.01304	0.05730	873.0	61.4	934.4	1.0329	0.0527	1.0856	705.0
704.5	3187.8	0.04233	0.01822	0.06055	861.9	85.3	947.2	1.0234	0.0732	1.0967	704.5
704.0	3177.2	0.04108	0.02192	0.06300	854.2	102.0	956.2	1.0169	0.0876	1.1046	704.0
703.5	3166.8	0.04015	0.02492	0.06504	848.2	115.2	963.5	1.0118	0.0991	1.1109	703.5
703.0	3156.3	0.03940	0.02744	0.06684	843.2	126.4	969.6	1.0076	0.1087	1.1163	703.0
702.5	3145.9	0.03878	0.02969	0.06847	838.9	136.1	974.9	1.0039	0.1171	1.1210	702.5
702.0	3135.5	0.03824	0.03173	0.06997	835.0	144.7	979.7	1.0006	0.1246	1.1252	702.0
701.5	3125.2	0.03777	0.03361	0.07138	831.5	152.6	984.0	0.9977	0.1314	1.1291	701.5
701.0	3114.9	0.03735	0.03536	0.07271	828.2	159.8	988.0	0.9949	0.1377	1.1326	701.0
700.5	3104.6	0.03697	0.03701	0.07397	825.2	166.5	991.7	0.9924	0.1435	1.1359	700.5
700.0	3094.3	0.03662	0.03857	0.07519	822.4	172.7	995.2	0.9901	0.1490	1.1390	700.0
699.0	3073.9	0.03600	0.04149	0.07749	817.3	184.2	1001.5	0.9858	0.1590	1.1447	699.0
698.0	3053.6	0.03546	0.04420	0.07966	812.6	194.6	1007.2	0.9818	0.1681	1.1499	698.0
697.0	3033.5	0.03498	0.04674	0.08172	808.4	204.0	1012.4	0.9783	0.1764	1.1547	697.0
696.0	3013.4	0.03455	0.04916	0.08371	804.4	212.8	1017.2	0.9749	0.1841	1.1591	696.0
695.0	2993.5	0.03415	0.05147	0.08563	800.6	221.0	1021.7	0.9718	0.1914	1.1632	695.0
694.0	2973.7	0.03379	0.05370	0.08749	797.1	228.8	1025.9	0.9689	0.1982	1.1671	694.0
693.0	2954.0	0.03345	0.05587	0.08931	793.8	236.1	1029.9	0.9660	0.2048	1.1708	693.0
692.0	2934.5	0.03313	0.05797	0.09110	790.5	243.1	1033.6	0.9634	0.2110	1.1744	692.0
690.0	2895.7	0.03256	0.06203	0.09459	784.5	256.1	1040.6	0.9583	0.2227	1.1810	690.0
688.0	2857.4	0.03204	0.06595	0.09799	778.8	268.2	1047.0	0.9535	0.2337	1.1872	688.0
686.0	2819.5	0.03157	0.06976	0.10133	773.4	279.5	1052.9	0.9490	0.2439	1.1930	686.0
684.0	2782.1	0.03114	0.07349	0.10463	768.2	290.2	1058.4	0.9447	0.2537	1.1984	684.0
682.0	2745.1	0.03074	0.07716	0.10790	763.3	300.4	1063.6	0.9406	0.2631	1.2036	682.0
680.0	2708.6	0.03037	0.08080	0.11117	758.5	310.0	1068.5	0.9365	0.2720	1.2086	680.0
678.0	2672.5	0.03002	0.08440	0.11442	753.8	319.4	1073.2	0.9326	0.2807	1.2133	678.0
676.0	2636.8	0.02970	0.08799	0.11769	749.2	328.5	1077.6	0.9287	0.2892	1.2179	676.0
674.0	2601.5	0.02939	0.09156	0.12096	744.7	337.2	1081.9	0.9249	0.2974	1.2223	674.0
672.0	2566.6	0.02911	0.09514	0.12424	740.2	345.7	1085.9	0.9212	0.3054	1.2266	672.0
670.0	2532.2	0.02884	0.09871	0.12755	735.8	354.0	1089.8	0.9174	0.3133	1.2307	670.0
668.0	2498.1	0.02858	0.10229	0.13087	731.5	362.1	1093.5	0.9137	0.3210	1.2347	668.0
666.0	2464.4	0.02834	0.10588	0.13421	727.1	370.0	1097.1	0.9100	0.3286	1.2387	666.0
664.0	2431.1	0.02811	0.10947	0.13757	722.9	377.7	1100.6	0.9064	0.3361	1.2425	664.0
662.0	2398.2	0.02789	0.11306	0.14095	718.8	385.1	1103.9	0.9028	0.3434	1.2462	662.0

660.0	1.2498	0.3502	0.8995	1107.0	392.1	714.9	0.14431	0.11663	0.02768	2365.7	660.0
658.0	1.2533	0.3570	0.8963	1110.1	399.0	711.1	0.14771	0.12023	0.02748	2333.5	658.0
656.0	1.2567	0.3637	0.8931	1113.1	405.7	707.4	0.15115	0.12387	0.02728	2301.7	656.0
654.0	1.2601	0.3702	0.8899	1115.9	412.2	703.7	0.15463	0.12754	0.02709	2270.3	654.0
652.0	1.2634	0.3767	0.8868	1118.7	418.7	700.0	0.15816	0.13124	0.02691	2239.2	652.0
650.0	1.2667	0.3830	0.8837	1121.4	425.0	696.4	0.16173	0.13499	0.02674	2208.4	650.0
648.0	1.2699	0.3893	0.8806	1124.0	431.1	692.9	0.16534	0.13876	0.02657	2178.1	648.0
646.0	1.2730	0.3954	0.8776	1126.6	437.2	689.4	0.16899	0.14258	0.02641	2148.0	646.0
644.0	1.2761	0.4015	0.8746	1129.0	443.1	685.9	0.17269	0.14644	0.02625	2118.3	644.0
642.0	1.2791	0.4075	0.8716	1131.4	448.9	682.5	0.17643	0.15033	0.02610	2088.9	642.0
640.0	1.2821	0.4134	0.8686	1133.7	454.6	679.1	0.18021	0.15427	0.02595	2059.9	640.0
638.0	1.2850	0.4193	0.8657	1136.0	460.2	675.8	0.18405	0.15824	0.02580	2031.2	638.0
636.0	1.2879	0.4251	0.8628	1138.1	465.7	672.4	0.18792	0.16226	0.02566	2002.8	636.0
634.0	1.2907	0.4307	0.8599	1140.2	471.1	669.1	0.19185	0.16633	0.02553	1974.7	634.0
632.0	1.2934	0.4364	0.8571	1142.2	476.4	665.9	0.19583	0.17044	0.02539	1947.0	632.0
630.0	1.2962	0.4419	0.8542	1144.2	481.6	662.7	0.19986	0.17459	0.02526	1919.5	630.0
628.0	1.2988	0.4474	0.8514	1146.1	486.7	659.5	0.20394	0.17880	0.02514	1892.4	628.0
626.0	1.3015	0.4529	0.8486	1148.0	491.7	656.3	0.20807	0.18306	0.02501	1865.6	626.0
624.0	1.3041	0.4583	0.8458	1149.8	496.6	653.1	0.21226	0.18737	0.02489	1839.0	624.0
622.0	1.3066	0.4636	0.8430	1151.5	501.5	650.0	0.21650	0.19173	0.02477	1812.8	622.0
620.0	1.3092	0.4689	0.8403	1153.2	506.3	646.9	0.22081	0.19615	0.02466	1786.9	620.0
618.0	1.3117	0.4742	0.8375	1154.8	511.0	643.8	0.22517	0.20063	0.02455	1761.2	618.0
616.0	1.3141	0.4794	0.8348	1156.4	515.6	640.8	0.22960	0.20516	0.02444	1735.9	616.0
614.0	1.3166	0.4845	0.8321	1158.0	520.2	637.8	0.23409	0.20976	0.02433	1710.8	614.0
612.0	1.3190	0.4896	0.8294	1159.5	524.7	634.8	0.23865	0.21442	0.02422	1686.1	612.0
610.0	1.3214	0.4947	0.8267	1160.9	529.2	631.8	0.24327	0.21915	0.02412	1661.6	610.0
608.0	1.3238	0.4997	0.8240	1162.4	533.6	628.8	0.24796	0.22394	0.02402	1637.3	608.0
606.0	1.3261	0.5048	0.8214	1163.8	537.9	625.9	0.25273	0.22881	0.02392	1613.4	606.0
604.0	1.3284	0.5097	0.8187	1165.1	542.2	622.9	0.25757	0.23374	0.02382	1589.7	604.0
602.0	1.3307	0.5147	0.8161	1166.4	546.4	620.0	0.26248	0.23875	0.02373	1566.3	602.0
600.0	1.3330	0.5196	0.8134	1167.7	550.6	617.1	0.26747	0.24384	0.02364	1543.2	600.0
598.0	1.3353	0.5245	0.8108	1169.0	554.7	614.3	0.27255	0.24900	0.02354	1520.4	598.0
596.0	1.3375	0.5293	0.8082	1170.2	558.8	611.4	0.27770	0.25425	0.02345	1497.8	596.0
594.0	1.3398	0.5342	0.8056	1171.4	562.8	608.6	0.28294	0.25958	0.02337	1475.4	594.0
592.0	1.3420	0.5390	0.8030	1172.6	566.8	605.7	0.28827	0.26499	0.02328	1453.3	592.0
590.0	1.3442	0.5437	0.8004	1173.7	570.8	602.9	0.29368	0.27049	0.02319	1431.5	590.0
588.0	1.3464	0.5485	0.7978	1174.8	574.7	600.1	0.29919	0.27608	0.02311	1410.0	588.0
586.0	1.3485	0.5532	0.7953	1175.9	578.5	597.3	0.30478	0.28176	0.02303	1388.6	586.0
584.0	1.3507	0.5580	0.7927	1176.9	582.4	594.6	0.31048	0.28753	0.02295	1367.6	584.0
582.0	1.3528	0.5627	0.7902	1178.0	586.1	591.8	0.31627	0.29340	0.02287	1346.7	582.0
580.0	1.3550	0.5673	0.7876	1179.0	589.9	589.1	0.32216	0.29937	0.02279	1326.2	580.0

Table D-1. Properties of Saturated Steam and Saturated Water (Temperature) (*Continued*)

Temp. F	Press. psia	Volume, ft³/lbm			Enthalpy, Btu/lbm			Entropy, Btu/lbm × R			Temp. F
		Water v_f	Evap. v_{fg}	Steam v_g	Water h_f	Evap. h_{fg}	Steam h_g	Water s_f	Evap. s_{fg}	Steam s_g	
580.0	1332.17	0.02279	0.29937	0.32216	589.1	589.9	1179.0	0.7876	0.5673	1.3559	580.0
578.0	1305.84	0.02271	0.30544	0.32816	586.4	593.6	1179.9	0.7851	0.5720	1.3571	578.0
576.0	1285.74	0.02264	0.31162	0.33426	583.7	597.2	1180.9	0.7825	0.5766	1.3592	576.0
574.0	1265.89	0.02256	0.31790	0.34046	581.0	600.9	1181.8	0.7800	0.5813	1.3613	574.0
572.0	1246.26	0.02249	0.32429	0.34678	578.3	604.5	1182.7	0.7775	0.5859	1.3634	572.0
570.0	1226.88	0.02242	0.33079	0.35321	575.6	608.0	1183.6	0.7750	0.5905	1.3654	570.0
568.0	1207.72	0.02235	0.33741	0.35975	572.9	611.5	1184.5	0.7725	0.5950	1.3675	568.0
566.0	1188.80	0.02228	0.34414	0.36642	570.3	615.0	1185.3	0.7699	0.5996	1.3696	566.0
564.0	1170.10	0.02221	0.35099	0.37320	567.6	618.5	1186.1	0.7674	0.6041	1.3716	564.0
562.0	1151.63	0.02214	0.35797	0.38011	565.0	621.9	1186.9	0.7650	0.6087	1.3736	562.0
560.0	1135.38	0.02207	0.36507	0.38714	562.4	625.3	1187.7	0.7625	0.6132	1.3757	560.0
558.0	1115.36	0.02201	0.37230	0.39431	559.8	628.6	1188.4	0.7600	0.6177	1.3777	558.0
556.0	1097.55	0.02194	0.37966	0.40160	557.2	632.0	1189.2	0.7575	0.6222	1.3297	556.0
554.0	1079.96	0.02188	0.38715	0.40903	554.6	635.3	1189.9	0.7550	0.6267	1.3817	554.0
552.0	1062.59	0.02182	0.39479	0.41660	552.0	638.5	1190.6	0.7525	0.6311	1.3837	552.0
550.0	1045.43	0.02176	0.40256	0.42432	549.5	641.8	1191.2	0.7501	0.6356	1.3856	550.0
548.0	1028.49	0.02169	0.41048	0.43217	546.9	645.0	1191.9	0.7476	0.6400	1.3876	548.0
546.0	1011.75	0.02163	0.41855	0.44018	544.4	648.1	1192.5	0.7451	0.6445	1.3896	546.0
544.0	995.22	0.02157	0.42677	0.44834	541.8	651.3	1193.1	0.7427	0.6489	1.3915	544.0
542.0	978.90	0.02151	0.43514	0.45665	539.3	654.4	1193.7	0.7402	0.6533	1.3935	542.0
540.0	962.79	0.02146	0.44367	0.46513	536.8	657.5	1194.3	0.7378	0.6577	1.3954	540.0
538.0	946.88	0.02140	0.45237	0.47377	534.2	660.6	1194.8	0.7353	0.6621	1.3974	538.0
536.0	931.17	0.02134	0.46123	0.48257	531.7	663.6	1195.4	0.7329	0.6665	1.3993	536.0
534.0	915.66	0.02129	0.47026	0.49155	529.2	666.6	1195.9	0.7304	0.6708	1.4012	534.0
532.0	900.34	0.02123	0.47947	0.50070	526.8	669.6	1196.4	0.7280	0.6752	1.4032	532.0
530.0	885.23	0.02118	0.48886	0.51004	524.3	672.6	1196.9	0.7255	0.6796	1.4051	530.0
528.0	870.31	0.02112	0.49843	0.51955	521.8	675.5	1197.3	0.7231	0.6839	1.4070	528.0
526.0	855.58	0.02107	0.50819	0.52926	519.3	678.4	1197.8	0.7206	0.6883	1.4089	526.0
524.0	841.04	0.02102	0.51814	0.53916	516.9	681.3	1198.2	0.7182	0.6926	1.4108	524.0
522.0	826.69	0.02097	0.52829	0.54926	514.4	684.2	1198.6	0.7158	0.6969	1.4127	522.0
520.0	812.53	0.02091	0.53864	0.55956	512.0	687.0	1199.0	0.7133	0.7013	1.4146	520.0
518.0	798.55	0.02086	0.54920	0.57006	509.6	689.9	1199.4	0.7109	0.7056	1.4165	518.0
516.0	784.76	0.02081	0.55997	0.58079	507.1	692.7	1199.8	0.7085	0.7099	1.4183	516.0
514.0	771.15	0.02076	0.57096	0.59173	504.7	695.4	1200.2	0.7060	0.7142	1.4202	514.0
512.0	757.72	0.02072	0.58218	0.60289	502.3	698.2	1200.5	0.7036	0.7185	1.4221	512.0

Temp °F	s_g	s_{fg}	s_f	h_g	h_{fg}	h_f	v_g	v_{fg}	v_f	Press.	Temp °F
510.0	1.4240	0.7228	0.7012	1200.8	700.9	499.9	0.61429	0.59362	0.02067	744.47	510.0
508.0	1.4258	0.7271	0.6987	1201.1	703.7	497.5	0.62592	0.60530	0.02062	731.40	508.0
506.0	1.4277	0.7314	0.6963	1201.4	706.3	495.1	0.63779	0.61722	0.02057	718.50	506.0
504.0	1.4296	0.7357	0.6939	1201.7	709.0	492.7	0.64991	0.62938	0.02053	705.78	504.0
502.0	1.4314	0.7400	0.6915	1202.0	711.7	490.3	0.66228	0.64180	0.02048	693.23	502.0
500.0	1.4333	0.7443	0.6890	1202.2	714.3	487.9	0.67492	0.65448	0.02043	680.86	500.0
498.0	1.4352	0.7486	0.6866	1202.5	716.9	485.5	0.68782	0.66743	0.02039	668.65	498.0
496.0	1.4370	0.7528	0.6842	1202.7	719.5	483.2	0.70100	0.68065	0.02034	656.61	496.0
494.0	1.4389	0.7571	0.6818	1202.7	722.1	480.8	0.71445	0.69415	0.02030	644.73	494.0
492.0	1.4407	0.7614	0.6793	1203.1	724.6	478.5	0.72820	0.70794	0.02026	633.03	492.0
490.0	1.4426	0.7657	0.6769	1203.3	727.2	476.1	0.74224	0.72203	0.02021	621.48	490.0
488.0	1.4444	0.7700	0.6745	1203.5	729.7	473.8	0.75658	0.73641	0.02017	610.10	488.0
486.0	1.4463	0.7742	0.6721	1203.7	732.2	471.5	0.77124	0.75111	0.02013	598.87	486.0
484.0	1.4481	0.7785	0.6696	1203.8	734.7	469.1	0.78622	0.76613	0.02009	587.81	484.0
482.0	1.4500	0.7828	0.6672	1204.0	737.2	466.8	0.80152	0.78148	0.02004	576.90	482.0
480.0	1.4518	0.7871	0.6648	1204.1	739.6	464.5	0.81717	0.79716	0.02000	566.15	480.0
478.0	1.4537	0.7913	0.6624	1204.2	742.1	462.2	0.83315	0.81319	0.01996	555.55	478.0
476.0	1.4555	0.7956	0.6599	1204.4	744.5	459.9	0.84950	0.82958	0.01992	545.11	476.0
474.0	1.4574	0.7999	0.6575	1204.4	746.9	457.5	0.86621	0.84632	0.01988	534.81	474.0
472.0	1.4592	0.8042	0.6551	1204.5	749.3	455.2	0.88329	0.86345	0.01984	524.67	472.0
470.0	1.4611	0.8084	0.6527	1204.6	751.6	452.9	0.90076	0.88095	0.01980	514.67	470.0
468.0	1.4629	0.8127	0.6502	1204.6	754.0	450.7	0.91862	0.89885	0.01976	504.83	468.0
466.0	1.4648	0.8170	0.6478	1204.7	756.3	448.4	0.93689	0.91716	0.01973	495.12	466.0
464.0	1.4667	0.8213	0.6454	1204.7	758.6	446.1	0.95557	0.93588	0.01969	485.56	464.0
462.0	1.4685	0.8256	0.6429	1204.8	761.0	443.8	0.97469	0.95504	0.01965	476.14	462.0
460.0	1.4704	0.8299	0.6405	1204.8	763.2	441.5	0.99424	0.97463	0.01961	466.87	460.0
458.0	1.4722	0.8342	0.6381	1204.8	765.5	439.3	1.01425	0.99467	0.01958	457.73	458.0
456.0	1.4741	0.8385	0.6356	1204.8	767.8	437.0	1.03472	1.01518	0.01954	448.73	456.0
454.0	1.4759	0.8428	0.6332	1204.8	770.0	434.7	1.05567	1.03616	0.01950	439.87	454.0
452.0	1.4778	0.8471	0.6308	1204.8	772.3	432.5	1.07711	1.05764	0.01947	431.14	452.0
450.0	1.4797	0.8514	0.6283	1204.7	774.5	430.2	1.09905	1.07962	0.01943	422.55	450.0
448.0	1.4815	0.8557	0.6259	1204.7	776.7	428.0	1.12152	1.10212	0.01940	414.09	448.0
446.0	1.4834	0.8600	0.6234	1204.6	778.9	425.7	1.14452	1.12515	0.01936	405.76	446.0
444.0	1.4853	0.8643	0.6210	1204.6	781.1	423.5	1.16806	1.14874	0.01933	397.56	444.0
442.0	1.4872	0.8686	0.6185	1204.5	783.2	421.3	1.19217	1.17288	0.01929	389.49	442.0
440.0	1.4890	0.8729	0.6161	1204.4	785.4	419.0	1.21687	1.19761	0.01926	381.54	440.0
438.0	1.4909	0.8773	0.6136	1204.3	787.5	416.8	1.24216	1.22293	0.01923	373.72	438.0
436.0	1.4928	0.8816	0.6112	1204.2	789.7	414.6	1.26806	1.24887	0.01919	366.03	436.0
434.0	1.4947	0.8859	0.6087	1204.1	791.8	412.4	1.29460	1.27544	0.01916	358.46	434.0
432.0	1.4966	0.8903	0.6063	1204.0	793.9	410.1	1.32179	1.30266	0.01913	351.00	432.0
430.0	1.4985	0.8946	0.6038	1203.9	796.0	407.9	1.34965	1.33055	0.01909	343.67	430.0

Table D-1. Properties of Saturated Steam and Saturated Water (Temperature) (Continued)

Temp. F	Press. psia	Volume, ft³/lbm			Enthalpy, Btu/lbm			Entropy, Btu/lbm × F			Temp. F
		Water v_f	Evap. v_{fg}	Steam v_g	Water h_f	Evap. h_{fg}	Steam h_g	Water s_f	Evap. s_{fg}	Steam s_g	
430.0	343.674	0.01909	1.3306	1.3496	407.9	796.0	1203.9	0.6038	0.8946	1.4985	430.0
428.0	336.463	0.01906	1.3591	1.3782	405.7	798.0	1203.7	0.6014	0.8990	1.5004	428.0
426.0	329.369	0.01903	1.3884	1.4075	403.5	800.1	1203.6	0.5989	0.9034	1.5023	426.0
424.0	322.391	0.01900	1.4184	1.4374	401.3	802.2	1203.5	0.5964	0.9077	1.5042	424.0
422.0	315.529	0.01897	1.4492	1.4682	399.1	804.2	1203.3	0.5940	0.9121	1.5061	422.0
420.0	308.780	0.01894	1.4808	1.4997	396.9	806.2	1203.1	0.5915	0.9165	1.5080	420.0
418.0	302.143	0.01890	1.5131	1.5320	394.7	808.2	1202.9	0.5890	0.9209	1.5099	418.0
416.0	295.617	0.01887	1.5463	1.5651	392.5	810.2	1202.8	0.5866	0.9253	1.5118	416.0
414.0	289.201	0.01884	1.5803	1.5991	390.3	812.2	1202.6	0.5841	0.9297	1.5137	414.0
412.0	282.894	0.01881	1.6152	1.6340	388.1	814.2	1202.4	0.5816	0.9341	1.5157	412.0
410.0	276.694	0.01878	1.6510	1.6697	386.0	816.2	1202.1	0.5791	0.9385	1.5176	410.0
408.0	270.600	0.01875	1.6877	1.7064	383.8	818.2	1201.9	0.5766	0.9429	1.5195	408.0
406.0	264.611	0.01872	1.7253	1.7441	381.6	820.1	1201.7	0.5742	0.9473	1.5215	406.0
404.0	258.725	0.01870	1.7640	1.7827	379.4	822.1	1201.5	0.5717	0.9518	1.5234	404.0
402.0	252.942	0.01867	1.8037	1.8223	377.3	824.0	1201.2	0.5692	0.9562	1.5254	402.0
400.0	247.259	0.01864	1.8444	1.8630	375.1	825.9	1201.0	0.5667	0.9607	1.5274	400.0
398.0	241.677	0.01861	1.8862	1.9048	372.9	827.8	1200.7	0.5642	0.9651	1.5293	398.0
396.0	236.193	0.01858	1.9291	1.9477	370.8	829.7	1200.4	0.5617	0.9696	1.5313	396.0
394.0	230.807	0.01855	1.9731	1.9917	368.6	831.6	1200.2	0.5592	0.9741	1.5333	394.0
392.0	225.516	0.01853	2.0184	2.0369	366.5	833.4	1199.9	0.5567	0.9786	1.5352	392.0
390.0	220.321	0.01850	2.0649	2.0833	364.3	835.3	1199.6	0.5542	0.9831	1.5372	390.0
388.0	215.220	0.01847	2.1126	2.1311	362.2	837.2	1199.3	0.5516	0.9876	1.5392	388.0
386.0	210.211	0.01844	2.1616	2.1801	360.0	839.0	1199.0	0.5491	0.9921	1.5412	386.0
384.0	205.294	0.01842	2.2120	2.2304	357.9	840.8	1198.7	0.5466	0.9966	1.5432	384.0
382.0	200.467	0.01839	2.2638	2.2821	355.7	842.7	1198.4	0.5441	1.0012	1.5452	382.0
380.0	195.729	0.01836	2.3170	2.3353	353.6	844.5	1198.0	0.5416	1.0057	1.5473	380.0
378.0	191.080	0.01834	2.3716	2.3900	351.4	846.3	1197.7	0.5390	1.0103	1.5493	378.0
376.0	186.517	0.01831	2.4279	2.4462	349.3	848.1	1197.4	0.5365	1.0148	1.5513	376.0
374.0	182.040	0.01829	2.4857	2.5039	347.2	849.8	1197.0	0.5340	1.0194	1.5534	374.0
372.0	177.648	0.01826	2.5451	2.5633	345.0	851.6	1196.7	0.5314	1.0240	1.5554	372.0
370.0	173.339	0.01823	2.6062	2.6244	342.9	853.4	1196.3	0.5289	1.0286	1.5575	370.0
368.0	169.113	0.01821	2.6691	2.6873	340.8	855.1	1195.9	0.5263	1.0332	1.5595	368.0
366.0	164.968	0.01818	2.7337	2.7519	338.7	856.9	1195.6	0.5238	1.0378	1.5616	366.0
364.0	160.903	0.01816	2.8002	2.8184	336.5	858.6	1195.2	0.5212	1.0424	1.5637	364.0
362.0	156.917	0.01813	2.8687	2.8868	334.4	860.4	1194.8	0.5187	1.0471	1.5658	362.0

Temp											Temp
360.0	1.5678	1.0517	0.5161	1194.4	862.1	332.3	2.9573	2.9392	0.01811	153.010	360.0
358.0	1.5699	1.0564	0.5135	1194.0	863.8	330.2	3.0298	3.0117	0.01809	149.179	358.0
356.0	1.5721	1.0611	0.5110	1193.6	865.5	328.1	3.1044	3.0863	0.01806	145.424	356.0
354.0	1.5742	1.0658	0.5084	1193.2	867.2	326.0	3.1812	3.1632	0.01804	141.744	354.0
352.0	1.5763	1.0705	0.5058	1192.7	868.9	323.9	3.2603	3.2423	0.01801	138.138	352.0
350.0	1.5784	1.0752	0.5032	1192.3	870.6	321.8	3.3418	3.3238	0.01799	134.604	350.0
348.0	1.5806	1.0799	0.5006	1191.9	872.2	319.7	3.4258	3.4078	0.01797	131.142	348.0
346.0	1.5827	1.0847	0.4980	1191.4	873.9	317.6	3.5122	3.4943	0.01794	127.751	346.0
344.0	1.5849	1.0894	0.4954	1191.0	875.5	315.5	3.6013	3.5834	0.01792	124.430	344.0
342.0	1.5871	1.0942	0.4928	1190.5	877.2	313.4	3.6931	3.6752	0.01790	121.177	342.0
340.0	1.5892	1.0990	0.4902	1190.1	878.8	311.3	3.7878	3.7699	0.01787	117.992	340.0
338.0	1.5914	1.1038	0.4876	1189.6	880.5	309.2	3.8853	3.8675	0.01785	114.873	338.0
336.0	1.5936	1.1086	0.4850	1189.1	882.1	307.1	3.9859	3.9681	0.01783	111.820	336.0
334.0	1.5958	1.1134	0.4824	1188.7	883.7	305.0	4.0896	4.0718	0.01781	108.832	334.0
332.0	1.5981	1.1183	0.4798	1188.2	885.3	302.9	4.1966	4.1788	0.01779	105.907	332.0
330.0	1.6003	1.1231	0.4772	1187.7	886.9	300.8	4.3069	4.2892	0.01776	103.045	330.0
328.0	1.6025	1.1280	0.4745	1187.2	888.5	298.7	4.4208	4.4030	0.01774	100.245	328.0
326.0	1.6048	1.1329	0.4719	1186.7	890.1	296.6	4.5382	4.5205	0.01772	97.506	326.0
324.0	1.6071	1.1378	0.4692	1186.2	891.6	294.6	4.6595	4.6418	0.01770	94.826	324.0
322.0	1.6093	1.1427	0.4666	1185.7	893.2	292.5	4.7846	4.7669	0.01768	92.205	322.0
320.0	1.6116	1.1477	0.4640	1185.2	894.8	290.4	4.9138	4.8961	0.01766	89.643	320.0
318.0	1.6139	1.1526	0.4613	1184.7	896.3	288.3	5.0471	5.0295	0.01764	87.137	318.0
316.0	1.6162	1.1576	0.4586	1184.1	897.9	286.3	5.1849	5.1673	0.01761	84.688	316.0
314.0	1.6185	1.1626	0.4560	1183.6	899.4	284.2	5.3272	5.3096	0.01759	82.293	314.0
312.0	1.6209	1.1676	0.4533	1183.1	901.0	282.1	5.4742	5.4566	0.01757	79.953	312.0
310.0	1.6232	1.1726	0.4506	1182.5	902.5	280.0	5.6260	5.6085	0.01755	77.667	310.0
308.0	1.6256	1.1776	0.4479	1182.0	904.0	278.0	5.7830	5.7655	0.01753	75.433	308.0
306.0	1.6279	1.1827	0.4453	1181.4	905.5	275.9	5.9452	5.9277	0.01751	73.251	306.0
304.0	1.6303	1.1877	0.4426	1180.9	907.0	273.8	6.1130	6.0955	0.01749	71.119	304.0
302.0	1.6327	1.1928	0.4399	1180.3	908.5	271.8	6.2864	6.2689	0.01747	69.038	302.0
300.0	1.6351	1.1979	0.4372	1179.7	910.0	269.7	6.4650	6.4483	0.01745	67.005	300.0
298.0	1.6375	1.2031	0.4345	1179.2	911.5	267.7	6.6513	6.6339	0.01743	65.021	298.0
296.0	1.6400	1.2082	0.4317	1178.6	913.0	265.6	6.8433	6.8259	0.01741	63.084	296.0
294.0	1.6424	1.2134	0.4290	1178.0	914.5	263.5	7.0419	7.0245	0.01739	61.194	294.0
292.0	1.6449	1.2186	0.4263	1177.4	915.9	261.5	7.2475	7.2301	0.01738	59.350	292.0
290.0	1.6473	1.2238	0.4236	1176.8	917.4	259.4	7.4603	7.4430	0.01736	57.550	290.0
288.0	1.6498	1.2290	0.4208	1176.2	918.8	257.4	7.6807	7.6634	0.01734	55.795	288.0
286.0	1.6523	1.2342	0.4181	1175.6	920.3	255.3	7.9089	7.8916	0.01732	54.083	286.0
284.0	1.6548	1.2395	0.4154	1175.0	921.7	253.3	8.1453	8.1280	0.01730	52.414	284.0
282.0	1.6574	1.2448	0.4126	1174.4	923.2	251.2	8.3902	8.3729	0.01728	50.786	282.0
280.0	1.6599	1.2501	0.4098	1173.8	924.6	249.2	8.6439	8.6267	0.01726	49.200	280.0

Table D-1. Properties of Saturated Steam and Saturated Water (Temperature) (*Continued*)

Temp. F	Press. psia	Volume, ft³/lbm			Enthalpy, Btu/lbm			Entropy, Btu/lbm × R			Temp. F
		Water v_f	Evap. v_{fg}	Steam v_g	Water h_f	Evap. h_{fg}	Steam h_g	Water s_f	Evap. s_{fg}	Steam s_g	
280.0	49.200	0.017264	8.627	8.644	249.17	924.6	1173.8	0.4098	1.2501	1.6599	280.0
278.0	47.653	0.017246	8.890	8.907	247.13	926.0	1173.2	0.4071	1.2554	1.6625	278.0
276.0	46.147	0.017228	9.162	9.180	245.08	927.5	1172.5	0.4043	1.2607	1.6650	276.0
274.0	44.678	0.017210	9.445	9.462	243.03	928.9	1171.9	0.4015	1.2661	1.6676	274.0
272.0	43.249	0.017193	9.738	9.755	240.99	930.3	1171.3	0.3987	1.2715	1.6702	272.0
270.0	41.856	0.017175	10.042	10.060	238.95	931.7	1170.6	0.3960	1.2769	1.6729	270.0
268.0	40.500	0.017157	10.358	10.375	236.91	933.1	1170.0	0.3932	1.2823	1.6755	268.0
266.0	39.179	0.017140	10.685	10.703	234.87	934.5	1169.3	0.3904	1.2878	1.6781	266.0
264.0	37.894	0.017123	11.025	11.042	232.83	935.9	1168.7	0.3876	1.2933	1.6808	264.0
262.0	36.644	0.017106	11.378	11.395	230.79	937.3	1168.0	0.3847	1.2988	1.6835	262.0
260.0	35.427	0.017089	11.745	11.762	228.76	938.6	1167.4	0.3819	1.3043	1.6862	260.0
258.0	34.243	0.017072	12.125	12.142	226.72	940.0	1166.7	0.3791	1.3098	1.6889	258.0
256.0	33.091	0.017055	12.521	12.538	224.69	941.4	1166.1	0.3763	1.3154	1.6917	256.0
254.0	31.972	0.017039	12.931	12.948	222.65	942.7	1165.4	0.3734	1.3210	1.6944	254.0
252.0	30.883	0.017022	13.358	13.375	220.62	944.1	1164.7	0.3706	1.3266	1.6972	252.0
250.0	29.825	0.017006	13.802	13.819	218.59	945.4	1164.0	0.3677	1.3323	1.7000	250.0
248.0	28.796	0.016990	14.264	14.281	216.56	946.8	1163.4	0.3649	1.3379	1.7028	248.0
246.0	27.797	0.016974	14.744	14.761	214.53	948.1	1162.7	0.3620	1.3436	1.7056	246.0
244.0	26.826	0.016958	15.243	15.260	212.50	949.5	1162.0	0.3591	1.3494	1.7085	244.0
242.0	25.883	0.016942	15.763	15.780	210.48	950.8	1161.3	0.3562	1.3551	1.7113	242.0
240.0	24.968	0.016926	16.304	16.321	208.45	952.1	1160.6	0.3533	1.3609	1.7142	240.0
238.0	24.079	0.016910	16.867	16.884	206.42	953.5	1159.9	0.3505	1.3667	1.7171	238.0
236.0	23.216	0.016895	17.454	17.471	204.40	954.8	1159.2	0.3476	1.3725	1.7201	236.0
234.0	22.379	0.016880	18.065	18.082	202.38	956.1	1158.5	0.3446	1.3784	1.7230	234.0
232.0	21.567	0.016864	18.701	18.718	200.35	957.4	1157.8	0.3417	1.3842	1.7260	232.0
230.0	20.779	0.016849	19.364	19.381	198.33	958.7	1157.1	0.3388	1.3902	1.7290	230.0
229.0	20.394	0.016842	19.707	19.723	197.32	959.4	1156.7	0.3373	1.3931	1.7305	229.0
228.0	20.015	0.016834	20.056	20.073	196.31	960.0	1156.3	0.3359	1.3961	1.7320	228.0
227.0	19.642	0.016827	20.413	20.429	195.30	960.7	1156.0	0.3344	1.3991	1.7335	227.0
226.0	19.274	0.016819	20.777	20.794	194.29	961.3	1155.6	0.3329	1.4021	1.7350	226.0
225.0	18.912	0.016812	21.149	21.166	193.28	962.0	1155.3	0.3315	1.4051	1.7365	225.0
224.0	18.556	0.016805	21.529	21.545	192.27	962.6	1154.9	0.3300	1.4081	1.7380	224.0
223.0	18.206	0.016797	21.917	21.933	191.26	963.3	1154.5	0.3285	1.4111	1.7396	223.0
222.0	17.860	0.016790	22.313	22.330	190.25	963.9	1154.2	0.3270	1.4141	1.7411	222.0
221.0	17.521	0.016783	22.718	22.735	189.24	964.6	1153.8	0.3255	1.4171	1.7427	221.0

Temp											Temp
220.0	1.7442	1.4201	0.3241	1153.4	965.2	188.23	23.148	23.131	0.016775	17.186	220.0
219.0	1.7458	1.4232	0.3226	1153.1	965.8	187.22	23.571	23.554	0.016768	16.857	219.0
218.0	1.7473	1.4262	0.3211	1152.7	966.5	186.21	24.002	23.986	0.016761	16.533	218.0
217.0	1.7489	1.4293	0.3196	1152.3	967.1	185.21	24.444	24.427	0.016754	16.214	217.0
216.0	1.7505	1.4323	0.3181	1152.0	967.8	184.20	24.894	24.878	0.016747	15.901	216.0
215.0	1.7520	1.4354	0.3166	1151.6	968.4	183.19	25.355	25.338	0.016740	15.592	215.0
214.0	1.7536	1.4385	0.3151	1151.2	969.0	182.18	25.826	25.809	0.016733	15.289	214.0
213.0	1.7552	1.4416	0.3136	1150.8	969.7	181.17	26.307	26.290	0.016726	14.990	213.0
212.0	1.7568	1.4447	0.3121	1150.5	970.3	180.17	26.799	26.782	0.016719	14.696	212.0
211.0	1.7584	1.4478	0.3106	1150.1	970.9	179.16	27.302	27.285	0.016712	14.407	211.0
210.0	1.7600	1.4509	0.3091	1149.7	971.6	178.15	27.816	27.799	0.016705	14.123	210.0
209.0	1.7616	1.4540	0.3076	1149.4	972.2	177.14	28.341	28.324	0.016698	13.843	209.0
208.0	1.7632	1.4571	0.3061	1149.0	972.8	176.14	28.878	28.862	0.016691	13.568	208.0
207.0	1.7649	1.4602	0.3046	1148.6	973.5	175.13	29.428	29.411	0.016684	13.297	207.0
206.0	1.7665	1.4634	0.3031	1148.2	974.1	174.12	29.989	29.973	0.016677	13.031	206.0
205.0	1.7681	1.4665	0.3016	1147.9	974.7	173.12	30.564	30.547	0.016670	12.770	205.0
204.0	1.7698	1.4697	0.3001	1147.5	975.4	172.11	31.151	31.135	0.016664	12.512	204.0
203.0	1.7714	1.4728	0.2986	1147.1	976.0	171.10	31.752	31.736	0.016657	12.259	203.0
202.0	1.7731	1.4760	0.2971	1146.7	976.7	170.10	32.367	32.350	0.016650	12.011	202.0
201.0	1.7747	1.4792	0.2955	1146.3	977.2	169.09	32.996	32.979	0.016643	11.766	201.0
200.0	1.7764	1.4824	0.2940	1146.0	977.9	168.09	33.639	33.622	0.016637	11.526	200.0
199.0	1.7781	1.4856	0.2925	1145.6	978.5	167.08	34.297	34.280	0.016630	11.290	199.0
198.0	1.7798	1.4888	0.2910	1145.2	979.1	166.08	34.970	34.954	0.016624	11.058	198.0
197.0	1.7814	1.4920	0.2894	1144.8	979.7	165.07	35.659	35.643	0.016617	10.830	197.0
196.0	1.7831	1.4952	0.2879	1144.4	980.4	164.06	36.364	36.348	0.016611	10.605	196.0
195.0	1.7848	1.4985	0.2864	1144.0	981.0	163.06	37.086	37.069	0.016604	10.385	195.0
194.0	1.7865	1.5017	0.2848	1143.7	981.6	162.05	37.824	37.808	0.016598	10.168	194.0
193.0	1.7882	1.5050	0.2833	1143.3	982.2	161.05	38.580	38.564	0.016591	9.956	193.0
192.0	1.7900	1.5082	0.2818	1142.9	982.8	160.05	39.354	39.337	0.016585	9.747	192.0
191.0	1.7917	1.5115	0.2802	1142.5	983.5	159.04	40.146	40.130	0.016578	9.541	191.0
190.0	1.7934	1.5148	0.2787	1142.1	984.1	158.04	40.957	40.941	0.016572	9.340	190.0
189.0	1.7952	1.5180	0.2771	1141.7	984.7	157.03	41.787	41.771	0.016566	9.141	189.0
188.0	1.7969	1.5213	0.2756	1141.3	985.3	156.03	42.638	42.621	0.016559	8.947	188.0
187.0	1.7987	1.5246	0.2740	1140.9	985.9	155.02	43.508	43.492	0.016553	8.756	187.0
186.0	1.8004	1.5279	0.2725	1140.5	986.5	154.02	44.400	44.383	0.016547	8.568	186.0
185.0	1.8022	1.5313	0.2709	1140.2	987.1	153.02	45.313	45.297	0.016541	8.384	185.0
184.0	1.8040	1.5346	0.2694	1139.8	987.8	152.01	46.249	46.232	0.016534	8.203	184.0
183.0	1.8057	1.5379	0.2678	1139.4	988.4	151.01	47.207	47.190	0.016528	8.025	183.0
182.0	1.8075	1.5413	0.2662	1139.0	989.0	150.01	48.189	48.172	0.016522	7.850	182.0
181.0	1.8093	1.5446	0.2647	1138.6	989.6	149.00	49.194	49.178	0.016516	7.679	181.0
180.0	1.8111	1.5480	0.2631	1138.2	990.2	148.00	50.225	50.208	0.016510	7.511	180.0

Table D-1. Properties of Saturated Steam and Saturated Water (Temperature) (Continued)

Temp. F	Press. psia	Volume, ft³/lbm			Enthalpy, Btu/lbm			Entropy, Btu/lbm × R			Temp. F
		Water v_f	Evap. v_{fg}	Steam v_g	Water h_f	Evap. h_{fg}	Steam h_g	Water s_f	Evap. s_{fg}	Steam s_g	
180.0	7.5110	0.016510	50.21	50.22	148.00	990.2	1138.2	0.2631	1.5480	1.8111	180.0
179.0	7.3460	0.016504	51.26	51.28	147.00	990.8	1137.8	0.2615	1.5514	1.8129	179.0
178.0	7.1840	0.016498	52.35	52.36	145.99	991.4	1137.4	0.2600	1.5548	1.8147	178.0
177.0	7.0250	0.016492	53.46	53.47	144.99	992.0	1137.0	0.2584	1.5582	1.8166	177.0
176.0	6.8690	0.016486	54.59	54.61	143.99	992.6	1136.6	0.2568	1.5616	1.8184	176.0
175.0	6.7159	0.016480	55.76	55.77	142.99	993.2	1136.2	0.2552	1.5650	1.8202	175.0
174.0	6.5656	0.016474	56.95	56.97	141.98	993.8	1135.8	0.2537	1.5684	1.8221	174.0
173.0	6.4182	0.016468	58.18	58.19	140.98	994.4	1135.4	0.2521	1.5718	1.8239	173.0
172.0	6.2736	0.016463	59.43	59.45	139.98	995.0	1135.0	0.2505	1.5753	1.8258	172.0
171.0	6.1318	0.016457	60.72	60.74	138.98	995.6	1134.6	0.2489	1.5787	1.8276	171.0
170.0	5.9926	0.016451	62.04	62.06	137.97	996.2	1134.2	0.2473	1.5822	1.8295	170.0
169.0	5.8562	0.016445	63.39	63.41	136.97	996.8	1133.8	0.2457	1.5857	1.8314	169.0
168.0	5.7223	0.016440	64.78	64.80	135.97	997.4	1133.4	0.2441	1.5892	1.8333	168.0
167.0	5.5911	0.016434	66.21	66.22	134.97	998.0	1133.0	0.2425	1.5926	1.8352	167.0
166.0	5.4623	0.016428	67.67	67.68	133.97	998.6	1132.6	0.2409	1.5961	1.8371	166.0
165.0	5.3361	0.016423	69.17	69.18	132.96	999.2	1132.2	0.2393	1.5997	1.8390	165.0
164.0	5.2124	0.016417	70.70	70.72	131.96	999.8	1131.8	0.2377	1.6032	1.8409	164.0
163.0	5.0911	0.016412	72.28	72.30	130.96	1000.4	1131.4	0.2361	1.6067	1.8428	163.0
162.0	4.9722	0.016406	73.90	73.92	129.96	1001.0	1131.0	0.2345	1.6103	1.8448	162.0
161.0	4.8556	0.016401	75.56	75.58	128.96	1001.6	1130.6	0.2329	1.6138	1.8467	161.0
160.0	4.7414	0.016395	77.27	77.29	127.96	1002.2	1130.2	0.2313	1.6174	1.8487	160.0
159.0	4.6294	0.016390	79.02	79.04	126.96	1002.8	1129.8	0.2297	1.6210	1.8506	159.0
158.0	4.5197	0.016384	80.82	80.83	125.96	1003.4	1129.4	0.2281	1.6245	1.8526	158.0
157.0	4.4122	0.016379	82.66	82.68	124.95	1004.0	1129.0	0.2264	1.6281	1.8546	157.0
156.0	4.3068	0.016374	84.56	84.57	123.95	1004.6	1128.6	0.2248	1.6318	1.8566	156.0
155.0	4.2036	0.016369	86.50	86.52	122.95	1005.2	1128.2	0.2232	1.6354	1.8586	155.0
154.0	4.1025	0.016363	88.50	88.52	121.95	1005.8	1127.7	0.2216	1.6390	1.8606	154.0
153.0	4.0035	0.016358	90.55	90.57	120.95	1006.4	1127.3	0.2199	1.6426	1.8626	153.0
152.0	3.9065	0.016353	92.66	92.68	119.95	1007.0	1126.9	0.2183	1.6463	1.8646	152.0
151.0	3.8114	0.016348	94.83	94.84	118.95	1007.6	1126.5	0.2167	1.6500	1.8666	151.0
150.0	3.7184	0.016343	97.05	97.07	117.95	1008.2	1126.1	0.2150	1.6536	1.8686	150.0
149.0	3.6273	0.016337	99.33	99.35	116.95	1008.7	1125.7	0.2134	1.6573	1.8707	149.0
148.0	3.5381	0.016332	101.68	101.70	115.95	1009.3	1125.3	0.2117	1.6610	1.8727	148.0
147.0	3.4508	0.016327	104.10	104.11	114.95	1009.9	1124.9	0.2101	1.6647	1.8748	147.0
146.0	3.3653	0.016322	106.58	106.59	113.95	1010.5	1124.5	0.2084	1.6684	1.8769	146.0

145.0	1.8789	1.6722	0.2068	1124.0	1011.1	112.95	109.14	109.12	0.016317	3.2816	145.0
144.0	1.8810	1.6759	0.2051	1123.6	1011.7	111.95	111.76	111.74	0.016312	3.1997	144.0
143.0	1.8831	1.6797	0.2035	1123.2	1012.3	110.95	114.45	114.44	0.016308	3.1195	143.0
142.0	1.8852	1.6834	0.2018	1122.8	1012.9	109.95	117.22	117.21	0.016303	3.0411	142.0
141.0	1.8873	1.6872	0.2001	1122.4	1013.4	108.95	120.07	120.05	0.016298	2.9643	141.0
140.0	1.8895	1.6910	0.1985	1122.0	1014.0	107.95	123.00	122.98	0.016293	2.8892	140.0
139.0	1.8916	1.6948	0.1968	1121.6	1014.6	106.95	126.01	125.99	0.016288	2.8157	139.0
138.0	1.8937	1.6986	0.1951	1121.1	1015.2	105.95	129.11	129.09	0.016284	2.7438	138.0
137.0	1.8959	1.7024	0.1935	1120.7	1015.8	104.95	132.29	132.28	0.016279	2.6735	137.0
136.0	1.8980	1.7063	0.1918	1120.3	1016.4	103.95	135.57	135.55	0.016274	2.6047	136.0
135.0	1.9002	1.7101	0.1901	1119.9	1016.9	102.95	138.94	138.93	0.016270	2.5375	135.0
134.0	1.9024	1.7140	0.1884	1119.5	1017.5	101.95	142.41	142.40	0.016265	2.4717	134.0
133.0	1.9046	1.7178	0.1867	1119.1	1018.1	100.95	145.98	145.97	0.016260	2.4074	133.0
132.0	1.9068	1.7217	0.1851	1118.6	1018.7	99.95	149.66	149.64	0.016256	2.3445	132.0
131.0	1.9090	1.7256	0.1834	1118.2	1019.3	98.95	153.44	153.42	0.016251	2.2830	131.0
130.0	1.9112	1.7295	0.1817	1117.8	1019.9	97.96	157.33	157.32	0.016247	2.2230	130.0
129.0	1.9134	1.7335	0.1800	1117.4	1020.4	96.96	161.32	161.32	0.016243	2.1642	129.0
128.0	1.9157	1.7374	0.1783	1117.0	1021.0	95.96	165.47	165.45	0.016238	2.1068	128.0
127.0	1.9179	1.7413	0.1766	1116.5	1021.6	94.96	169.72	169.70	0.016234	2.0507	127.0
126.0	1.9202	1.7453	0.1749	1116.1	1022.2	93.96	174.09	174.08	0.016229	1.9959	126.0
125.0	1.9224	1.7493	0.1732	1115.7	1022.7	92.96	178.60	178.58	0.016225	1.9424	125.0
124.0	1.9247	1.7533	0.1715	1115.3	1023.3	91.96	183.24	183.23	0.016221	1.8901	124.0
123.0	1.9270	1.7573	0.1697	1114.9	1023.9	90.96	188.03	188.01	0.016217	1.8390	123.0
122.0	1.9293	1.7613	0.1680	1114.4	1024.5	89.96	192.95	192.94	0.016213	1.7891	122.0
121.0	1.9316	1.7653	0.1663	1114.0	1025.0	88.96	198.03	198.01	0.016208	1.7403	121.0
120.0	1.9339	1.7693	0.1646	1113.6	1025.6	87.97	203.26	203.25	0.016204	1.6927	120.0
119.0	1.9362	1.7734	0.1629	1113.2	1026.2	86.97	208.66	208.64	0.016200	1.6463	119.0
118.0	1.9386	1.7774	0.1611	1112.7	1026.8	85.97	214.21	214.20	0.016196	1.6009	118.0
117.0	1.9409	1.7815	0.1594	1112.3	1027.3	84.97	219.94	219.93	0.016192	1.5566	117.0
116.0	1.9433	1.7856	0.1577	1111.9	1027.9	83.97	225.85	225.84	0.016188	1.5133	116.0
115.0	1.9457	1.7897	0.1559	1111.5	1028.5	82.97	231.94	231.93	0.016184	1.4711	115.0
114.0	1.9480	1.7938	0.1542	1111.0	1029.1	81.97	238.22	238.21	0.016180	1.4299	114.0
113.0	1.9504	1.7980	0.1525	1110.6	1029.6	80.98	244.70	244.69	0.016177	1.3898	113.0
112.0	1.9528	1.8021	0.1507	1110.2	1030.2	79.98	251.38	251.37	0.016173	1.3505	112.0
111.0	1.9552	1.8063	0.1490	1109.8	1030.8	78.98	258.28	258.26	0.016169	1.3123	111.0
110.0	1.9577	1.8105	0.1472	1109.3	1031.4	77.98	265.39	265.37	0.016165	1.2750	110.0
109.0	1.9601	1.8146	0.1455	1108.9	1031.9	76.98	272.72	272.71	0.016162	1.2385	109.0
108.0	1.9626	1.8188	0.1437	1108.5	1032.5	75.98	280.30	280.28	0.016158	1.2030	108.0
107.0	1.9650	1.8231	0.1419	1108.1	1033.1	74.99	288.11	288.09	0.016154	1.1684	107.0
106.0	1.9675	1.8273	0.1402	1107.6	1033.6	73.99	296.18	296.16	0.016151	1.1347	106.0
105.0	1.9700	1.8315	0.1384	1107.2	1034.2	72.99	304.50	304.49	0.016147	1.1017	105.0

Table D-1. Properties of Saturated Steam and Saturated Water (Temperature) *(Continued)*

Temp. F	Press. psia	Volume, ft³/lbm			Enthalpy, Btu/lbm			Entropy, Btu/lbm × R			Temp. F
		Water v_f	Evap. v_{fg}	Steam v_g	Water h_f	Evap. h_{fg}	Steam h_g	Water s_f	Evap. s_{fg}	Steam s_g	
105.0	1.10174	0.016147	304.5	304.5	72.990	1034.2	1107.2	0.1384	1.8315	1.9700	105.0
104.0	1.06965	0.016144	313.1	313.1	71.992	1034.8	1106.8	0.1366	1.8358	1.9725	104.0
103.0	1.03838	0.016140	322.0	322.0	70.993	1035.4	1106.3	0.1349	1.8401	1.9750	103.0
102.0	1.00789	0.016137	331.1	331.1	69.995	1035.9	1105.9	0.1331	1.8444	1.9775	102.0
101.0	0.97818	0.016133	340.6	340.6	68.997	1036.5	1105.5	0.1313	1.8487	1.9800	101.0
100.0	0.94924	0.016130	350.4	350.4	67.999	1037.1	1105.1	0.1295	1.8530	1.9825	100.0
99.0	0.92103	0.016127	360.5	360.5	67.001	1037.6	1104.6	0.1278	1.8573	1.9851	99.0
98.0	0.89356	0.016123	370.9	370.9	66.003	1038.2	1104.2	0.1260	1.8617	1.9876	98.0
97.0	0.86679	0.016120	381.7	381.7	65.005	1038.8	1103.8	0.1242	1.8660	1.9902	97.0
96.0	0.84072	0.016117	392.8	392.8	64.006	1039.3	1103.3	0.1224	1.8704	1.9928	96.0
95.0	0.81534	0.016114	404.4	404.4	63.008	1039.9	1102.9	0.1206	1.8748	1.9954	95.0
94.0	0.79062	0.016111	416.3	416.3	62.010	1040.5	1102.5	0.1188	1.8792	1.9980	94.0
93.0	0.76655	0.016108	428.6	428.6	61.012	1041.0	1102.1	0.1170	1.8837	2.0006	93.0
92.0	0.74313	0.016105	441.3	441.3	60.014	1041.6	1101.6	0.1152	1.8881	2.0033	92.0
91.0	0.72032	0.016102	454.5	454.5	59.016	1042.2	1101.2	0.1134	1.8926	2.0059	91.0
90.0	0.69813	0.016099	468.1	468.1	58.018	1042.7	1100.8	0.1115	1.8970	2.0086	90.0
89.0	0.67653	0.016096	482.2	482.2	57.020	1043.3	1100.3	0.1097	1.9015	2.0112	89.0
88.0	0.65551	0.016093	496.8	496.8	56.022	1043.9	1099.9	0.1079	1.9060	2.0139	88.0
87.0	0.63507	0.016090	511.9	511.9	55.024	1044.4	1099.5	0.1061	1.9105	2.0166	87.0
86.0	0.61518	0.016087	527.5	527.5	54.026	1045.0	1099.0	0.1043	1.9151	2.0193	86.0
85.0	0.59583	0.016085	543.6	543.6	53.027	1045.6	1098.6	0.1024	1.9196	2.0221	85.0
84.0	0.57702	0.016082	560.3	560.3	52.029	1046.1	1098.2	0.1006	1.9242	2.0248	84.0
83.0	0.55872	0.016079	577.6	577.6	51.031	1046.7	1097.7	0.0988	1.9288	2.0275	83.0
82.0	0.54093	0.016077	595.6	595.6	50.033	1047.3	1097.3	0.0969	1.9334	2.0303	82.0
81.0	0.52364	0.016074	614.1	614.1	49.035	1047.8	1096.9	0.0951	1.9380	2.0331	81.0
80.0	0.50683	0.016072	633.3	633.3	48.037	1048.4	1096.4	0.0932	1.9426	2.0359	80.0
79.0	0.49049	0.016070	653.2	653.2	47.038	1049.0	1096.0	0.0914	1.9473	2.0387	79.0
78.0	0.47461	0.016067	673.8	673.8	46.040	1049.5	1095.6	0.0895	1.9520	2.0415	78.0
77.0	0.45919	0.016065	695.2	695.2	45.042	1050.1	1095.1	0.0877	1.9567	2.0443	77.0
76.0	0.44420	0.016063	717.4	717.4	44.043	1050.7	1094.7	0.0858	1.9614	2.0472	76.0
75.0	0.42964	0.016060	740.3	740.3	43.045	1051.2	1094.3	0.0839	1.9661	2.0500	75.0
74.0	0.41550	0.016058	764.1	764.1	42.046	1051.8	1093.8	0.0821	1.9708	2.0529	74.0
73.0	0.40177	0.016056	788.8	788.8	41.048	1052.4	1093.4	0.0802	1.9756	2.0558	73.0
72.0	0.38844	0.016054	814.3	814.3	40.049	1052.9	1093.0	0.0783	1.9804	2.0587	72.0
71.0	0.37549	0.016052	840.9	840.9	39.050	1053.5	1092.5	0.0764	1.9852	2.0616	71.0

70.0	2.0645	1.9900	0.0745	1092.1	1054.0	38.052	868.4	868.3	0.016050	0.36292	70.0
69.0	2.0675	1.9948	0.0727	1091.7	1054.6	37.053	896.9	896.9	0.016048	0.35073	69.0
68.0	2.0704	1.9996	0.0708	1091.2	1055.1	36.054	926.5	926.5	0.016046	0.33889	68.0
67.0	2.0734	2.0045	0.0689	1090.8	1055.7	35.055	957.2	957.2	0.016044	0.32740	67.0
66.0	2.0764	2.0094	0.0670	1090.4	1056.3	34.056	989.1	989.0	0.016043	0.31626	66.0
65.0	2.0794	2.0143	0.0651	1089.9	1056.9	33.057	1022.1	1022.1	0.016041	0.30545	65.0
64.0	2.0824	2.0192	0.0632	1089.5	1057.4	32.058	1056.5	1056.5	0.016039	0.29497	64.0
63.0	2.0854	2.0242	0.0613	1089.0	1058.0	31.058	1092.1	1092.1	0.016038	0.28480	63.0
62.0	2.0885	2.0291	0.0593	1088.6	1058.5	30.059	1129.2	1129.2	0.016036	0.27494	62.0
61.0	2.0915	2.0341	0.0574	1088.2	1059.1	29.059	1167.6	1167.6	0.016035	0.26538	61.0
60.0	2.0946	2.0391	0.0555	1087.7	1059.7	28.060	1207.6	1207.6	0.016033	0.25611	60.0
59.0	2.0977	2.0441	0.0536	1087.3	1060.2	27.060	1249.1	1249.1	0.016032	0.24713	59.0
58.0	2.1008	2.0491	0.0516	1086.9	1060.8	26.060	1292.2	1292.2	0.016031	0.23843	58.0
57.0	2.1039	2.0542	0.0497	1086.4	1061.3	25.060	1337.0	1337.0	0.016029	0.23000	57.0
56.0	2.1070	2.0593	0.0478	1086.0	1061.9	24.059	1383.6	1383.6	0.016028	0.22183	56.0
55.0	2.1102	2.0644	0.0458	1085.6	1062.5	23.059	1432.0	1432.0	0.016027	0.21392	55.0
54.0	2.1134	2.0695	0.0439	1085.1	1063.1	22.058	1482.4	1482.4	0.016026	0.20625	54.0
53.0	2.1165	2.0746	0.0419	1084.7	1063.6	21.058	1534.8	1534.7	0.016025	0.19883	53.0
52.0	2.1197	2.0798	0.0400	1084.2	1064.2	20.057	1589.2	1589.2	0.016024	0.19165	52.0
51.0	2.1230	2.0849	0.0380	1083.8	1064.7	19.056	1645.9	1645.9	0.016023	0.18469	51.0
50.0	2.1262	2.0901	0.0361	1083.4	1065.3	18.054	1704.8	1704.8	0.016023	0.17796	50.0
49.0	2.1294	2.0953	0.0341	1082.9	1065.9	17.053	1766.2	1766.2	0.016022	0.17144	49.0
48.0	2.1327	2.1006	0.0321	1082.5	1066.4	16.051	1830.0	1830.0	0.016021	0.16514	48.0
47.0	2.1360	2.1058	0.0301	1082.1	1067.0	15.049	1896.5	1896.5	0.016021	0.15904	47.0
46.0	2.1393	2.1111	0.0282	1081.6	1067.6	14.047	1965.7	1965.7	0.016020	0.15314	46.0
45.0	2.1426	2.1164	0.0262	1081.2	1068.1	13.044	2037.8	2037.7	0.016020	0.14744	45.0
44.0	2.1459	2.1217	0.0242	1080.7	1068.7	12.041	2112.8	2112.8	0.016019	0.14192	44.0
43.0	2.1493	2.1271	0.0222	1080.3	1069.3	11.038	2191.0	2191.0	0.016019	0.13659	43.0
42.0	2.1527	2.1325	0.0202	1079.9	1069.8	10.035	2272.4	2272.4	0.016019	0.13143	42.0
41.0	2.1560	2.1378	0.0182	1079.4	1070.4	9.031	2357.3	2357.3	0.016019	0.12645	41.0
40.0	2.1594	2.1432	0.0162	1079.0	1071.0	8.027	2445.8	2445.8	0.016019	0.12163	40.0
39.0	2.1629	2.1487	0.0142	1078.5	1071.5	7.023	2538.2	2538.1	0.016019	0.11698	39.0
38.0	2.1663	2.1541	0.0122	1078.1	1072.1	6.018	2634.2	2634.1	0.016019	0.11249	38.0
37.0	2.1697	2.1596	0.0101	1077.7	1072.7	5.013	2734.4	2734.4	0.016019	0.10815	37.0
36.0	2.1732	2.1651	0.0081	1077.2	1073.2	4.008	2839.0	2839.0	0.016020	0.10395	36.0
35.0	2.1767	2.1706	0.0061	1076.8	1073.8	3.002	2948.1	2948.1	0.016020	0.09991	35.0
34.0	2.1802	2.1762	0.0041	1076.4	1074.4	1.996	3061.9	3061.9	0.016021	0.09600	34.0
33.0	2.1837	2.1817	0.0020	1075.9	1074.9	0.989	3180.7	3180.7	0.016021	0.09223	33.0
32.018	2.1872	2.1872	0.0000	1075.5	1075.5	0.0003	3302.4	3302.4	0.016022	0.08865	32.018
32.0	2.1873	2.1873	-0.0000	1075.5	1075.5	-0.0179	3304.7	3304.7	0.016022	0.08859	*32.0

* THE STATES HERE SHOWN ARE METASTABLE

Table D-2. Properties of Saturated Steam and Saturated Water (Pressure, inches H_g)

Pressure in.H_g, psia	Temp. F	Volume, ft³/lbm Water v_f	Evap. v_{fg}	Steam v_g	Enthalpy, Btu/lbm Water h_f	Evap. h_{fg}	Steam h_g	Entropy, Btu/lbm×R Water s_f	Evap. s_{fg}	Steam s_g	Energy, Btu/lbm Water u_f	Steam u_g
30.0	14.735	212.13 0.016720	26.716	26.733	180.30	970.2	1150.5	0.3123	1.4443	1.7566	180.25	1077.6
29.0	14.243	210.43 0.016708	27.578	27.595	178.58	971.3	1149.9	0.3098	1.4495	1.7593	178.54	1077.2
28.0	13.752	208.67 0.016695	28.499	28.516	176.81	972.4	1149.2	0.3071	1.4550	1.7621	176.77	1076.7
27.0	13.261	206.87 0.016683	29.486	29.502	174.99	973.6	1148.6	0.3044	1.4607	1.7651	174.95	1076.2
26.0	12.770	205.00 0.016670	30.546	30.563	173.12	974.7	1147.9	0.3016	1.4665	1.7681	173.08	1075.6
25.0	12.279	203.08 0.016657	31.689	31.705	171.18	975.9	1147.1	0.2987	1.4726	1.7713	171.14	1075.1
24.0	11.788	201.09 0.016644	32.923	32.940	169.18	977.2	1146.4	0.2957	1.4789	1.7746	169.14	1074.5
23.0	11.297	199.03 0.016630	34.261	34.278	167.11	978.5	1145.6	0.2925	1.4855	1.7780	167.07	1073.9
22.0	10.805	196.89 0.016616	35.717	35.734	164.96	979.8	1144.8	0.2893	1.4924	1.7816	164.93	1073.3
21.0	10.314	194.68 0.016602	37.307	37.324	162.73	981.2	1143.9	0.2859	1.4995	1.7854	162.70	1072.7
20.0	9.823	192.37 0.016587	39.051	39.067	160.41	982.6	1143.0	0.2823	1.5070	1.7893	160.38	1072.0
19.0	9.332	189.96 0.016572	40.972	40.989	158.00	984.1	1142.1	0.2786	1.5149	1.7935	157.97	1071.3
18.0	8.841	187.45 0.016556	43.100	43.116	155.47	985.6	1141.1	0.2747	1.5232	1.7979	155.45	1070.6
17.0	8.350	184.81 0.016539	45.470	45.486	152.83	987.3	1140.1	0.2706	1.5319	1.8025	152.80	1069.8
16.0	7.858	182.05 0.016522	48.126	48.143	150.05	988.9	1139.0	0.2663	1.5411	1.8074	150.03	1069.0
15.0	7.3673	179.13 0.016505	51.13	51.14	147.13	990.7	1137.8	0.2617	1.5509	1.8127	147.10	1068.1
14.0	6.8762	176.05 0.016486	54.54	54.55	144.03	992.6	1136.6	0.2569	1.5614	1.8183	144.01	1067.2
13.0	6.3850	172.77 0.016467	58.46	58.48	140.75	994.6	1135.3	0.2517	1.5726	1.8243	140.73	1066.2
12.0	5.8938	169.28 0.016447	63.01	63.03	137.25	996.7	1133.9	0.2462	1.5847	1.8309	137.23	1065.2
11.0	5.4027	165.53 0.016426	68.37	68.38	133.50	998.9	1132.4	0.2402	1.5978	1.8380	133.48	1064.1
10.0	4.9115	161.48 0.016403	74.76	74.77	129.44	1001.3	1130.8	0.2337	1.6121	1.8458	129.43	1062.8
9.5	4.6660	159.33 0.016392	78.44	78.46	127.29	1002.6	1129.9	0.2302	1.6198	1.8500	127.27	1062.2
9.0	4.4204	157.08 0.016380	82.52	82.53	125.03	1004.0	1129.0	0.2266	1.6279	1.8544	125.02	1061.5
8.5	4.1748	154.72 0.016367	87.06	87.08	122.67	1005.4	1128.0	0.2227	1.6364	1.8591	122.66	1060.8
8.0	3.9292	152.24 0.016354	92.16	92.17	120.19	1006.8	1127.0	0.2187	1.6454	1.8641	120.18	1060.0
7.5	3.6837	149.62 0.016341	97.91	97.92	117.57	1008.4	1125.9	0.2144	1.6550	1.8694	117.56	1059.2
7.0	3.4381	146.85 0.016327	104.46	104.47	114.80	1010.0	1124.8	0.2098	1.6653	1.8751	114.79	1058.3
6.5	3.1925	143.91 0.016312	111.98	112.00	111.86	1011.7	1123.6	0.2050	1.6762	1.8812	111.85	1057.4
6.0	2.9469	140.77 0.016297	120.72	120.73	108.72	1013.6	1122.3	0.1998	1.6881	1.8878	108.71	1056.5
5.5	2.7013	137.40 0.016281	131.00	131.01	105.35	1015.5	1120.9	0.1941	1.7009	1.8950	105.34	1055.4
5.0	2.4558	133.75 0.016264	143.26	143.28	101.71	1017.7	1119.4	0.1880	1.7149	1.9029	101.70	1054.3
4.5	2.2102	129.78 0.016246	158.17	158.19	97.74	1020.0	1117.7	0.1813	1.7304	1.9117	97.73	1053.0
4.0	1.9646	125.42 0.016227	176.68	176.70	93.38	1022.5	1115.8	0.1739	1.7476	1.9215	93.37	1051.6
3.5	1.7190	120.56 0.016207	200.32	200.34	88.52	1025.3	1113.8	0.1655	1.7671	1.9326	88.52	1050.1
3.4	1.6699	119.51 0.016202	205.86	205.87	87.48	1025.9	1113.4	0.1638	1.7713	1.9351	87.47	1049.8

3.3	1.6208	118.44	0.016198	211.72	211.74	86.41	1026.5	1112.9	0.1619	1.7756	1.9375	86.41	1049.4
3.2	1.5717	117.34	0.016194	217.94	217.95	85.31	1027.1	1112.5	0.1600	1.7801	1.9401	85.31	1049.1
3.1	1.5226	116.22	0.016189	224.55	224.56	84.19	1027.8	1112.0	0.1581	1.7847	1.9428	84.18	1048.7
3.0	1.4735	115.06	0.016184	231.58	231.60	83.03	1028.5	1111.5	0.1560	1.7895	1.9455	83.02	1048.3
2.9	1.4243	113.86	0.016180	239.09	239.11	81.84	1029.1	1111.0	0.1540	1.7944	1.9484	81.83	1048.0
2.8	1.3752	112.63	0.016175	247.12	247.13	80.61	1029.9	1110.5	0.1518	1.7995	1.9513	80.61	1047.6
2.7	1.3261	111.36	0.016170	255.72	255.74	79.34	1030.6	1109.9	0.1496	1.8048	1.9544	79.34	1047.2
2.6	1.2770	110.06	0.016165	264.97	264.99	78.04	1031.3	1109.4	0.1473	1.8102	1.9575	78.03	1046.7
2.5	1.2279	108.70	0.016160	274.94	274.95	76.69	1032.1	1108.8	0.1449	1.8159	1.9608	76.68	1046.3
2.4	1.1788	107.30	0.016155	285.71	285.73	75.29	1032.9	1108.2	0.1425	1.8218	1.9643	75.28	1045.9
2.3	1.12965	105.85	0.016150	297.40	297.41	73.838	1033.7	1107.6	0.1399	1.8279	1.9679	73.83	1045.4
2.2	1.08054	104.34	0.016145	310.11	310.13	72.333	1034.6	1106.9	0.1375	1.8343	1.9716	72.33	1044.9
2.1	1.03142	102.77	0.016139	324.00	324.02	70.768	1035.5	1106.3	0.1345	1.8410	1.9755	70.76	1044.4
2.0	0.98231	101.14	0.016134	339.25	339.26	69.137	1036.4	1105.6	0.1316	1.8481	1.9796	69.13	1043.9
1.9	0.93319	99.43	0.016128	356.05	356.06	67.434	1037.4	1104.8	0.1285	1.8554	1.9840	67.43	1043.3
1.8	0.88408	97.65	0.016122	374.66	374.68	65.652	1038.4	1104.1	0.1253	1.8632	1.9885	65.65	1042.8
1.7	0.83496	95.78	0.016116	395.40	395.42	63.782	1039.5	1103.2	0.1220	1.8714	1.9934	63.78	1042.2
1.6	0.78585	93.80	0.016110	418.66	418.68	61.814	1040.6	1102.4	0.1184	1.8801	1.9985	61.81	1041.5
1.5	0.73673	91.72	0.016104	444.94	444.95	59.737	1041.8	1101.5	0.1147	1.8893	2.0040	59.73	1040.8
1.4	0.68762	89.52	0.016097	474.86	474.87	57.536	1043.0	1100.6	0.1107	1.8992	2.0099	57.53	1040.1
1.30	0.63850	87.17	0.016091	509.21	509.27	55.193	1044.3	1099.5	0.1064	1.9098	2.0162	55.19	1039.4
1.20	0.58838	84.66	0.016084	549.21	549.23	52.689	1045.8	1098.5	0.1018	1.9212	2.0230	52.69	1038.6
1.10	0.54027	81.96	0.016077	596.24	596.25	49.995	1047.3	1097.3	0.0968	1.9336	2.0304	49.99	1037.7
1.00	0.49115	79.04	0.016070	652.39	652.41	47.079	1048.9	1096.0	0.0914	1.9471	2.0386	47.08	1036.7
0.95	0.46660	77.48	0.016066	684.78	684.80	45.525	1049.8	1095.3	0.0886	1.9544	2.0429	45.52	1036.2
0.90	0.44204	75.85	0.016062	720.67	720.69	43.897	1050.7	1094.6	0.0855	1.9621	2.0476	43.90	1035.7
0.85	0.41748	74.14	0.016058	760.67	760.68	42.188	1051.7	1093.9	0.0823	1.9702	2.0525	42.19	1035.1
0.80	0.39292	72.34	0.016055	805.53	805.54	40.388	1052.7	1093.1	0.0789	1.9787	2.0577	40.39	1034.5
0.75	0.36637	70.44	0.016051	856.21	856.22	38.488	1053.8	1092.3	0.0754	1.9879	2.0632	38.49	1033.9
0.70	0.34381	68.42	0.016047	913.93	913.94	36.473	1054.9	1091.4	0.0716	1.9976	2.0692	36.47	1033.3
0.65	0.31925	66.27	0.016043	980.29	980.30	34.327	1056.1	1090.5	0.0675	2.0081	2.0756	34.33	1032.6
0.60	0.29469	63.97	0.016039	1057.40	1057.42	32.031	1057.4	1089.5	0.0631	2.0193	2.0825	32.03	1031.8
0.55	0.27013	61.50	0.016035	1148.16	1148.16	29.560	1058.6	1088.4	0.0584	2.0316	2.0900	29.56	1031.0
0.50	0.24558	58.82	0.016032	1256.57	1256.58	26.883	1060.0	1087.2	0.0532	2.0450	2.0982	26.88	1030.1
0.45	0.22102	55.90	0.016028	1388.40	1388.42	23.958	1062.0	1085.9	0.0476	2.0598	2.1074	23.96	1029.2
0.40	0.19646	52.67	0.016025	1552.29	1552.30	20.731	1063.8	1084.5	0.0413	2.0763	2.1176	20.73	1028.1
0.35	0.17190	49.07	0.016022	1761.71	1761.71	17.124	1065.8	1083.2	0.0342	2.0950	2.1292	17.12	1026.9
0.30	0.14735	44.98	0.016020	2038.95	2038.97	13.028	1068.1	1081.2	0.0262	2.1165	2.1427	13.03	1025.6
0.25	0.12279	40.24	0.016019	2423.54	2423.96	8.271	1070.8	1079.1	0.0167	2.1419	2.1586	8.27	1024.0
0.20	0.09823	34.58	0.016020	2995.82	2995.84	2.575	1074.0	1076.6	0.0052	2.1730	2.1782	2.57	1022.1

Table D-2. Properties of Saturated Steam and Saturated Water (Pressure, inches H_g) (Continued)

Press. psia	Temp. F	Volume, ft³/lbm			Enthalpy, Btu/lbm			Entropy, Btu/lbm × R			Energy, Btu/lb	
		Water v_f	Evap. v_{fg}	Steam v_g	Water h_f	Evap. h_{fg}	Steam h_g	Water s_f	Evap. s_{fg}	Steam s_g	Water u_f	Steam u_g
3208.2	705.47	0.05078	0.00000	0.05078	906.0	0.0	906.0	1.0612	0.0000	1.0612	875.9	875.9
3200.0	705.08	0.04472	0.01191	0.05663	875.5	56.1	931.6	1.0351	0.0482	1.0832	849.1	898.1
3190.0	704.61	0.04266	0.01729	0.05995	863.9	81.0	944.8	1.0251	0.0696	1.0947	838.7	909.5
3180.0	704.13	0.04137	0.02103	0.06240	856.0	98.0	954.1	1.0185	0.0842	1.1027	831.7	917.3
3170.0	703.65	0.04041	0.02403	0.06444	850.0	111.4	961.4	1.0133	0.0958	1.1091	826.3	923.6
3160.0	703.18	0.03965	0.02658	0.06623	844.9	122.6	967.5	1.0090	0.1055	1.1145	821.7	928.8
3150.0	702.70	0.03901	0.02883	0.06785	840.5	132.4	972.9	1.0053	0.1139	1.1192	817.8	933.4
3140.0	702.22	0.03847	0.03087	0.06934	836.6	141.1	977.7	1.0020	0.1214	1.1234	814.3	937.4
3130.0	701.73	0.03798	0.03275	0.07073	833.1	149.0	982.1	0.9990	0.1283	1.1273	811.1	941.1
3120.0	701.25	0.03755	0.03450	0.07206	829.8	156.3	986.1	0.9963	0.1346	1.1309	808.1	944.5
3100.0	700.28	0.03681	0.03771	0.07452	824.0	169.3	993.3	0.9914	0.1460	1.1373	802.9	950.6
3080.0	699.30	0.03617	0.04064	0.07682	818.8	180.9	999.7	0.9870	0.1561	1.1431	798.2	955.9
3060.0	698.31	0.03562	0.04337	0.07899	814.1	191.4	1005.5	0.9830	0.1653	1.1483	793.9	960.7
3040.0	697.32	0.03513	0.04593	0.08106	809.7	201.1	1010.8	0.9794	0.1738	1.1532	790.0	965.2
3020.0	696.33	0.03468	0.04838	0.08306	805.7	210.0	1015.7	0.9760	0.1817	1.1577	786.3	969.3
3000.0	695.33	0.03428	0.05073	0.08500	801.8	218.4	1020.3	0.9728	0.1891	1.1619	782.8	973.1
2980.0	694.32	0.03390	0.05300	0.08690	798.2	226.4	1024.6	0.9698	0.1961	1.1659	779.5	976.7
2960.0	693.30	0.03355	0.05521	0.08876	794.8	233.9	1028.7	0.9669	0.2029	1.1697	776.4	980.1
2940.0	692.28	0.03322	0.05738	0.09060	791.4	241.1	1032.6	0.9641	0.2093	1.1734	773.4	983.3
2920.0	691.26	0.03291	0.05950	0.09241	788.2	248.0	1036.3	0.9614	0.2155	1.1769	770.4	986.3
2900.0	690.22	0.03262	0.06158	0.09420	785.1	254.7	1039.8	0.9588	0.2215	1.1803	767.6	989.3
2880.0	689.18	0.03234	0.06364	0.09598	782.1	261.2	1043.3	0.9563	0.2273	1.1836	764.9	992.1
2860.0	688.14	0.03208	0.06568	0.09776	779.2	267.4	1046.6	0.9539	0.2329	1.1868	762.2	994.8
2840.0	687.09	0.03182	0.06770	0.09952	776.3	273.5	1049.7	0.9515	0.2384	1.1899	759.6	997.4
2820.0	686.03	0.03158	0.06971	0.10129	773.5	279.4	1052.8	0.9491	0.2438	1.1929	757.0	1000.0
2800.0	684.96	0.03134	0.07171	0.10305	770.7	285.1	1055.8	0.9468	0.2491	1.1958	754.4	1002.4
2780.0	683.89	0.03112	0.07370	0.10481	768.0	290.8	1058.7	0.9445	0.2543	1.1987	751.9	1004.8
2760.0	682.81	0.03090	0.07569	0.10658	765.3	296.3	1061.6	0.9422	0.2593	1.2016	749.5	1007.1
2740.0	681.72	0.03069	0.07767	0.10836	762.6	301.8	1064.3	0.9400	0.2643	1.2043	747.0	1009.4
2720.0	680.63	0.03048	0.07966	0.11014	760.0	307.1	1067.0	0.9378	0.2693	1.2070	744.6	1011.6
2700.0	679.53	0.03029	0.08165	0.11194	757.3	312.3	1069.7	0.9356	0.2741	1.2097	742.2	1013.7
2680.0	678.42	0.03009	0.08365	0.11374	754.7	317.5	1072.2	0.9334	0.2789	1.2123	739.8	1015.8
2660.0	677.30	0.02991	0.08565	0.11556	752.2	322.6	1074.8	0.9312	0.2837	1.2149	737.4	1017.9
2640.0	676.18	0.02973	0.08766	0.11739	749.6	327.7	1077.2	0.9291	0.2884	1.2175	735.1	1019.9
2620.0	675.05	0.02955	0.08968	0.11924	747.0	332.6	1079.7	0.9269	0.2931	1.2200	732.7	1021.9

2600.0	673.91	0.02938	0.09172	0.12110	744.5	337.6	1082.0	0.9247	0.2977	1.2225	730.3	1023.8
2580.0	672.77	0.02922	0.09376	0.12298	741.9	342.3	1084.4	0.9226	0.3024	1.2254	728.6	1025.7
2560.0	671.62	0.02905	0.09582	0.12488	739.4	347.5	1086.7	0.9204	0.3069	1.2274	725.6	1027.5
2540.0	670.46	0.02890	0.09789	0.12679	736.8	352.1	1088.9	0.9183	0.3115	1.2298	723.2	1029.3
2520.0	669.29	0.02874	0.09998	0.12873	734.3	356.9	1091.1	0.9161	0.3161	1.2322	720.9	1031.1
2500.0	668.11	0.02859	0.10209	0.13068	731.7	361.6	1093.3	0.9139	0.3206	1.2345	718.5	1032.9
2480.0	666.93	0.02845	0.10421	0.13266	729.1	366.3	1095.5	0.9117	0.3251	1.2369	716.1	1034.6
2460.0	665.74	0.02831	0.10635	0.13466	726.6	371.0	1097.6	0.9096	0.3296	1.2392	713.7	1036.3
2440.0	664.54	0.02817	0.10851	0.13667	724.0	375.7	1099.7	0.9074	0.3341	1.2415	711.3	1038.0
2420.0	663.33	0.02803	0.11068	0.13871	721.4	380.3	1101.7	0.9052	0.3386	1.2438	708.9	1039.6
2400.0	662.11	0.02790	0.11287	0.14076	719.0	384.8	1103.7	0.9031	0.3430	1.2460	706.6	1041.2
2380.0	660.88	0.02777	0.11505	0.14282	716.4	389.3	1105.7	0.9010	0.3472	1.2482	704.4	1042.8
2360.0	659.65	0.02764	0.11726	0.14490	714.3	393.3	1107.6	0.8990	0.3514	1.2504	702.2	1044.3
2340.0	658.41	0.02752	0.11949	0.14701	711.9	397.5	1109.5	0.8969	0.3556	1.2526	700.0	1045.8
2320.0	657.15	0.02739	0.12176	0.14916	709.5	401.8	1111.4	0.8949	0.3598	1.2548	697.8	1047.3
2300.0	655.89	0.02727	0.12406	0.15133	707.2	406.0	1113.2	0.8929	0.3640	1.2569	695.6	1048.8
2280.0	654.62	0.02715	0.12639	0.15354	704.8	410.2	1115.0	0.8909	0.3682	1.2591	693.4	1050.3
2260.0	653.34	0.02703	0.12875	0.15579	702.5	414.4	1116.6	0.8889	0.3723	1.2612	691.2	1051.7
2240.0	652.05	0.02692	0.13114	0.15806	700.8	418.5	1118.6	0.8868	0.3765	1.2633	689.0	1053.1
2220.0	650.75	0.02680	0.13357	0.16037	697.8	422.6	1120.4	0.8848	0.3806	1.2655	686.8	1054.5
2200.0	649.45	0.02669	0.13603	0.16272	695.5	426.7	1122.2	0.8828	0.3848	1.2676	684.6	1055.9
2180.0	648.13	0.02658	0.13852	0.16510	693.1	430.8	1123.9	0.8808	0.3889	1.2697	682.2	1057.3
2160.0	646.80	0.02647	0.14105	0.16752	690.8	434.8	1125.6	0.8788	0.3930	1.2718	680.2	1058.6
2140.0	645.46	0.02636	0.14361	0.16998	688.5	438.8	1127.3	0.8768	0.3971	1.2738	678.0	1059.9
2120.0	644.12	0.02626	0.14621	0.17247	686.1	442.8	1128.9	0.8747	0.4012	1.2759	675.8	1061.2
2100.0	642.76	0.02615	0.14885	0.17501	683.8	446.7	1130.5	0.8727	0.4053	1.2780	673.6	1062.5
2080.0	641.39	0.02605	0.15153	0.17758	681.5	450.7	1132.1	0.8707	0.4093	1.2800	671.4	1063.8
2060.0	640.01	0.02595	0.15425	0.18020	679.1	454.6	1133.7	0.8686	0.4134	1.2821	669.2	1065.0
2040.0	638.62	0.02585	0.15701	0.18286	676.8	458.5	1135.3	0.8666	0.4175	1.2841	667.0	1066.3
2020.0	637.22	0.02575	0.15981	0.18556	674.5	462.4	1136.8	0.8646	0.4216	1.2861	664.8	1067.5
2000.0	635.80	0.02565	0.16266	0.18831	672.1	466.2	1138.3	0.8625	0.4256	1.2881	662.6	1068.6
1980.0	634.38	0.02555	0.16555	0.19111	669.8	470.1	1139.8	0.8605	0.4297	1.2901	660.4	1069.8
1960.0	632.94	0.02545	0.16849	0.19395	667.4	473.9	1141.3	0.8584	0.4337	1.2921	658.2	1071.0
1940.0	631.50	0.02536	0.17148	0.19684	665.1	477.7	1142.7	0.8563	0.4378	1.2941	656.0	1072.1
1920.0	630.04	0.02526	0.17452	0.19978	662.7	481.5	1144.2	0.8543	0.4418	1.2961	653.7	1073.2
1900.0	628.56	0.02517	0.17761	0.20278	660.4	485.2	1145.6	0.8522	0.4459	1.2981	651.5	1074.3
1880.0	627.08	0.02508	0.18075	0.20583	658.0	489.0	1147.0	0.8501	0.4500	1.3001	649.3	1075.4
1860.0	625.58	0.02499	0.18395	0.20894	655.6	492.7	1148.3	0.8480	0.4540	1.3020	647.0	1076.4
1840.0	624.07	0.02490	0.18721	0.21210	653.3	496.4	1149.7	0.8459	0.4581	1.3040	644.8	1077.5
1820.0	622.55	0.02481	0.19052	0.21533	650.9	500.1	1151.0	0.8438	0.4622	1.3059	642.5	1078.5
1800.0	621.02	0.02472	0.19390	0.21861	648.5	503.8	1152.3	0.8417	0.4662	1.3079	640.3	1079.5

Table D-2. Properties of Saturated Steam and Saturated Water (Pressure, inches H$_g$) (Continued)

Press. psia	Temp. F	Volume, ft³/lbm			Enthalpy, Btu/lbm			Entropy, Btu/lbm × R			Energy, Btu/lbm	
		Water v_f	Evap. v_{fg}	Steam v_g	Water h_f	Evap. h_{fg}	Steam h_g	Water s_f	Evap. s_{fg}	Steam s_g	Water u_f	Steam u_g
1800.0	621.02	0.02472	0.19390	0.21861	648.5	503.8	1152.3	0.8417	0.4662	1.3079	640.3	1079.5
1780.0	619.47	0.02463	0.19734	0.22197	646.1	507.5	1153.6	0.8395	0.4703	1.3098	638.0	1080.5
1760.0	617.90	0.02454	0.20084	0.22538	643.7	511.2	1154.9	0.8374	0.4744	1.3118	635.7	1081.5
1740.0	616.33	0.02445	0.20442	0.22887	641.3	514.9	1156.2	0.8352	0.4785	1.3137	633.4	1082.5
1720.0	614.73	0.02437	0.20806	0.23243	638.9	518.5	1157.4	0.8331	0.4826	1.3157	631.1	1083.4
1700.0	613.13	0.02428	0.21178	0.23607	636.5	522.2	1158.6	0.8309	0.4867	1.3176	628.8	1084.4
1680.0	611.51	0.02420	0.21558	0.23978	634.0	525.8	1159.8	0.8287	0.4909	1.3196	626.5	1085.3
1660.0	609.87	0.02411	0.21945	0.24357	631.6	529.5	1161.0	0.8265	0.4950	1.3215	624.2	1086.2
1640.0	608.22	0.02403	0.22341	0.24744	629.1	533.1	1162.2	0.8243	0.4992	1.3235	621.8	1087.1
1620.0	606.55	0.02395	0.22745	0.25140	626.7	536.7	1163.4	0.8221	0.5034	1.3255	619.5	1088.0
1600.0	604.87	0.02387	0.23159	0.25545	624.2	540.3	1164.5	0.8199	0.5076	1.3274	617.1	1088.9
1580.0	603.17	0.02378	0.23581	0.25960	621.7	543.9	1165.7	0.8176	0.5118	1.3294	614.8	1089.8
1560.0	601.45	0.02370	0.24014	0.26384	619.2	547.6	1166.8	0.8153	0.5160	1.3314	612.4	1090.6
1540.0	599.72	0.02362	0.24456	0.26818	616.7	551.2	1167.9	0.8131	0.5203	1.3333	610.0	1091.5
1520.0	597.97	0.02354	0.24909	0.27263	614.2	554.8	1169.0	0.8108	0.5245	1.3353	607.6	1092.3
1500.0	596.20	0.02346	0.25372	0.27719	611.7	558.4	1170.1	0.8085	0.5288	1.3373	605.2	1093.1
1480.0	594.41	0.02338	0.25847	0.28186	609.1	562.0	1171.2	0.8061	0.5332	1.3393	602.7	1094.0
1460.0	592.61	0.02331	0.26334	0.28665	606.6	565.6	1172.2	0.8038	0.5375	1.3413	600.3	1094.8
1440.0	590.78	0.02323	0.26833	0.29156	604.0	569.2	1173.3	0.8014	0.5419	1.3433	597.8	1095.6
1420.0	588.93	0.02315	0.27345	0.29660	601.4	572.9	1174.3	0.7990	0.5463	1.3453	595.3	1096.3
1400.0	587.07	0.02307	0.27871	0.30178	598.8	576.5	1175.3	0.7966	0.5507	1.3474	592.9	1097.1
1380.0	585.18	0.02300	0.28410	0.30710	596.2	580.1	1176.3	0.7942	0.5552	1.3494	590.3	1097.9
1360.0	583.28	0.02292	0.28965	0.31256	593.6	583.7	1177.3	0.7918	0.5597	1.3515	587.8	1098.6
1340.0	581.35	0.02284	0.29534	0.31818	590.9	587.4	1178.3	0.7893	0.5642	1.3535	585.3	1099.4
1320.0	579.40	0.02277	0.30119	0.32396	588.3	591.0	1179.3	0.7868	0.5688	1.3556	582.7	1100.1
1300.0	577.42	0.02269	0.30722	0.32991	585.6	594.6	1180.2	0.7843	0.5733	1.3577	580.1	1100.9
1290.0	576.43	0.02265	0.31029	0.33295	584.2	596.5	1180.7	0.7831	0.5757	1.3587	578.8	1101.2
1280.0	575.42	0.02262	0.31341	0.33603	582.9	598.3	1181.2	0.7818	0.5780	1.3598	577.5	1101.6
1270.0	574.42	0.02258	0.31658	0.33916	581.5	600.1	1181.6	0.7805	0.5803	1.3608	576.2	1101.9
1260.0	573.40	0.02254	0.31980	0.34234	580.2	601.9	1182.1	0.7793	0.5826	1.3619	574.9	1102.3
1250.0	572.38	0.02250	0.32306	0.34556	578.8	603.8	1182.6	0.7780	0.5850	1.3630	573.6	1102.6
1240.0	571.36	0.02247	0.32637	0.34884	577.4	605.6	1183.0	0.7767	0.5874	1.3640	572.3	1103.0
1230.0	570.32	0.02243	0.32973	0.35216	576.0	607.4	1183.5	0.7754	0.5897	1.3651	570.9	1103.3
1220.0	569.28	0.02239	0.33314	0.35554	574.6	609.3	1183.9	0.7741	0.5921	1.3662	569.6	1103.7
1210.0	568.24	0.02236	0.33661	0.35897	573.2	611.1	1184.4	0.7728	0.5945	1.3673	568.2	1104.0

1104.3	566.9	1.3683	0.5969	0.7714	1184.3	613.0	571.9	0.36245	0.34013	0.02232	567.19	1200.0
1104.7	565.5	1.3694	0.5993	0.7701	1185.3	614.8	570.2	0.36599	0.34371	0.02228	566.13	1190.0
1105.0	564.2	1.3705	0.6017	0.7688	1185.7	616.6	569.0	0.36958	0.34734	0.02225	565.06	1180.0
1105.3	562.8	1.3716	0.6042	0.7674	1186.1	618.5	567.6	0.37324	0.35103	0.02221	563.99	1170.0
1105.6	561.4	1.3727	0.6066	0.7661	1186.6	620.3	566.2	0.37695	0.35478	0.02217	562.91	1160.0
1106.0	560.1	1.3738	0.6091	0.7647	1187.0	622.2	564.8	0.38073	0.35859	0.02214	561.82	1150.0
1106.3	558.7	1.3749	0.6115	0.7634	1187.4	624.1	563.3	0.38457	0.36247	0.02210	560.73	1140.0
1106.6	557.3	1.3760	0.6140	0.7620	1187.8	625.9	561.9	0.38847	0.36641	0.02206	559.63	1130.0
1106.9	555.9	1.3771	0.6165	0.7606	1188.2	627.8	560.5	0.39244	0.37041	0.02203	558.52	1120.0
1107.2	554.5	1.3783	0.6190	0.7592	1188.7	629.6	559.0	0.39648	0.37449	0.02199	557.40	1110.0
1107.5	553.1	1.3794	0.6216	0.7578	1189.1	631.5	557.5	0.40058	0.37863	0.02195	556.28	1100.0
1107.8	551.7	1.3805	0.6241	0.7564	1189.5	633.4	556.1	0.40476	0.38285	0.02192	555.14	1090.0
1108.1	550.2	1.3817	0.6266	0.7550	1189.9	635.3	554.6	0.40902	0.38714	0.02188	554.00	1080.0
1108.4	548.8	1.3828	0.6292	0.7536	1190.3	637.1	553.1	0.41335	0.39150	0.02184	552.86	1070.0
1108.7	547.4	1.3840	0.6318	0.7522	1190.7	639.0	551.6	0.41775	0.39594	0.02181	551.70	1060.0
1109.1	545.9	1.3851	0.6344	0.7507	1191.0	640.9	550.1	0.42224	0.40047	0.02177	550.53	1050.0
1109.3	544.5	1.3863	0.6370	0.7493	1191.4	642.8	548.6	0.42681	0.40507	0.02174	549.36	1040.0
1109.6	543.0	1.3874	0.6396	0.7478	1191.8	644.7	547.1	0.43146	0.40976	0.02170	548.18	1030.0
1109.9	541.5	1.3886	0.6423	0.7463	1192.2	646.6	545.6	0.43620	0.41454	0.02166	546.99	1020.0
1110.1	540.0	1.3898	0.6449	0.7449	1192.6	648.5	544.1	0.44103	0.41941	0.02163	545.79	1010.0
1110.4	538.6	1.3910	0.6476	0.7434	1192.9	650.4	542.6	0.44596	0.42436	0.02159	544.58	1000.0
1110.7	537.1	1.3922	0.6503	0.7419	1193.3	652.3	541.0	0.45097	0.42942	0.02155	543.36	990.0
1111.0	535.6	1.3934	0.6530	0.7404	1193.7	654.2	539.5	0.45609	0.43457	0.02152	542.14	980.0
1111.2	534.0	1.3946	0.6557	0.7389	1194.0	656.1	537.9	0.46130	0.43982	0.02148	540.90	970.0
1111.5	532.5	1.3958	0.6584	0.7373	1194.4	658.0	536.3	0.46662	0.44518	0.02145	539.65	960.0
1111.7	531.0	1.3970	0.6612	0.7358	1194.7	660.0	534.7	0.47205	0.45064	0.02141	538.39	950.0
1112.0	529.4	1.3982	0.6640	0.7342	1195.1	661.9	533.2	0.47759	0.45621	0.02137	537.13	940.0
1112.2	527.9	1.3995	0.6668	0.7327	1195.4	663.8	531.6	0.48324	0.46190	0.02134	535.85	930.0
1112.5	526.3	1.4007	0.6696	0.7311	1195.7	665.8	530.0	0.48901	0.46770	0.02130	534.56	920.0
1112.7	524.8	1.4019	0.6724	0.7295	1196.1	667.7	528.3	0.49490	0.47363	0.02127	533.26	910.0
1113.0	523.2	1.4032	0.6753	0.7279	1196.4	669.7	526.7	0.50091	0.47968	0.02123	531.95	900.0
1113.2	521.6	1.4045	0.6782	0.7263	1196.7	671.6	525.1	0.50706	0.48586	0.02119	530.63	890.0
1113.4	520.0	1.4057	0.6811	0.7247	1197.0	673.6	523.4	0.51333	0.49218	0.02116	529.30	880.0
1113.7	518.4	1.4070	0.6840	0.7230	1197.3	675.6	521.8	0.51975	0.49863	0.02112	527.96	870.0
1113.9	516.7	1.4083	0.6869	0.7214	1197.7	677.6	520.1	0.52631	0.50522	0.02109	526.60	860.0
1114.1	515.1	1.4096	0.6899	0.7197	1198.0	679.5	518.4	0.53302	0.51197	0.02105	525.24	850.0
1114.3	513.4	1.4109	0.6929	0.7180	1198.2	681.5	516.7	0.53988	0.51886	0.02101	523.86	840.0
1114.5	511.8	1.4122	0.6959	0.7163	1198.5	683.5	515.0	0.54689	0.52592	0.02098	522.46	830.0
1114.8	510.1	1.4136	0.6990	0.7146	1198.8	685.5	513.3	0.55408	0.53314	0.02094	521.06	820.0
1115.0	508.4	1.4149	0.7020	0.7129	1199.1	687.6	511.6	0.56143	0.54052	0.02091	519.64	810.0
1115.2	506.7	1.4163	0.7051	0.7111	1199.4	689.6	509.8	0.56896	0.54809	0.02087	518.21	800.0

Table D-2. Properties of Saturated Steam and Saturated Water (Pressure, inches H$_g$) (Continued)

Press. psia	Temp. F	Volume, ft³/lbm Water v_f	Evap. v_{fg}	Steam v_g	Enthalpy, Btu/lbm Water h_f	Evap. h_{fg}	Steam h_g	Entropy, Btu/lbm × R Water s_f	Evap. s_{fg}	Steam s_g	Energy, Btu/lb Water u_f	Steam u_g
800.0	518.21	0.02087	0.54809	0.56896	509.8	689.6	1199.4	0.7111	0.7051	1.4163	506.7	1115.2
790.0	516.76	0.02083	0.55584	0.57667	508.1	691.6	1199.7	0.7094	0.7082	1.4176	505.0	1115.4
780.0	515.30	0.02080	0.56377	0.58457	506.3	693.6	1199.9	0.7076	0.7114	1.4190	503.3	1115.5
770.0	513.83	0.02076	0.57191	0.59267	504.5	695.7	1200.2	0.7058	0.7146	1.4204	501.5	1115.7
760.0	512.34	0.02072	0.58025	0.60097	502.7	697.7	1200.4	0.7040	0.7178	1.4218	499.8	1115.9
750.0	510.84	0.02069	0.58880	0.60949	500.9	699.8	1200.7	0.7022	0.7210	1.4232	498.0	1116.1
740.0	509.32	0.02065	0.59757	0.61822	499.1	701.9	1200.9	0.7003	0.7243	1.4246	496.2	1116.3
730.0	507.78	0.02061	0.60657	0.62719	497.2	703.9	1201.2	0.6985	0.7276	1.4260	494.4	1116.4
720.0	506.23	0.02058	0.61581	0.63639	495.4	706.0	1201.4	0.6966	0.7309	1.4275	492.6	1116.6
710.0	504.67	0.02054	0.62530	0.64585	493.5	708.1	1201.6	0.6947	0.7343	1.4290	490.8	1116.8
700.0	503.08	0.02050	0.63505	0.65556	491.6	710.2	1201.8	0.6928	0.7377	1.4304	488.9	1116.9
690.0	501.48	0.02047	0.64507	0.66554	489.7	712.4	1202.1	0.6908	0.7411	1.4319	487.1	1117.1
680.0	499.86	0.02043	0.65538	0.67581	487.8	714.5	1202.3	0.6889	0.7446	1.4334	485.2	1117.2
670.0	498.22	0.02039	0.66598	0.68637	485.8	716.6	1202.5	0.6869	0.7481	1.4350	483.3	1117.4
660.0	496.57	0.02036	0.67688	0.69724	483.9	718.8	1202.7	0.6849	0.7516	1.4365	481.4	1117.5
650.0	494.89	0.02032	0.68811	0.70843	481.9	720.9	1202.8	0.6828	0.7552	1.4381	479.4	1117.6
640.0	493.19	0.02028	0.69967	0.71995	479.9	723.1	1203.0	0.6808	0.7588	1.4396	477.5	1117.8
630.0	491.48	0.02024	0.71159	0.73183	477.9	725.3	1203.2	0.6787	0.7625	1.4412	475.5	1117.9
620.0	489.74	0.02021	0.72387	0.74408	475.8	727.5	1203.4	0.6766	0.7662	1.4428	473.5	1118.0
610.0	487.98	0.02017	0.73654	0.75671	473.8	729.7	1203.5	0.6745	0.7700	1.4445	471.5	1118.1
600.0	486.20	0.02013	0.74962	0.76975	471.7	732.0	1203.7	0.6723	0.7738	1.4461	469.5	1118.2
590.0	484.40	0.02009	0.76312	0.78321	469.6	734.2	1203.8	0.6701	0.7777	1.4478	467.4	1118.3
580.0	482.57	0.02006	0.77706	0.79712	467.5	736.5	1203.9	0.6679	0.7816	1.4495	465.3	1118.4
570.0	480.72	0.02002	0.79148	0.81150	465.3	738.7	1204.1	0.6657	0.7855	1.4512	463.2	1118.5
560.0	478.84	0.01998	0.80639	0.82637	463.1	741.0	1204.2	0.6634	0.7895	1.4529	461.1	1118.5
550.0	476.94	0.01994	0.82183	0.84177	460.9	743.3	1204.3	0.6611	0.7936	1.4547	458.9	1118.6
540.0	475.01	0.01990	0.83781	0.85771	458.7	745.7	1204.4	0.6587	0.7977	1.4565	456.7	1118.7
530.0	473.05	0.01986	0.85437	0.87423	456.5	748.0	1204.5	0.6564	0.8019	1.4583	454.5	1118.7
520.0	471.07	0.01982	0.87155	0.89137	454.2	750.4	1204.5	0.6540	0.8062	1.4601	452.3	1118.8
510.0	469.05	0.01978	0.88937	0.90915	451.9	752.7	1204.6	0.6515	0.8105	1.4620	450.0	1118.8
500.0	467.01	0.01975	0.90787	0.92762	449.5	755.1	1204.7	0.6490	0.8148	1.4639	447.7	1118.8
490.0	464.93	0.01971	0.92710	0.94681	447.2	757.6	1204.7	0.6465	0.8193	1.4658	445.4	1118.9
480.0	462.82	0.01967	0.94711	0.96677	444.7	760.0	1204.8	0.6439	0.8238	1.4677	443.0	1118.9
470.0	460.68	0.01963	0.96793	0.98756	442.3	762.5	1204.8	0.6413	0.8284	1.4697	440.6	1118.9
460.0	458.50	0.01959	0.98962	1.00921	439.8	765.0	1204.8	0.6387	0.8331	1.4718	438.2	1118.9

450.0	456.28	0.01954	1.01224	1.03179	437.3	767.5	1204.8	0.6360	0.8378	1.4738	435.7	1118.9
440.0	454.03	0.01950	1.03585	1.05535	434.8	770.0	1204.8	0.6332	0.8427	1.4759	433.2	1118.8
430.0	451.74	0.01946	1.06051	1.07998	432.2	772.6	1204.8	0.6304	0.8476	1.4781	430.6	1118.8
420.0	449.40	0.01942	1.08631	1.10573	429.6	775.2	1204.7	0.6276	0.8527	1.4802	428.0	1118.8
410.0	447.02	0.01938	1.11331	1.13269	426.9	777.8	1204.7	0.6247	0.8578	1.4825	425.4	1118.7
400.0	444.60	0.01934	1.14162	1.16095	424.2	780.4	1204.6	0.6217	0.8630	1.4847	422.7	1118.7
390.0	442.13	0.01930	1.17132	1.19061	421.4	783.1	1204.5	0.6187	0.8683	1.4870	420.0	1118.6
380.0	439.61	0.01925	1.20252	1.22177	418.6	785.8	1204.4	0.6156	0.8738	1.4894	417.2	1118.5
370.0	437.04	0.01921	1.23535	1.25456	415.7	788.6	1204.3	0.6125	0.8794	1.4918	414.4	1118.4
360.0	434.41	0.01917	1.26993	1.28910	412.8	791.3	1204.1	0.6092	0.8851	1.4943	411.5	1118.3
350.0	431.73	0.01912	1.30642	1.32554	409.8	794.2	1204.0	0.6059	0.8909	1.4968	408.6	1118.1
345.0	430.36	0.01910	1.32542	1.34452	408.3	795.6	1203.9	0.6043	0.8939	1.4981	407.1	1118.1
340.0	428.99	0.01908	1.34497	1.36405	406.8	797.0	1203.8	0.6026	0.8969	1.4994	405.6	1118.0
335.0	427.59	0.01905	1.36507	1.38413	405.3	798.5	1203.7	0.6009	0.8999	1.5008	404.1	1117.9
330.0	426.18	0.01903	1.38577	1.40480	403.7	799.9	1203.6	0.5991	0.9030	1.5021	402.5	1117.8
325.0	424.75	0.01901	1.40707	1.42608	402.1	801.4	1203.5	0.5974	0.9061	1.5035	401.0	1117.7
320.0	423.31	0.01899	1.42902	1.44801	400.5	802.9	1203.4	0.5956	0.9092	1.5048	399.4	1117.7
315.0	421.84	0.01896	1.45164	1.47060	398.9	804.4	1203.3	0.5938	0.9124	1.5062	397.8	1117.6
310.0	420.36	0.01894	1.47496	1.49390	397.3	805.9	1203.2	0.5920	0.9157	1.5076	396.2	1117.5
305.0	418.87	0.01892	1.49902	1.51793	395.7	807.4	1203.0	0.5901	0.9190	1.5091	394.6	1117.4
300.0	417.35	0.01889	1.52384	1.54274	394.0	808.9	1202.9	0.5882	0.9223	1.5105	392.9	1117.2
295.0	415.81	0.01887	1.54948	1.56835	392.3	810.4	1202.7	0.5863	0.9257	1.5120	391.3	1117.1
290.0	414.25	0.01885	1.57597	1.59482	390.6	812.0	1202.6	0.5844	0.9291	1.5135	389.6	1117.0
285.0	412.67	0.01882	1.60336	1.62218	388.9	813.6	1202.4	0.5824	0.9326	1.5150	387.8	1116.9
280.0	411.07	0.01880	1.63169	1.65049	387.1	815.1	1202.3	0.5805	0.9361	1.5166	386.1	1116.7
275.0	409.45	0.01878	1.66101	1.67978	385.4	816.7	1202.1	0.5784	0.9397	1.5181	384.4	1116.6
270.0	407.80	0.01875	1.69137	1.71013	383.6	818.3	1201.9	0.5764	0.9433	1.5197	382.6	1116.5
265.0	406.13	0.01873	1.72284	1.74157	381.7	820.0	1201.7	0.5743	0.9470	1.5214	380.8	1116.3
260.0	404.44	0.01870	1.75548	1.77418	379.9	821.6	1201.5	0.5722	0.9508	1.5230	379.0	1116.2
255.0	402.72	0.01868	1.78935	1.80802	378.0	823.3	1201.3	0.5701	0.9546	1.5247	377.2	1116.0
250.0	400.97	0.01865	1.82452	1.84317	376.1	825.0	1201.1	0.5679	0.9585	1.5264	375.3	1115.8
245.0	399.19	0.01863	1.86107	1.87970	374.2	826.6	1200.9	0.5657	0.9625	1.5281	373.4	1115.6
240.0	397.39	0.01860	1.89909	1.91769	372.3	828.4	1200.6	0.5634	0.9665	1.5299	371.4	1115.5
235.0	395.56	0.01857	1.93867	1.95725	370.3	830.1	1200.4	0.5611	0.9706	1.5317	369.5	1115.3
230.0	393.70	0.01855	1.97991	1.99846	368.3	831.8	1200.1	0.5588	0.9748	1.5336	367.5	1115.1
225.0	391.80	0.01852	2.02291	2.04143	366.2	833.6	1199.9	0.5564	0.9790	1.5354	365.5	1114.9
220.0	389.88	0.01850	2.06779	2.08629	364.2	835.4	1199.6	0.5540	0.9834	1.5374	363.4	1114.6
215.0	387.91	0.01847	2.11469	2.13315	362.1	837.2	1199.3	0.5515	0.9878	1.5393	361.3	1114.4
210.0	385.91	0.01844	2.16373	2.18217	359.9	839.1	1199.0	0.5490	0.9923	1.5413	359.2	1114.2
205.0	383.88	0.01841	2.21508	2.23349	357.7	840.9	1198.7	0.5465	0.9969	1.5434	357.0	1113.9
200.0	381.80	0.01839	2.26890	2.28728	355.5	842.8	1198.3	0.5438	1.0016	1.5454	354.8	1113.7

Table D-2. Properties of Saturated Steam and Saturated Water (Pressure, inches H_g) *(Continued)*

Press. psia	Temp. F	Volume, ft³/lbm			Enthalpy, Btu/lbm			Entropy, Btu/lbm × R			Energy, Btu/lbm	
		Water v_f	Evap. v_{fg}	Steam v_g	Water h_f	Evap. h_{fg}	Steam h_g	Water s_f	Evap. s_{fg}	Steam s_g	Water u_f	Steam u_g
200.0	381.80	0.01839	2.2689	2.2873	355.5	842.8	1198.3	0.5438	1.0016	1.5454	354.8	1113.7
198.0	380.96	0.01838	2.2912	2.3095	354.6	843.6	1198.2	0.5428	1.0035	1.5463	353.9	1113.6
196.0	380.12	0.01836	2.3139	2.3322	353.7	844.4	1198.1	0.5417	1.0054	1.5471	353.0	1113.5
194.0	379.26	0.01835	2.3370	2.3554	352.8	845.1	1197.9	0.5406	1.0074	1.5480	352.1	1113.4
192.0	378.40	0.01834	2.3606	2.3790	351.9	845.9	1197.8	0.5395	1.0094	1.5489	351.2	1113.2
190.0	377.53	0.01833	2.3847	2.4030	350.9	846.7	1197.6	0.5384	1.0113	1.5498	350.3	1113.1
188.0	376.65	0.01832	2.4093	2.4276	350.0	847.5	1197.5	0.5373	1.0133	1.5507	349.4	1113.0
186.0	375.77	0.01831	2.4344	2.4527	349.1	848.3	1197.3	0.5362	1.0153	1.5516	348.4	1112.9
184.0	374.88	0.01830	2.4600	2.4783	348.1	849.1	1197.2	0.5351	1.0174	1.5525	347.5	1112.8
182.0	373.98	0.01828	2.4862	2.5045	347.2	849.9	1197.0	0.5339	1.0194	1.5534	346.5	1112.7
180.0	373.08	0.01827	2.5129	2.5312	346.2	850.7	1196.9	0.5328	1.0215	1.5543	345.6	1112.5
178.0	372.16	0.01826	2.5402	2.5585	345.2	851.5	1196.7	0.5316	1.0236	1.5552	344.6	1112.4
176.0	371.24	0.01825	2.5681	2.5864	344.2	852.3	1196.5	0.5305	1.0257	1.5562	343.6	1112.3
174.0	370.31	0.01824	2.5966	2.6149	343.2	853.1	1196.4	0.5293	1.0279	1.5571	342.7	1112.2
172.0	369.37	0.01823	2.6258	2.6440	342.2	853.9	1196.2	0.5281	1.0300	1.5581	341.7	1112.0
170.0	368.42	0.01821	2.6556	2.6738	341.2	854.8	1196.0	0.5269	1.0322	1.5591	340.7	1111.9
168.0	367.47	0.01820	2.6861	2.7043	340.2	855.6	1195.8	0.5256	1.0344	1.5601	339.7	1111.8
166.0	366.50	0.01819	2.7173	2.7355	339.2	856.5	1195.7	0.5244	1.0367	1.5611	338.6	1111.6
164.0	365.53	0.01818	2.7493	2.7674	338.2	857.3	1195.5	0.5232	1.0389	1.5621	337.6	1111.5
162.0	364.54	0.01817	2.7820	2.8001	337.1	858.2	1195.3	0.5219	1.0412	1.5631	336.6	1111.3
160.0	363.55	0.01815	2.8155	2.8336	336.1	859.0	1195.1	0.5206	1.0435	1.5641	335.5	1111.2
158.0	362.55	0.01814	2.8498	2.8679	335.0	859.9	1194.9	0.5194	1.0458	1.5652	334.4	1111.0
156.0	361.53	0.01813	2.8849	2.9031	333.9	860.8	1194.7	0.5181	1.0482	1.5662	333.4	1110.9
154.0	360.51	0.01812	2.9210	2.9391	332.8	861.6	1194.5	0.5168	1.0506	1.5673	332.3	1110.7
152.0	359.48	0.01810	2.9579	2.9760	331.8	862.5	1194.3	0.5154	1.0530	1.5684	331.2	1110.6
150.0	358.43	0.01809	2.9958	3.0139	330.6	863.4	1194.1	0.5141	1.0554	1.5695	330.1	1110.4
149.0	357.91	0.01808	3.0151	3.0332	330.1	863.9	1194.0	0.5134	1.0566	1.5700	329.6	1110.3
148.0	357.38	0.01808	3.0347	3.0528	329.5	864.3	1193.9	0.5127	1.0579	1.5706	329.0	1110.3
147.0	356.84	0.01807	3.0545	3.0726	329.0	864.8	1193.8	0.5120	1.0591	1.5712	328.5	1110.2
146.0	356.31	0.01806	3.0746	3.0927	328.4	865.2	1193.6	0.5114	1.0604	1.5717	327.9	1110.1
145.0	355.77	0.01806	3.0950	3.1130	327.8	865.7	1193.5	0.5107	1.0616	1.5723	327.4	1110.0
144.0	355.23	0.01805	3.1156	3.1337	327.3	866.2	1193.4	0.5100	1.0629	1.5729	326.8	1109.9
143.0	354.69	0.01805	3.1365	3.1546	326.7	866.6	1193.3	0.5093	1.0642	1.5734	326.2	1109.8
142.0	354.14	0.01804	3.1577	3.1757	326.1	867.1	1193.2	0.5086	1.0655	1.5740	325.6	1109.7
141.0	353.59	0.01803	3.1792	3.1972	325.5	867.5	1193.1	0.5079	1.0668	1.5746	325.1	1109.7

Abs. Press.	Temp. °F	v_f	v_{fg}	v_g	h_f	h_{fg}	h_g	s_f	s_{fg}	s_g	u_f	u_g
140.0	353.04	0.01803	3.2010	3.2190	325.0	868.0	1193.0	0.5071	1.0681	1.5752	324.5	1109.6
139.0	352.48	0.01802	3.2230	3.2411	324.4	868.5	1192.8	0.5064	1.0694	1.5758	323.9	1109.5
138.0	351.92	0.01801	3.2454	3.2634	323.8	868.9	1192.7	0.5057	1.0707	1.5764	323.3	1109.4
137.0	351.36	0.01801	3.2681	3.2861	323.2	869.4	1192.6	0.5050	1.0720	1.5770	322.7	1109.3
136.0	350.79	0.01800	3.2912	3.3091	322.6	869.9	1192.5	0.5043	1.0733	1.5776	322.2	1109.2
135.0	350.23	0.01799	3.3145	3.3325	322.0	870.4	1192.4	0.5035	1.0747	1.5782	321.5	1109.1
134.0	349.65	0.01799	3.3382	3.3562	321.4	870.8	1192.2	0.5028	1.0760	1.5788	320.9	1109.0
133.0	349.08	0.01798	3.3622	3.3802	320.8	871.3	1192.1	0.5020	1.0774	1.5794	320.3	1108.9
132.0	348.50	0.01798	3.3866	3.4046	320.2	871.8	1192.0	0.5013	1.0788	1.5800	319.7	1108.8
131.0	347.92	0.01797	3.4113	3.4293	319.6	872.3	1191.9	0.5005	1.0801	1.5807	319.1	1108.7
130.0	347.33	0.01796	3.4364	3.4544	319.0	872.8	1191.7	0.4998	1.0815	1.5813	318.5	1108.6
129.0	346.74	0.01795	3.4619	3.4799	318.3	873.3	1191.6	0.4990	1.0829	1.5819	317.9	1108.5
128.0	346.15	0.01794	3.4878	3.5057	317.7	873.8	1191.5	0.4982	1.0843	1.5826	317.3	1108.4
127.0	345.55	0.01794	3.5141	3.5320	317.1	874.3	1191.3	0.4975	1.0858	1.5832	316.7	1108.3
126.0	344.95	0.01793	3.5407	3.5586	316.4	874.8	1191.2	0.4967	1.0872	1.5839	316.0	1108.2
125.0	344.35	0.01792	3.5678	3.5857	315.8	875.3	1191.1	0.4959	1.0886	1.5845	315.4	1108.1
124.0	343.74	0.01792	3.5953	3.6132	315.2	875.8	1190.9	0.4951	1.0901	1.5852	314.8	1108.0
123.0	343.13	0.01791	3.6232	3.6411	314.5	876.3	1190.8	0.4943	1.0915	1.5858	314.1	1107.9
122.0	342.51	0.01790	3.6516	3.6695	313.9	876.8	1190.7	0.4935	1.0930	1.5865	313.5	1107.8
121.0	341.89	0.01790	3.6804	3.6983	313.2	877.3	1190.5	0.4927	1.0945	1.5872	312.8	1107.7
120.0	341.27	0.01789	3.7097	3.7275	312.6	877.8	1190.4	0.4919	1.0960	1.5879	312.2	1107.6
119.0	340.64	0.01788	3.7394	3.7573	311.9	878.3	1190.2	0.4911	1.0975	1.5885	311.5	1107.5
118.0	340.01	0.01787	3.7697	3.7875	311.3	878.8	1190.1	0.4903	1.0990	1.5892	310.9	1107.4
117.0	339.37	0.01787	3.8004	3.8183	310.6	879.3	1189.9	0.4894	1.1005	1.5899	310.2	1107.3
116.0	338.73	0.01786	3.8316	3.8495	309.9	879.9	1189.8	0.4886	1.1021	1.5906	309.5	1107.2
115.0	338.08	0.01785	3.8634	3.8813	309.3	880.4	1189.6	0.4877	1.1036	1.5913	308.9	1107.0
114.0	337.43	0.01785	3.8957	3.9136	308.6	880.9	1189.5	0.4869	1.1052	1.5921	308.2	1106.9
113.0	336.78	0.01784	3.9286	3.9464	307.9	881.4	1189.3	0.4860	1.1067	1.5928	307.5	1106.8
112.0	336.12	0.01783	3.9620	3.9798	307.2	882.0	1189.2	0.4852	1.1083	1.5935	306.8	1106.7
111.0	335.46	0.01782	3.9960	4.0138	306.6	882.5	1189.0	0.4843	1.1099	1.5942	306.1	1106.6
110.0	334.79	0.01782	4.0306	4.0484	305.8	883.1	1188.9	0.4834	1.1115	1.5950	305.4	1106.5
109.0	334.11	0.01781	4.0658	4.0837	305.1	883.6	1188.7	0.4826	1.1132	1.5957	304.7	1106.3
108.0	333.44	0.01780	4.1017	4.1195	304.4	884.1	1188.5	0.4817	1.1148	1.5965	304.0	1106.2
107.0	332.75	0.01780	4.1382	4.1560	303.7	884.7	1188.4	0.4808	1.1165	1.5972	303.3	1106.1
106.0	332.06	0.01779	4.1753	4.1931	303.0	885.2	1188.2	0.4799	1.1181	1.5980	302.6	1106.0
105.0	331.37	0.01778	4.2132	4.2309	302.2	885.8	1188.0	0.4790	1.1198	1.5988	301.9	1105.8
104.0	330.67	0.01777	4.2517	4.2695	301.5	886.4	1187.9	0.4781	1.1215	1.5995	301.2	1105.7
103.0	329.97	0.01776	4.2910	4.3087	300.8	886.9	1187.7	0.4771	1.1232	1.6003	300.4	1105.6
102.0	329.26	0.01776	4.3310	4.3487	300.0	887.5	1187.5	0.4762	1.1249	1.6011	299.7	1105.4
101.0	328.54	0.01775	4.3717	4.3895	299.3	888.1	1187.3	0.4752	1.1267	1.6019	299.0	1105.3
100.0	327.82	0.01774	4.4133	4.4310	298.5	888.6	1187.2	0.4743	1.1284	1.6027	298.2	1105.2

Table D-2. Properties of Saturated Steam and Saturated Water (Pressure, inches H_g) (Continued)

Press. psia	Temp. F	Volume, ft³/lbm			Enthalpy, Btu/lbm			Entropy, Btu/lbm × R			Energy, Btu/lbm	
		Water v_f	Evap. v_{fg}	Steam v_g	Water h_f	Evap. h_{fg}	Steam h_g	Water s_f	Evap. s_{fg}	Steam s_g	Water u_f	Steam u_g
100.0	327.82	0.017740	4.4133	4.4310	298.5	888.6	1187.2	0.4743	1.1284	1.6027	298.2	1105.2
99.0	327.10	0.017732	4.4556	4.4734	297.8	889.2	1187.0	0.4733	1.1302	1.6036	297.5	1105.0
98.0	326.36	0.017724	4.4988	4.5166	297.0	889.8	1186.8	0.4724	1.1320	1.6044	296.7	1104.9
97.0	325.63	0.017716	4.5429	4.5606	296.3	890.4	1186.6	0.4714	1.1338	1.6052	295.9	1104.8
96.0	324.88	0.017708	4.5878	4.6055	295.5	891.0	1186.4	0.4704	1.1356	1.6061	295.2	1104.6
95.0	324.13	0.017700	4.6337	4.6514	294.7	891.5	1186.2	0.4694	1.1375	1.6069	294.4	1104.5
94.0	323.37	0.017692	4.6805	4.6982	293.9	892.1	1186.0	0.4684	1.1393	1.6078	293.6	1104.3
93.0	322.61	0.017684	4.7282	4.7459	293.1	892.7	1185.9	0.4674	1.1412	1.6086	292.8	1104.2
92.0	321.84	0.017675	4.7770	4.7947	292.3	893.3	1185.7	0.4664	1.1431	1.6095	292.0	1104.0
91.0	321.06	0.017667	4.8268	4.8445	291.5	893.9	1185.5	0.4654	1.1450	1.6104	291.2	1103.9
90.0	320.28	0.017659	4.8777	4.8953	290.7	894.6	1185.3	0.4643	1.1470	1.6113	290.4	1103.7
89.0	319.49	0.017651	4.9296	4.9473	289.9	895.2	1185.0	0.4633	1.1489	1.6122	289.6	1103.6
88.0	318.69	0.017642	4.9827	5.0004	289.0	895.8	1184.8	0.4622	1.1509	1.6131	288.9	1103.4
87.0	317.89	0.017634	5.0370	5.0546	288.2	896.4	1184.6	0.4611	1.1529	1.6140	287.9	1103.3
86.0	317.08	0.017625	5.0925	5.1101	287.4	897.0	1184.4	0.4601	1.1549	1.6150	287.1	1103.1
85.0	316.26	0.017617	5.1493	5.1669	286.5	897.7	1184.2	0.4590	1.1569	1.6159	286.2	1102.9
84.0	315.43	0.017608	5.2073	5.2249	285.7	898.3	1184.0	0.4579	1.1590	1.6169	285.4	1102.8
83.0	314.60	0.017600	5.2667	5.2843	284.8	899.0	1183.8	0.4568	1.1611	1.6178	284.5	1102.6
82.0	313.75	0.017591	5.3276	5.3451	283.9	899.6	1183.5	0.4556	1.1632	1.6188	283.7	1102.4
81.0	312.90	0.017582	5.3898	5.4074	283.0	900.3	1183.3	0.4545	1.1653	1.6198	282.8	1102.3
80.0	312.04	0.017573	5.4536	5.4711	282.1	900.9	1183.1	0.4534	1.1675	1.6208	281.9	1102.1
79.0	311.17	0.017565	5.5189	5.5364	281.3	901.6	1182.8	0.4522	1.1696	1.6218	281.0	1101.9
78.0	310.29	0.017556	5.5858	5.6034	280.3	902.3	1182.6	0.4510	1.1718	1.6229	280.1	1101.7
77.0	309.41	0.017547	5.6544	5.6720	279.4	902.9	1182.4	0.4498	1.1741	1.6239	279.2	1101.5
76.0	308.51	0.017538	5.7248	5.7423	278.5	903.6	1182.1	0.4486	1.1763	1.6250	278.3	1101.4
75.0	307.61	0.017529	5.7969	5.8144	277.6	904.3	1181.9	0.4474	1.1786	1.6260	277.3	1101.2
74.0	306.69	0.017519	5.8710	5.8885	276.6	905.0	1181.6	0.4462	1.1809	1.6271	276.4	1101.0
73.0	305.77	0.017510	5.9469	5.9645	275.7	905.7	1181.4	0.4449	1.1833	1.6282	275.4	1100.8
72.0	304.83	0.017501	6.0250	6.0425	274.7	906.4	1181.1	0.4437	1.1856	1.6293	274.5	1100.6
71.0	303.89	0.017491	6.1051	6.1226	273.7	907.1	1180.8	0.4424	1.1880	1.6304	273.5	1100.4
70.0	302.93	0.017482	6.1875	6.2050	272.7	907.8	1180.6	0.4411	1.1905	1.6316	272.5	1100.2
69.0	301.96	0.017472	6.2721	6.2896	271.7	908.5	1180.3	0.4398	1.1929	1.6327	271.5	1100.0
68.0	300.99	0.017463	6.3592	6.3767	270.7	909.3	1180.0	0.4385	1.1954	1.6339	270.5	1099.8
67.0	299.99	0.017453	6.4488	6.4662	269.7	910.0	1179.7	0.4372	1.1979	1.6351	269.6	1099.6
66.0	298.99	0.017443	6.5410	6.5584	268.7	910.8	1179.4	0.4358	1.2005	1.6363	268.5	1099.3

65.0	297.98	0.017433	6.6359	6.6533	267.6	911.5	1179.1	0.4344	1.2031	1.6375	267.4	1099.1
64.0	296.95	0.017423	6.7337	6.7511	266.6	912.3	1178.9	0.4330	1.2058	1.6388	266.4	1098.9
63.0	295.91	0.017413	6.8384	6.8558	265.5	913.0	1178.6	0.4316	1.2084	1.6401	265.3	1098.7
62.0	294.86	0.017403	6.9384	6.9558	264.4	913.8	1178.2	0.4302	1.2112	1.6413	264.2	1098.4
61.0	293.79	0.017393	7.0456	7.0630	263.3	914.6	1177.9	0.4287	1.2139	1.6427	263.1	1098.2
60.0	292.71	0.017383	7.1562	7.1736	262.2	915.4	1177.6	0.4273	1.2167	1.6440	262.0	1098.0
59.0	291.62	0.017372	7.2705	7.2879	261.1	916.2	1177.3	0.4258	1.2196	1.6453	260.9	1097.7
58.0	290.50	0.017362	7.3886	7.4059	259.9	917.0	1177.0	0.4243	1.2224	1.6467	259.8	1097.5
57.0	289.38	0.017351	7.5106	7.5280	258.8	917.8	1176.6	0.4227	1.2254	1.6481	258.6	1097.2
56.0	288.24	0.017340	7.6369	7.6543	257.6	918.7	1176.3	0.4212	1.2284	1.6495	257.4	1097.0
55.0	287.08	0.017329	7.7676	7.7850	256.4	919.5	1175.9	0.4196	1.2314	1.6510	256.2	1096.7
54.0	285.90	0.017319	7.9030	7.9203	255.2	920.4	1175.6	0.4180	1.2345	1.6524	255.0	1096.4
53.0	284.71	0.017307	8.0433	8.0606	254.0	921.2	1175.2	0.4163	1.2376	1.6539	253.8	1096.1
52.0	283.50	0.017296	8.1888	8.2061	252.8	922.1	1174.9	0.4147	1.2408	1.6555	252.6	1095.9
51.0	282.27	0.017285	8.3398	8.3571	251.5	923.0	1174.5	0.4130	1.2441	1.6570	251.3	1095.6
50.0	281.02	0.017274	8.4967	8.5140	250.2	923.9	1174.1	0.4112	1.2474	1.6586	250.1	1095.3
49.0	279.74	0.017262	8.6597	8.6770	248.9	924.8	1173.7	0.4095	1.2507	1.6602	248.8	1095.0
48.0	278.45	0.017250	8.8293	8.8465	247.6	925.7	1173.3	0.4077	1.2542	1.6619	247.4	1094.7
47.0	277.14	0.017238	9.0058	9.0231	246.2	926.6	1172.9	0.4059	1.2577	1.6636	246.1	1094.4
46.0	275.80	0.017226	9.1898	9.2070	244.9	927.6	1172.5	0.4040	1.2613	1.6653	244.7	1094.1
45.0	274.44	0.017214	9.3816	9.3988	243.5	928.6	1172.0	0.4021	1.2649	1.6671	243.3	1093.8
44.0	273.06	0.017202	9.5819	9.5991	242.1	929.5	1171.6	0.4002	1.2686	1.6689	241.9	1093.5
43.0	271.65	0.017189	9.7911	9.8083	240.6	930.5	1171.2	0.3983	1.2724	1.6707	240.5	1093.1
42.0	270.21	0.017177	10.0100	10.0272	239.2	931.5	1170.7	0.3962	1.2763	1.6726	239.0	1092.8
41.0	268.74	0.017164	10.2392	10.2563	237.7	932.6	1170.2	0.3942	1.2803	1.6745	237.5	1092.4
40.0	267.25	0.017151	10.4794	10.4965	236.1	933.6	1169.8	0.3921	1.2844	1.6765	236.0	1092.1
39.0	265.72	0.017138	10.7315	10.7487	234.6	934.7	1169.3	0.3900	1.2885	1.6785	234.5	1091.7
38.0	264.17	0.017124	10.9964	11.0136	233.0	935.8	1168.8	0.3878	1.2928	1.6806	232.9	1091.3
37.0	262.58	0.017111	11.2752	11.2923	231.4	936.9	1168.2	0.3856	1.2972	1.6827	231.3	1090.9
36.0	260.95	0.017097	11.5689	11.5860	229.7	938.0	1167.7	0.3833	1.3017	1.6849	229.6	1090.5
35.0	259.29	0.017083	11.8788	11.8959	228.0	939.1	1167.1	0.3809	1.3063	1.6872	227.9	1090.1
34.0	257.58	0.017069	12.2063	12.2234	226.3	940.3	1166.6	0.3785	1.3110	1.6895	226.2	1089.7
33.0	255.84	0.017054	12.5529	12.5700	224.5	941.5	1166.0	0.3760	1.3159	1.6919	224.5	1089.2
32.0	254.05	0.017039	12.9205	12.9376	222.7	942.7	1165.4	0.3735	1.3209	1.6944	222.6	1088.8
31.0	252.22	0.017024	13.3110	13.3280	220.8	943.9	1164.8	0.3709	1.3260	1.6969	220.7	1088.3
30.0	250.34	0.017009	13.7266	13.7436	218.9	945.2	1164.1	0.3682	1.3313	1.6995	218.8	1087.9
29.0	248.40	0.016993	14.1699	14.1869	217.0	946.5	1163.5	0.3654	1.3368	1.7022	216.9	1087.4
28.0	246.41	0.016977	14.6437	14.6607	214.9	947.9	1162.8	0.3626	1.3425	1.7050	214.9	1086.8
27.0	244.36	0.016961	15.1515	15.1684	212.9	949.2	1162.1	0.3596	1.3483	1.7080	212.8	1086.3
26.0	242.25	0.016944	15.6969	15.7138	210.7	950.6	1161.4	0.3566	1.3544	1.7110	210.6	1085.8
25.0	240.07	0.016927	16.2845	16.3014	208.5	952.1	1160.6	0.3535	1.3607	1.7141	208.4	1085.2

Table D-2. Properties of Saturated Steam and Saturated Water (Pressure, inches H$_g$) (Continued)

Press. psia	Temp. F	Volume, ft³/lbm			Enthalpy, Btu/lbm			Entropy, Btu/lbm × R			Energy, Btu/lbm	
		Water v_f	Evap. v_{fg}	Steam v_g	Water h_f	Evap. h_{fg}	Steam h_g	Water s_f	Evap. s_{fg}	Steam s_g	Water u_f	Steam u_g
25.0	240.07	0.016927	16.284	16.301	208.52	952.1	1160.6	0.3535	1.3607	1.7141	208.44	1085.2
24.5	238.95	0.016918	16.596	16.613	207.39	952.8	1160.2	0.3518	1.3639	1.7157	207.31	1084.9
24.0	237.82	0.016909	16.919	16.936	206.24	953.6	1159.8	0.3502	1.3672	1.7174	206.17	1084.6
23.5	236.66	0.016900	17.256	17.273	205.07	954.3	1159.4	0.3485	1.3706	1.7191	205.00	1084.3
23.0	235.49	0.016891	17.608	17.624	203.88	955.1	1159.0	0.3468	1.3740	1.7208	203.81	1084.0
22.5	234.29	0.016882	17.974	17.991	202.67	955.9	1158.6	0.3451	1.3775	1.7226	202.60	1083.7
22.0	233.07	0.016873	18.356	18.373	201.44	956.7	1158.1	0.3433	1.3811	1.7244	201.37	1083.4
21.5	231.83	0.016863	18.756	18.772	200.18	957.5	1157.7	0.3415	1.3847	1.7262	200.12	1083.0
21.0	230.57	0.016854	19.174	19.190	198.90	958.4	1157.3	0.3396	1.3885	1.7281	198.84	1082.7
20.5	229.28	0.016844	19.611	19.628	197.60	959.2	1156.8	0.3377	1.3923	1.7300	197.54	1082.3
20.0	227.96	0.016834	20.070	20.087	196.27	960.1	1156.3	0.3358	1.3962	1.7320	196.21	1082.0
19.5	226.62	0.016824	20.551	20.568	194.91	960.9	1155.8	0.3338	1.4002	1.7341	194.85	1081.6
19.0	225.24	0.016814	21.057	21.074	193.52	961.8	1155.3	0.3318	1.4043	1.7361	193.47	1081.2
18.5	223.84	0.016803	21.590	21.607	192.11	962.7	1154.8	0.3297	1.4085	1.7383	192.05	1080.9
18.0	222.41	0.016793	22.151	22.168	190.66	963.7	1154.3	0.3276	1.4129	1.7405	190.60	1080.5
17.5	220.94	0.016782	22.743	22.760	189.18	964.6	1153.8	0.3255	1.4173	1.7428	189.12	1080.1
17.0	219.44	0.016771	23.369	23.385	187.66	965.6	1153.2	0.3232	1.4219	1.7451	187.61	1079.7
16.5	217.90	0.016760	24.031	24.048	186.11	966.6	1152.7	0.3209	1.4265	1.7475	186.06	1079.2
16.0	216.32	0.016749	24.733	24.750	184.52	967.6	1152.1	0.3186	1.4314	1.7500	184.47	1078.8
15.5	214.70	0.016738	25.479	25.496	182.88	968.6	1151.5	0.3162	1.4363	1.7525	182.84	1078.3
15.0	213.03	0.016726	26.274	26.290	181.21	969.7	1150.9	0.3137	1.4415	1.7552	181.16	1077.9
14.696	212.00	0.016719	26.782	26.799	180.17	970.3	1150.5	0.3121	1.4447	1.7568	180.12	1077.6
14.5	211.32	0.016714	27.121	27.138	179.48	970.7	1150.2	0.3111	1.4468	1.7579	179.44	1077.4
14.0	209.56	0.016702	28.027	28.043	177.71	971.9	1149.6	0.3085	1.4522	1.7607	177.67	1076.9
13.5	207.75	0.016689	28.997	29.014	175.89	973.0	1148.9	0.3058	1.4579	1.7636	175.84	1076.4
13.0	205.88	0.016676	30.040	30.057	174.00	974.2	1148.2	0.3029	1.4638	1.7667	173.96	1075.9
12.5	203.95	0.016663	31.163	31.180	172.06	975.4	1147.5	0.3000	1.4698	1.7699	172.02	1075.3
12.0	201.96	0.016650	32.377	32.394	170.05	976.6	1146.7	0.2970	1.4762	1.7731	170.02	1074.8
11.5	199.89	0.016636	33.693	33.710	167.98	977.9	1145.9	0.2938	1.4827	1.7766	167.94	1074.2
11.0	197.75	0.016622	35.125	35.142	165.82	979.3	1145.1	0.2906	1.4896	1.7802	165.79	1073.6
10.5	195.52	0.016607	36.689	36.705	163.59	980.7	1144.2	0.2872	1.4968	1.7839	163.55	1072.9
10.0	193.21	0.016592	38.404	38.420	161.26	982.1	1143.3	0.2836	1.5043	1.7879	161.23	1072.3
9.5	190.80	0.016577	40.293	40.310	158.84	983.6	1142.4	0.2799	1.5121	1.7920	158.81	1071.6
9.0	188.27	0.016561	42.385	42.402	156.30	985.1	1141.4	0.2760	1.5204	1.7964	156.28	1070.8
8.5	185.63	0.016545	44.716	44.733	153.65	986.8	1140.4	0.2719	1.5292	1.8011	153.63	1070.0

8.0	182.86	0.016527	47.328	47.345	150.87	988.5	1139.3	0.2676	1.5384	1.8060	150.84	1069.2
7.5	179.93	0.016510	50.277	50.294	147.93	990.2	1138.2	0.2630	1.5482	1.8112	147.91	1068.4
7.0	176.84	0.016491	53.634	53.650	144.83	992.1	1136.9	0.2581	1.5587	1.8168	144.81	1067.4
6.5	173.56	0.016472	57.490	57.506	141.54	994.1	1135.6	0.2530	1.5699	1.8229	141.52	1066.5
6.0	170.05	0.016451	61.967	61.984	138.03	996.2	1134.2	0.2474	1.5820	1.8294	138.01	1065.4
5.5	166.29	0.016430	67.232	67.249	134.26	998.5	1132.7	0.2414	1.5951	1.8365	134.24	1064.3
5.0	162.24	0.016407	73.515	73.532	130.20	1000.9	1131.1	0.2349	1.6094	1.8443	130.18	1063.1
4.8	160.52	0.016398	76.383	76.400	128.47	1001.9	1130.4	0.2321	1.6155	1.8477	128.46	1062.5
4.6	158.73	0.016388	79.493	79.509	126.69	1003.0	1129.7	0.2292	1.6219	1.8512	126.68	1062.0
4.4	156.89	0.016379	82.876	82.893	124.84	1004.1	1128.9	0.2262	1.6286	1.8548	124.83	1061.4
4.2	154.96	0.016368	86.57	86.59	122.92	1005.2	1128.1	0.2231	1.6355	1.8586	122.90	1060.8
4.0	152.96	0.016358	90.63	90.64	120.92	1006.4	1127.3	0.2199	1.6428	1.8626	120.90	1060.2
3.8	150.88	0.016347	95.09	95.11	118.83	1007.6	1126.5	0.2165	1.6504	1.8669	118.82	1059.6
3.6	148.70	0.016336	100.04	100.06	116.65	1008.9	1125.6	0.2129	1.6584	1.8713	116.63	1058.9
3.4	146.41	0.016324	105.55	105.57	114.36	1010.3	1124.6	0.2091	1.6669	1.8760	114.35	1058.2
3.2	144.00	0.016313	111.73	111.75	111.95	1011.7	1123.6	0.2051	1.6759	1.8810	111.94	1057.5
3.0	141.47	0.016300	118.71	118.73	109.42	1013.2	1122.6	0.2009	1.6854	1.8864	109.41	1056.7
2.8	138.78	0.016287	126.66	126.67	106.73	1014.7	1121.5	0.1964	1.6956	1.8921	106.73	1055.8
2.6	135.93	0.016274	135.78	135.80	103.88	1016.4	1120.3	0.1917	1.7065	1.8982	103.87	1054.9
2.4	132.88	0.016260	146.39	146.40	100.84	1018.2	1119.0	0.1865	1.7183	1.9048	100.83	1054.0
2.2	129.61	0.016245	158.86	158.87	97.57	1020.1	1117.6	0.1810	1.7311	1.9121	97.56	1053.0
2.0	126.07	0.016230	173.74	173.76	94.03	1022.1	1116.2	0.1750	1.7450	1.9200	94.03	1051.8
1.8	122.22	0.016213	191.84	191.85	90.18	1024.3	1114.5	0.1684	1.7604	1.9288	90.18	1050.6
1.6	117.98	0.016196	214.31	214.33	85.95	1026.8	1112.7	0.1611	1.7775	1.9386	85.94	1049.3
1.4	113.26	0.016178	243.00	243.02	81.23	1029.5	1110.7	0.1529	1.7969	1.9498	81.23	1047.8
1.2	107.91	0.016158	280.95	280.96	75.90	1032.6	1108.5	0.1436	1.8192	1.9628	75.89	1046.1
1.0	101.74	0.016136	333.59	333.60	69.73	1036.1	1105.8	0.1326	1.8455	1.9781	69.73	1044.1
0.9	98.24	0.016124	368.41	368.43	66.24	1038.1	1104.3	0.1264	1.8606	1.9870	66.24	1042.9
0.8	94.38	0.016112	411.67	411.69	62.39	1040.3	1102.6	0.1195	1.8775	1.9970	62.39	1041.7
0.7	90.09	0.016099	466.93	466.94	58.10	1042.7	1100.8	0.1117	1.8966	2.0083	58.10	1040.3
0.60	85.218	0.016085	540.0	540.1	53.245	1045.5	1098.7	0.1028	1.9186	2.0215	53.243	1038.7
0.50	79.586	0.016071	641.5	641.5	47.623	1048.6	1096.3	0.0925	1.9446	2.0370	47.621	1036.9
0.40	72.869	0.016056	792.0	792.1	40.917	1052.4	1093.3	0.0799	1.9762	2.0562	40.916	1034.7
0.35	68.939	0.016048	898.6	898.6	36.992	1054.6	1091.6	0.0725	1.9951	2.0676	36.991	1033.4
0.30	64.484	0.016040	1039.7	1039.7	32.541	1057.1	1089.7	0.0641	2.0168	2.0809	32.540	1032.0
0.25	59.323	0.016032	1235.5	1235.5	27.382	1060.1	1087.4	0.0542	2.0425	2.0967	27.382	1030.3
0.20	53.160	0.016025	1526.3	1526.3	21.217	1063.5	1084.7	0.0422	2.0738	2.1160	21.217	1028.3
0.15	45.453	0.016020	2004.7	2004.7	13.498	1067.9	1081.4	0.0271	2.1140	2.1411	13.498	1025.7
0.10	35.023	0.016020	2945.5	2945.5	3.026	1073.8	1076.8	0.0061	2.1705	2.1766	3.025	1022.3
0.08865	32.018	0.016022	3302.4	3302.4	0.0003	1075.5	1075.5	0.0000	2.1872	2.1872	0.000	1021.3

Table D-3. Specific Heat at Constant Pressure of Steam and of Water.

c_p, Btu/lbm × F — Temp. F

Press., psia	150	200	300	400	600	800	1000	1500	2000	3000	4000	6000	8000	10000	15000	Press., psia
Sat. Water	1.054	1.067	1.093	1.118	1.168	1.224	1.286	1.492	1.841	7.646	—	—	—	—	—	Sat. Water
Sat. Steam	0.624	0.661	0.729	0.792	0.915	1.046	1.191	1.667	2.557	13.66	—	—	—	—	—	Sat. Steam
1500	0.562	0.563	0.565	0.567	0.571	0.576	0.580	0.590	0.601	0.623	0.645	0.691	0.737	0.780	0.868	1500
1480	0.561	0.562	0.564	0.566	0.570	0.575	0.579	0.590	0.601	0.623	0.647	0.694	0.742	0.786	0.878	1480
1460	0.559	0.560	0.562	0.565	0.569	0.573	0.578	0.589	0.601	0.624	0.648	0.698	0.747	0.793	0.888	1460
1440	0.557	0.559	0.561	0.563	0.568	0.572	0.577	0.589	0.600	0.625	0.650	0.701	0.753	0.800	0.900	1440
1420	0.556	0.557	0.559	0.562	0.566	0.571	0.576	0.588	0.600	0.625	0.651	0.705	0.759	0.808	0.912	1420
1400	0.554	0.555	0.558	0.560	0.565	0.570	0.575	0.587	0.600	0.626	0.653	0.709	0.765	0.817	0.926	1400
1380	0.553	0.554	0.556	0.559	0.564	0.569	0.574	0.587	0.600	0.627	0.655	0.714	0.773	0.827	0.939	1380
1360	0.551	0.552	0.555	0.558	0.563	0.568	0.573	0.586	0.600	0.628	0.657	0.719	0.781	0.838	0.953	1360
1340	0.549	0.551	0.553	0.556	0.561	0.567	0.572	0.586	0.600	0.629	0.660	0.725	0.790	0.850	0.968	1340
1320	0.548	0.549	0.552	0.555	0.560	0.566	0.571	0.585	0.600	0.630	0.663	0.731	0.800	0.864	0.983	1320
1300	0.546	0.548	0.550	0.553	0.559	0.565	0.570	0.585	0.600	0.632	0.666	0.738	0.811	0.879	0.998	1300
1280	0.545	0.546	0.549	0.552	0.558	0.564	0.570	0.585	0.600	0.634	0.669	0.746	0.824	0.897	1.014	1280
1260	0.543	0.544	0.547	0.550	0.556	0.563	0.569	0.585	0.601	0.636	0.673	0.755	0.838	0.918	1.033	1260
1240	0.541	0.543	0.546	0.549	0.555	0.562	0.568	0.584	0.601	0.638	0.678	0.765	0.855	0.942	1.053	1240
1220	0.540	0.541	0.544	0.548	0.554	0.561	0.567	0.584	0.602	0.641	0.683	0.777	0.875	0.969	1.072	1220
1200	0.538	0.540	0.543	0.546	0.553	0.560	0.567	0.584	0.603	0.644	0.689	0.790	0.897	1.000	1.095	1200
1180	0.536	0.538	0.541	0.544	0.552	0.559	0.566	0.584	0.604	0.647	0.696	0.805	0.922	1.033	1.117	1180
1160	0.535	0.536	0.540	0.544	0.551	0.558	0.565	0.585	0.606	0.652	0.704	0.823	0.952	1.070	1.143	1160
1140	0.533	0.535	0.539	0.542	0.550	0.557	0.565	0.585	0.607	0.656	0.713	0.843	0.986	1.107	1.167	1140
1120	0.531	0.533	0.537	0.541	0.549	0.557	0.565	0.586	0.609	0.662	0.723	0.866	1.025	1.149	1.190	1120
1100	0.530	0.532	0.536	0.540	0.548	0.556	0.564	0.587	0.612	0.668	0.735	0.893	1.070	1.193	1.220	1100
1080	0.528	0.530	0.534	0.538	0.547	0.555	0.564	0.588	0.615	0.676	0.749	0.924	1.120	1.242	1.240	1080
1060	0.527	0.529	0.533	0.537	0.546	0.555	0.564	0.590	0.618	0.685	0.765	0.960	1.176	1.295	1.260	1060
1040	0.525	0.527	0.532	0.536	0.545	0.555	0.565	0.592	0.622	0.695	0.783	1.002	1.238	1.351	1.282	1040
1020	0.523	0.526	0.530	0.535	0.545	0.555	0.565	0.594	0.627	0.707	0.804	1.051	1.306	1.399	1.298	1020
1000	0.522	0.524	0.529	0.534	0.544	0.555	0.566	0.597	0.633	0.721	0.829	1.110	1.382	1.471	1.306	1000
980	0.520	0.523	0.528	0.533	0.544	0.555	0.567	0.601	0.640	0.737	0.858	1.180	1.475	1.531	1.312	980
960	0.519	0.521	0.527	0.532	0.543	0.556	0.568	0.605	0.648	0.756	0.893	1.267	1.598	1.595	1.310	960
940	0.517	0.520	0.526	0.531	0.544	0.556	0.570	0.610	0.658	0.778	0.934	1.376	1.708	1.639	1.299	940
920	0.516	0.519	0.525	0.531	0.544	0.558	0.573	0.617	0.669	0.803	0.984	1.520	1.819	1.667	1.281	920

Temp																Temp
900	1.259	1.660	1.932	1.716	1.048	0.834	0.683	0.624	0.576	0.559	0.544	0.530	0.524	0.518	0.515	900
880	1.232	1.633	2.000	1.993	1.130	0.872	0.699	0.633	0.580	0.561	0.545	0.530	0.523	0.516	0.513	880
860	1.212	1.593	2.019	2.316	1.240	0.918	0.718	0.644	0.584	0.564	0.546	0.530	0.523	0.515	0.512	860
840	1.192	1.547	1.978	2.653	1.395	0.977	0.740	0.657	0.590	0.568	0.548	0.530	0.522	0.514	0.511	840
820	1.175	1.503	1.888	2.886	1.620	1.054	0.767	0.672	0.597	0.572	0.550	0.531	0.522	0.514	0.510	820
800	1.157	1.459	1.768	2.872	1.967	1.160	0.800	0.690	0.605	0.577	0.553	0.532	0.522	0.513	0.509	800
780	1.142	1.416	1.670	2.547	2.550	1.312	0.840	0.712	0.615	0.584	0.557	0.533	0.522	0.513	0.508	780
760	1.126	1.370	1.576	2.156	4.462	1.542	0.892	0.738	0.628	0.592	0.561	0.535	0.523	0.512	0.507	760
740	1.114	1.332	1.493	1.886	8.119	1.913	0.960	0.770	0.642	0.602	0.567	0.537	0.524	0.512	0.506	740
720	1.100	1.290	1.421	1.696	3.458	2.584	1.052	0.811	0.660	0.613	0.574	0.540	0.525	0.512	0.506	720
700	1.089	1.250	1.358	1.557	2.237	6.145	1.181	0.861	0.681	0.627	0.582	0.544	0.528	0.513	0.506	700
680	1.079	1.217	1.303	1.450	1.789	2.469	1.365	0.927	0.707	0.644	0.592	0.549	0.530	0.514	0.506	680
660	1.071	1.187	1.256	1.369	1.587	1.851	1.639	1.015	0.738	0.665	0.604	0.555	0.534	0.515	0.507	660
640	1.063	1.157	1.216	1.303	1.454	1.601	2.219	1.135	0.777	0.690	0.619	0.562	0.538	0.517	0.507	640
620	1.056	1.136	1.184	1.252	1.362	1.455	1.614	1.308	0.826	0.720	0.637	0.571	0.543	0.519	0.509	620
600	1.052	1.118	1.157	1.211	1.295	1.358	1.453	1.586	0.888	0.757	0.659	0.582	0.550	0.522	0.510	600
580	1.046	1.102	1.134	1.178	1.243	1.289	1.351	1.393	0.969	0.804	0.685	0.595	0.558	0.526	0.513	580
560	1.039	1.087	1.115	1.151	1.202	1.237	1.281	1.309	1.079	0.862	0.717	0.611	0.568	0.531	0.516	560
540	1.031	1.074	1.098	1.128	1.169	1.196	1.229	1.249	1.272	0.937	0.756	0.630	0.580	0.538	0.519	540
520	1.024	1.062	1.083	1.109	1.142	1.164	1.189	1.204	1.221	1.035	0.804	0.653	0.594	0.545	0.524	520
500	1.017	1.051	1.069	1.092	1.120	1.137	1.157	1.169	1.181	1.187	0.865	0.680	0.611	0.554	0.530	500
480	1.010	1.041	1.057	1.077	1.101	1.115	1.131	1.140	1.150	1.154	1.159	0.714	0.632	0.565	0.537	480
460	1.004	1.033	1.047	1.064	1.084	1.096	1.110	1.117	1.125	1.128	1.132	0.755	0.657	0.578	0.545	460
440	0.999	1.025	1.038	1.052	1.070	1.080	1.092	1.098	1.104	1.107	1.110	1.113	0.687	0.594	0.556	440
420	0.994	1.018	1.029	1.042	1.058	1.067	1.076	1.081	1.087	1.089	1.091	1.091	0.724	0.614	0.568	420
400	0.990	1.011	1.022	1.034	1.047	1.055	1.063	1.067	1.072	1.074	1.076	1.078	1.079	0.636	0.583	400
380	0.986	1.006	1.015	1.026	1.038	1.044	1.052	1.056	1.059	1.061	1.063	1.065	1.065	1.066	0.601	380
360	0.982	1.001	1.009	1.019	1.030	1.036	1.042	1.045	1.049	1.050	1.052	1.053	1.054	1.054	0.822	360
340	0.979	0.996	1.004	1.013	1.022	1.028	1.033	1.036	1.039	1.040	1.042	1.043	1.044	1.044	1.045	340
320	0.976	0.992	0.999	1.007	1.016	1.021	1.026	1.028	1.031	1.032	1.033	1.034	1.035	1.036	1.036	320
300	0.973	0.988	0.995	1.002	1.010	1.015	1.019	1.022	1.024	1.025	1.026	1.027	1.028	1.028	1.028	300
280	0.971	0.985	0.991	0.998	1.005	1.009	1.014	1.016	1.018	1.019	1.020	1.021	1.021	1.022	1.022	280
260	0.968	0.982	0.988	0.994	1.001	1.005	1.009	1.011	1.013	1.013	1.014	1.015	1.016	1.016	1.016	260
240	0.966	0.979	0.985	0.991	0.997	1.001	1.004	1.006	1.008	1.009	1.010	1.011	1.011	1.011	1.012	240
220	0.964	0.977	0.982	0.988	0.994	0.998	1.001	1.003	1.005	1.005	1.006	1.007	1.007	1.008	1.008	220
200	0.963	0.975	0.980	0.986	0.992	0.995	0.996	1.000	1.002	1.002	1.003	1.004	1.004	1.004	1.005	200
180	0.961	0.973	0.978	0.983	0.987	0.993	0.994	0.998	0.999	1.000	1.001	1.001	1.002	1.002	1.002	180
160	0.959	0.971	0.976	0.981	0.991	0.991	0.992	0.994	0.996	0.998	0.999	0.999	1.000	1.000	1.000	160
140	0.958	0.969	0.974	0.980	0.989	0.989	0.993	0.994	0.995	0.997	0.997	0.998	0.999	0.999	0.999	140
120	0.957	0.967	0.972	0.978	0.988	0.988	0.990	0.993	0.995	0.995	0.996	0.997	0.997	0.998	1.006	120
100	0.955	0.965	0.970	0.976	0.983	0.986	0.990	0.992	0.994	0.995	0.996	0.996	0.997	0.997	0.997	100
80	0.951	0.962	0.968	0.974	0.981	0.985	0.989	0.991	0.994	0.994	0.995	0.996	0.997	0.997	0.998	80
60	0.942	0.956	0.963	0.970	0.979	0.984	0.989	0.991	0.994	0.995	0.996	0.997	0.998	0.999	0.999	60
40	0.920	0.945	0.954	0.965	0.976	0.983	0.989	0.993	0.997	0.998	1.000	1.001	1.002	1.003	1.003	40
32	0.904	0.937	0.949	0.962	0.975	0.983	0.990	0.994	0.999	1.000	1.002	1.004	1.005	1.006	1.006	32

Figure D-1. Reciprocal Constant—Pressure specific Heat I/c_p, LBM × F/Btu for Steam and Water.

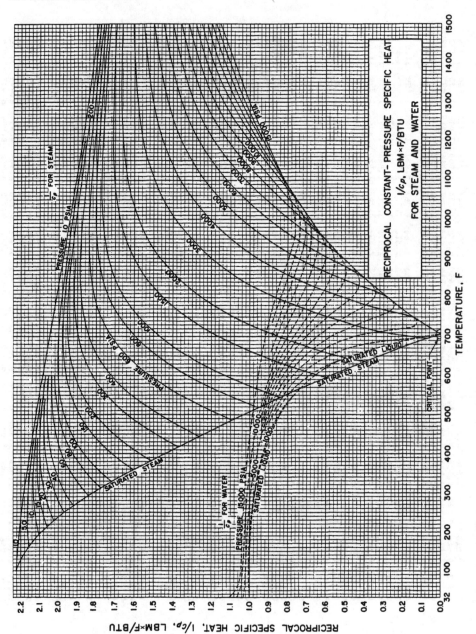

Figure D-2. Thermal Conductivity for Steam and Water.

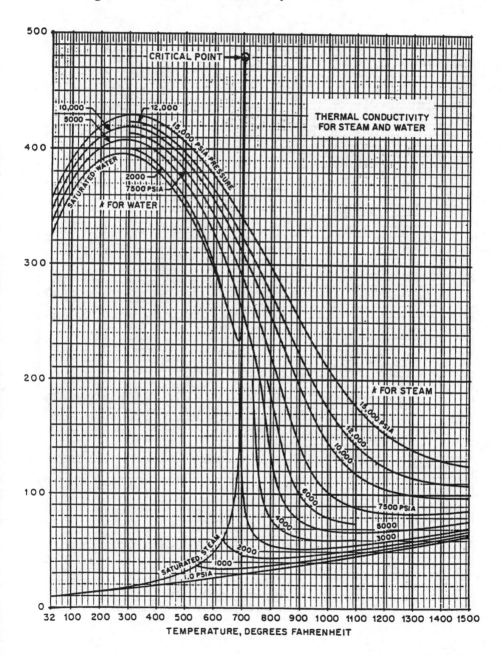

Figure D-3. Isentropic Work of Compression $(h-h_t)_s$, **for Water, (Ideal Pump Work).**

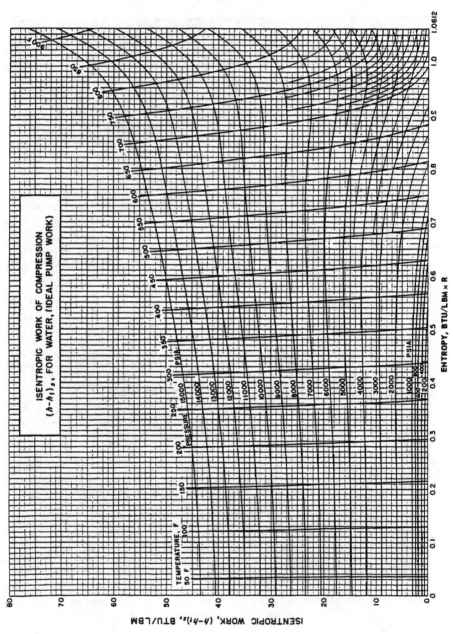

Figure D-4. (*pv*) Product for Low Temperature Steam.

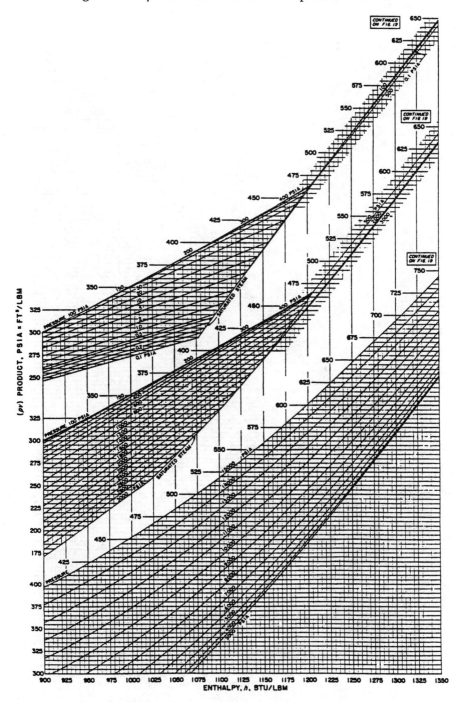

Figure D-5. (*pv*) Product for High Temperature Steam.

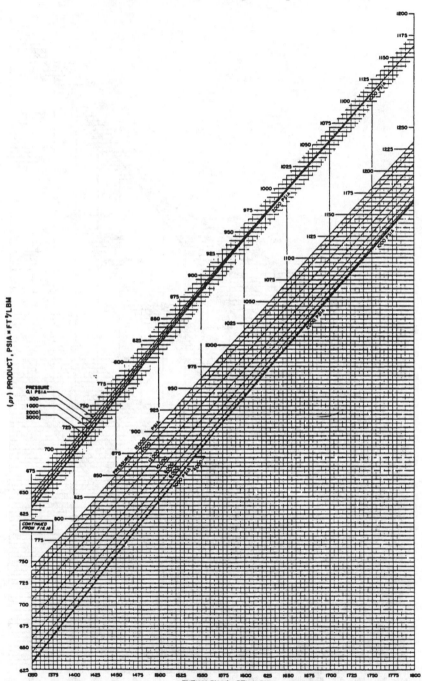

Figure D-6. Pressure-enthalpy Chart.

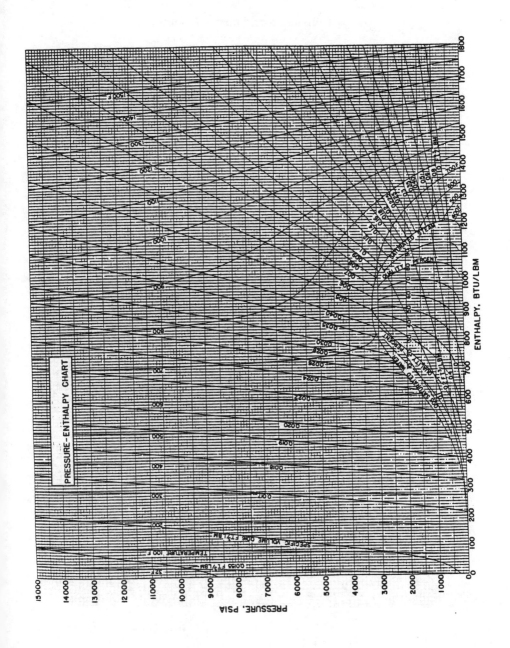

Figure D–7. Isentropic Exponent, T

$$T = \left(\frac{\partial p}{\partial v}\right)_s$$

$$pv^T = \text{Constant for a Short Expansion.}$$

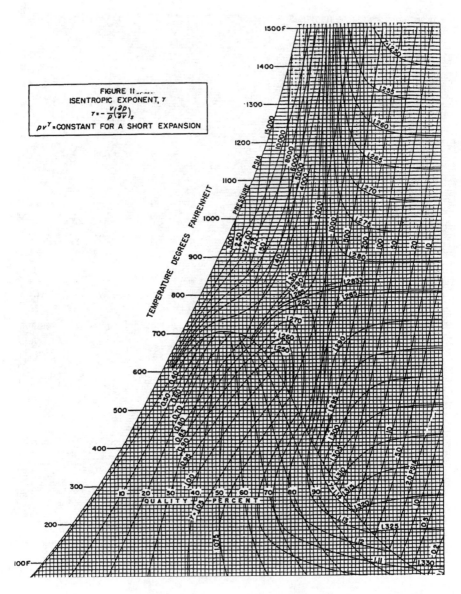

Figure D-8. (Heavy lines indicate divisions between sheers. Bold letters indicate sheer of A-Q. Overlap exists between all sheers.)

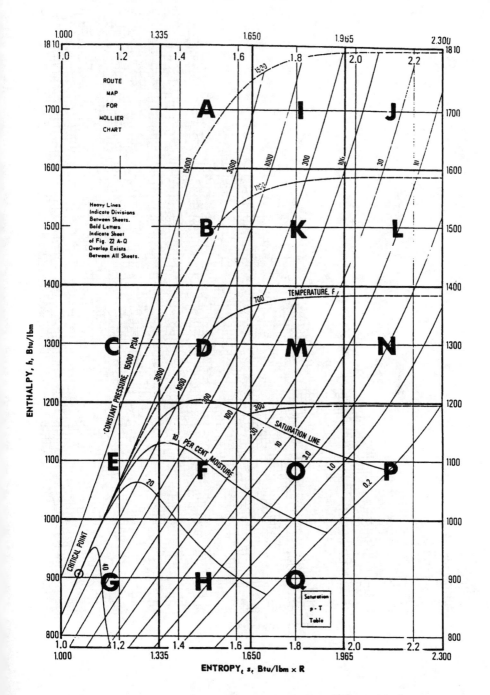

Figure D-9. Temperature Entropy Chart.

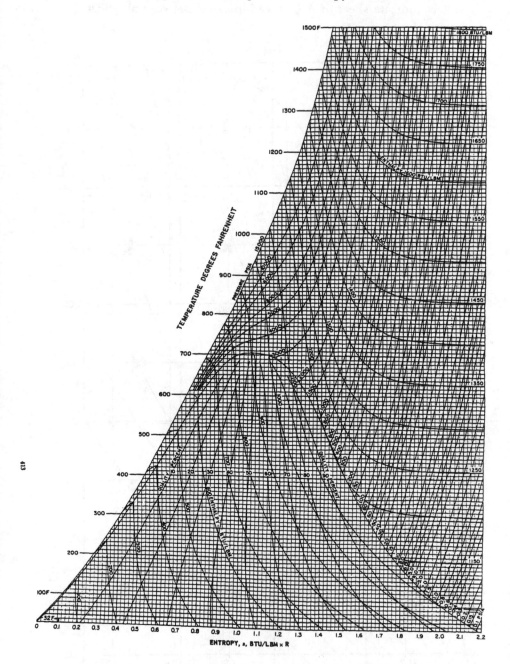

Figure D-10a.

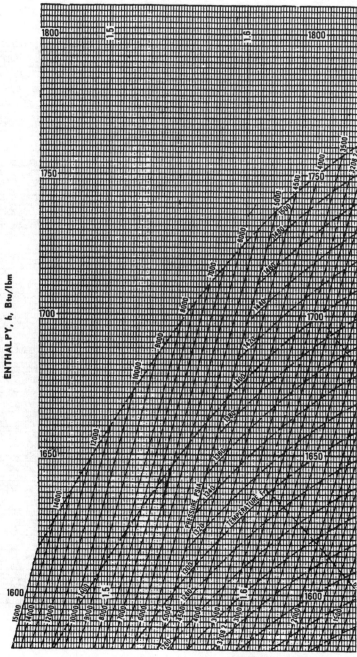

ENTROPY, s, Btu/lbm × R

Figure D-10b.

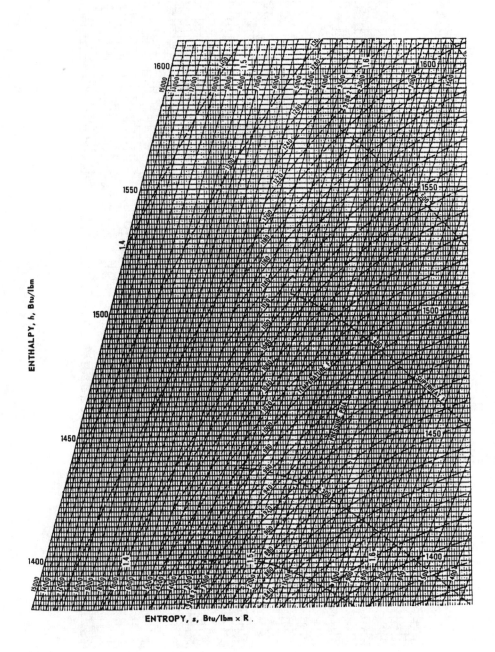

Figure D-10c.

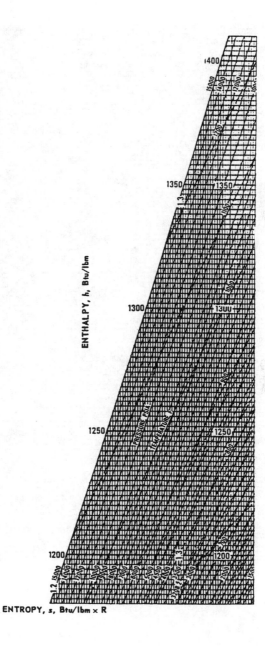

Figure D-10d.

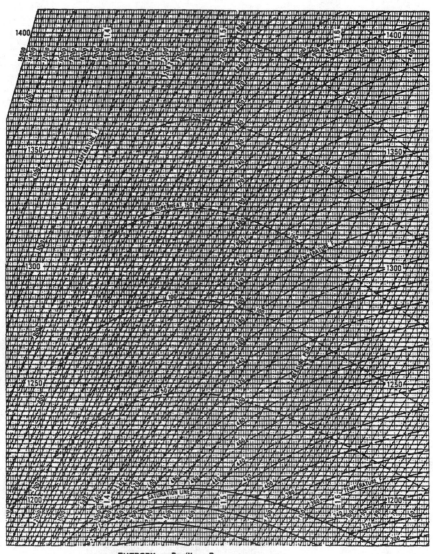

ENTHALPY, h, Btu/lbm

ENTROPY, s, Btu/lbm x R

Figure D-10e.

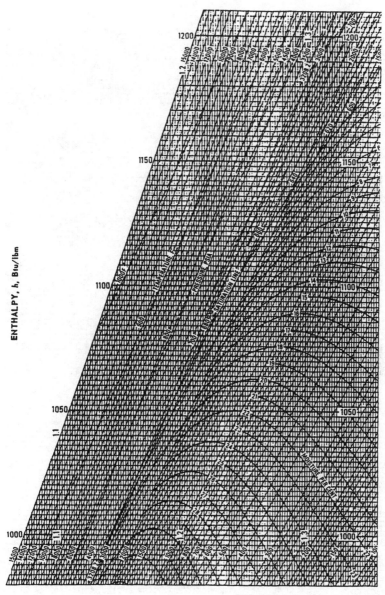

Figure D-10f.

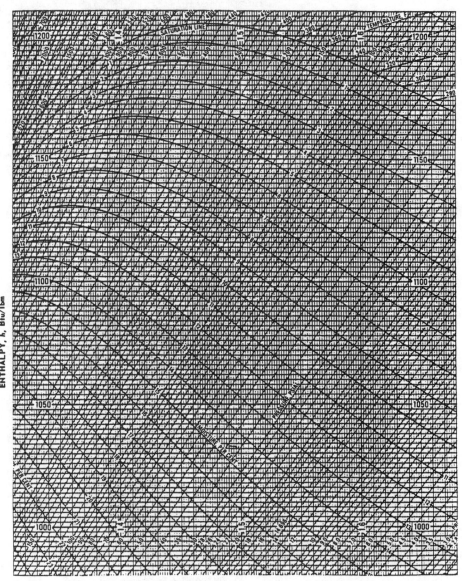

ENTHALPY, *h*, Btu/lbm

ENTROPY, *s*, Btu/lbm × R

Figure D-10g.

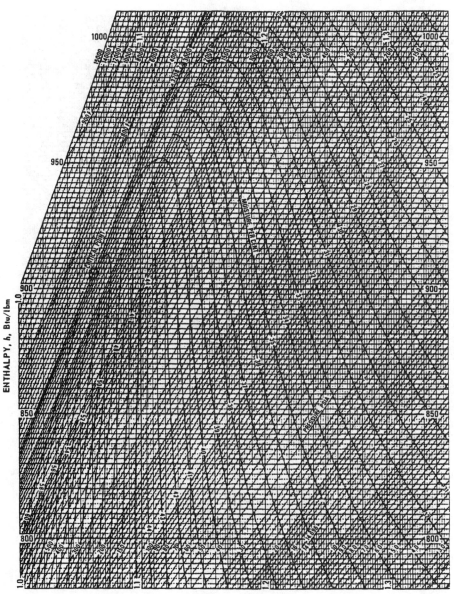

Figure D-10h.

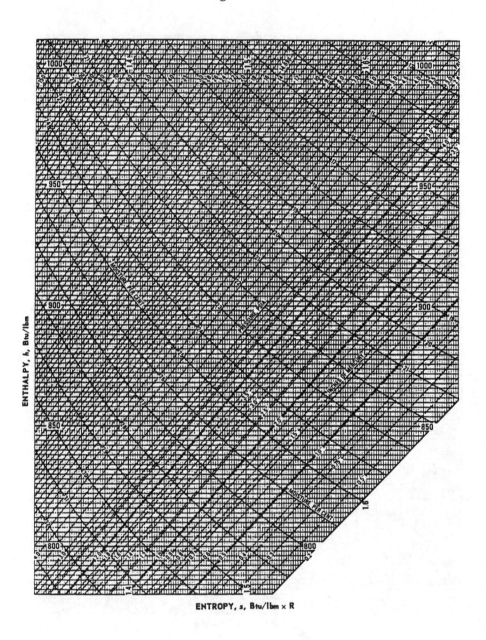

ENTROPY, *s*, Btu/lbm × R

Figure D-10i.

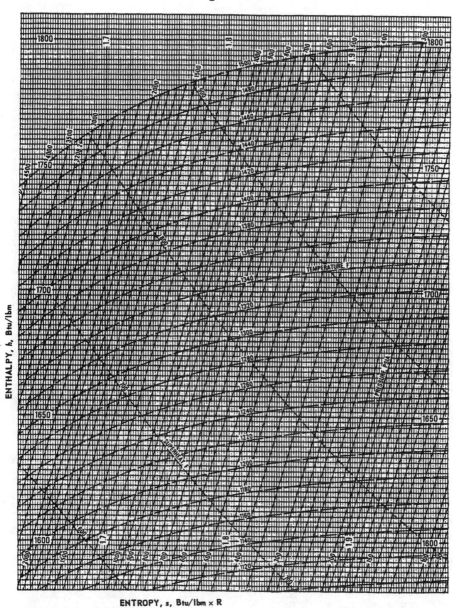

ENTROPY, s, Btu/lbm × R

Figure D-10j.

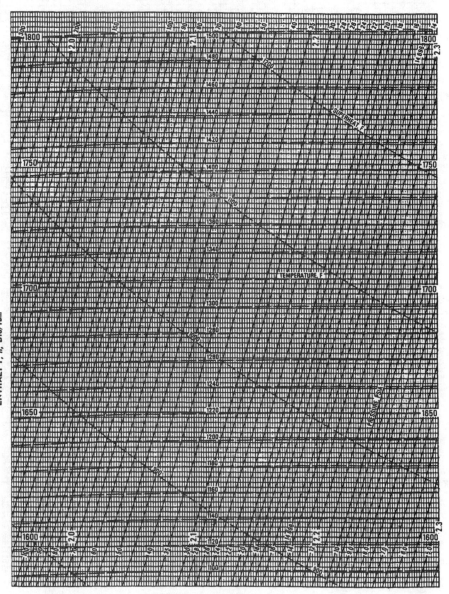

ENTHALPY, h, Btu/lbm

ENTROPY, s, Btu/lbm × R

Figure D-10k.

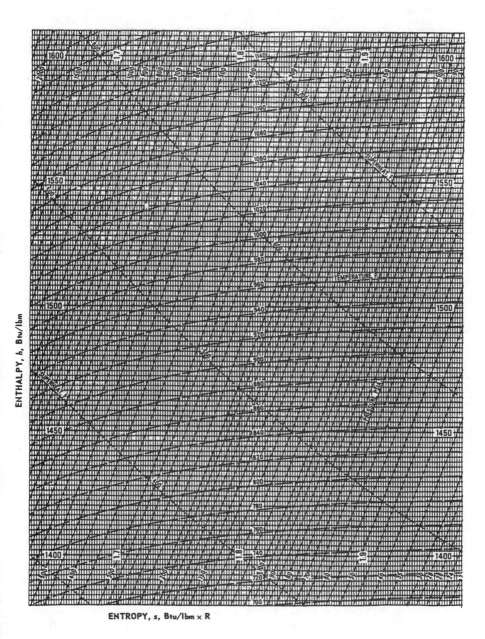

ENTROPY, s, Btu/lbm × R

Figure D-10l.

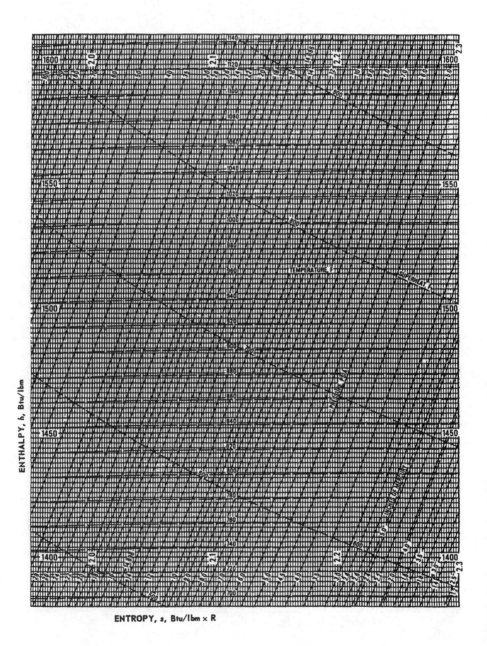

ENTHALPY, h, Btu/lbm

ENTROPY, s, Btu/lbm × R

Figure D-10m.

ENTHALPY, h, Btu/lbm

ENTROPY, s, Btu/lbm × R

Figure D-10n.

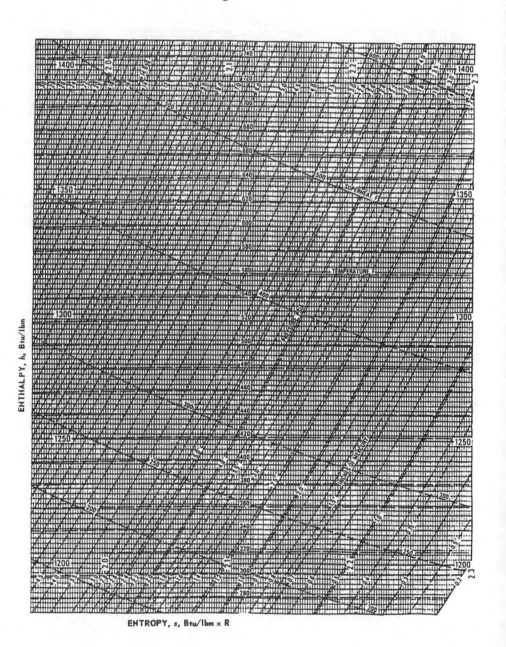

Figure D-10o.

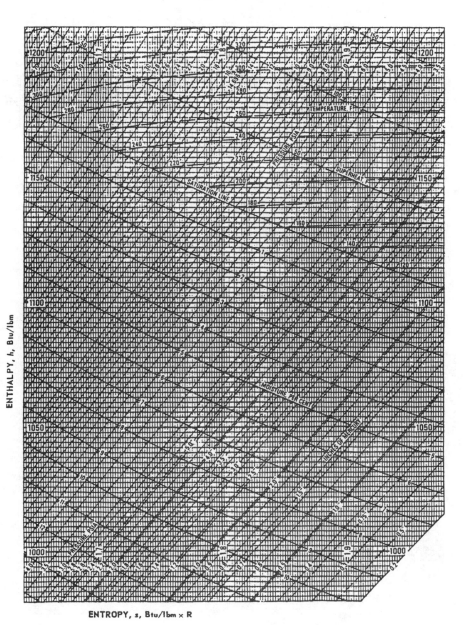

ENTROPY, s, Btu/lbm × R

Figure D-10p.

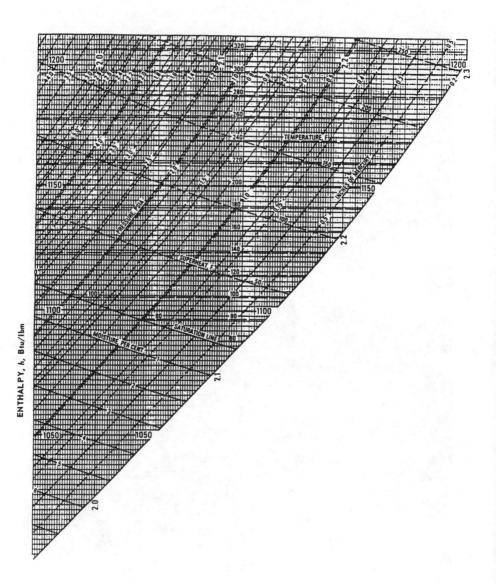

Figure D-10q.

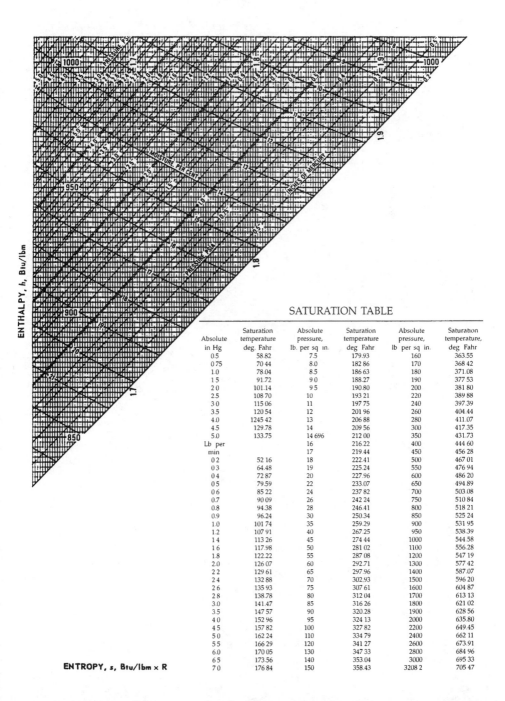

ENTHALPY, h, Btu/lbm

ENTROPY, s, Btu/lbm × R

SATURATION TABLE

Absolute in Hg	Saturation temperature deg. Fahr	Absolute pressure, lb. per sq in.	Saturation temperature deg. Fahr	Absolute pressure, lb per sq in.	Saturation temperature, deg Fahr
0.5	58.82	7.5	179.93	160	363.55
0.75	70.44	8.0	182.86	170	368.42
1.0	78.04	8.5	186.63	180	371.08
1.5	91.72	9.0	188.27	190	377.53
2.0	101.14	9.5	190.80	200	381.80
2.5	108.70	10	193.21	220	389.88
3.0	115.06	11	197.75	240	397.39
3.5	120.54	12	201.96	260	404.44
4.0	1245.42	13	206.88	280	411.07
4.5	129.78	14	209.56	300	417.35
5.0	133.75	14.696	212.00	350	431.73
Lb per min		16	216.22	400	444.60
		17	219.44	450	456.28
0.2	52.16	18	222.41	500	467.01
0.3	64.48	19	225.24	550	476.94
0.4	72.87	20	227.96	600	486.20
0.5	79.59	22	233.07	650	494.89
0.6	85.22	24	237.82	700	503.08
0.7	90.09	26	242.24	750	510.84
0.8	94.38	28	246.41	800	518.21
0.9	96.24	30	250.34	850	525.24
1.0	101.74	35	259.29	900	531.95
1.2	107.91	40	267.25	950	538.39
1.4	113.26	45	274.44	1000	544.58
1.6	117.98	50	281.02	1100	556.28
1.8	122.22	55	287.08	1200	547.19
2.0	126.07	60	292.71	1300	577.42
2.2	129.61	65	297.96	1400	587.07
2.4	132.88	70	302.93	1500	596.20
2.6	135.93	75	307.61	1600	604.87
2.8	138.78	80	312.04	1700	613.13
3.0	141.47	85	316.26	1800	621.02
3.5	147.57	90	320.28	1900	628.56
4.0	152.96	95	324.13	2000	635.80
4.5	157.82	100	327.82	2200	649.45
5.0	162.24	110	334.79	2400	662.11
5.5	166.29	120	341.27	2600	673.91
6.0	170.05	130	347.33	2800	684.96
6.5	173.56	140	353.04	3000	695.33
7.0	176.84	150	358.43	3208.2	705.47

Appendix E

The Orifice Steam Trap

*I*n 1964, the U.S. Navy investigated opportunities to improve steam trap performance in the Fleet. Their interest was primarily to reduce the use of steam in the main propulsion plants and steam driven auxiliary machinery in the Fleet. Among the concepts investigated was using a thin fixed size orifice plate engineered to pass only the condensate formed in a particular application

The researchers reasoned that, if a fixed orifice were sized to pass only the quantity of condensate formed, then steam emitted from the orifice, under conditions when there was no condensate present, would be minimal. For example, an orifice passing 1000 lbs/hr of condensate at 150 psi would theoretically pass only 5.9 lbs/hr of steam. When physically tested, the steam-to-condensate ratio was even greater.

This steam loss, although minute, was greater than encountered if a properly functioning conventional steam trap were in its place. However, the Navy's experience with conventional steam traps has been poor, even though maintenance was stressed and preventative maintenance systems and procedures employed. The conventional steam trap's failure rate was unacceptable.

Undoubtedly, the Navy was and is more concerned with dependability in steam traps than energy conservation. However, when opportunities are evident, the Navy takes advantage of them. Therefore, steam trap reliability also means steam energy savings.

A ship cannot depend on manufacturers' services when deployed at sea. Optimally, they want equipment that will function reliably during the ship's time at sea—away from her home port. Reliability, however, is not the only consideration in operating a vessel underweigh. The distance the ship can travel without replenishing its fuel supply is probably the second most important consideration to the Navy's compliment of engineers. Serious consideration is given to decisions involving steam utilization since it is directly proportional to fuel consumption.

NAVY EXPERIENCE

Therefore, the Navy was impressed with the researcher's work with the orifice steam trap. After a pilot program on a major combat ship, the decision was made to outfit the entire fossil-fueled fleet of more than 300 ships with this unique steam specialty device. Today, after more than 30 years of use the orifice steam trap is still used exclusively in the fleet. Results have been more than satisfactory.

INDUSTRIAL EXPERIENCE

The utility industry and the cogeneration (independent power) producers have also adapted the orifice steam trap, to the exclusion of conventional traps. Utility companies are equally as sensitive to the need for reliability and fuel economy for their power plants as the Navy is for its ships.

From the preceding it may be assumed that the days of the conventional steam trap are numbered, but oftentimes large industrial plants and production equipment have hundreds of steam traps in use. Except for very large steam consuming processes or machines, and where opportunities to install a number of equally sized orifice steam traps exist, the task of retrofitting can be overwhelming. Also, new genus of conventional steam traps are proving to be somewhat more reliable.

However, it would be appropriate for designers of new commercial and industrial facilities and expansion of older ones to give consideration to fitting new machinery with orifice traps.

Advantages

Some advantages of the fixed orifice steam trap are:
- No moving parts,
- Can accommodate load changes where the upstream pressure varies with the load,
- Performance can be easily calculated in advance, if the condensate flow and inlet and outlet pressures are known,
- Live steam losses are predictable and extremely small when the trap is properly sized,
- Maintenance is easy to accomplish and for most installations require only about 15 minutes,
- Requires only a minimum of inventory to support the necessary maintenance,

- Cannot fail open and insure large steam losses,
- The constant pressure drop across the orifice prevents over pressure of downstream condensate systems,
- Low initial cost,
- Ten-year or more expected life (this has been confirmed in the utility industry),
- Will not fail from thermal or hydraulic shock.

Disadvantages

Some disadvantaged of the orifice steam trap are:
- It must be supplemented with extra an automatic or manual drain valve to accommodate large than normal condensate loads, if they exist during machinery and for piping system warm-up,
- The orifice steam trap requires careful applications engineering

THE DRAIN FLANGE AND DRAIN UNION ASSEMBLIES

Diagrams of the common orifice steam trap, the Drain Flange Assembly (DFA) and the Drain Union Assembly (DUA), as used in the Navy, are shown in Figures E-1 and E-2.

There are several versions of the orifice steam trap on the market, all of which have merit in state-of-the-art steam trapping; however, the DFA and DUA are in most common use. Nicholson Steam Trap Company and most major manufacturers of steam traps and some newly formed companies specializing in versions of the orifice trap are quite good.

OTHER ORIFICE STEAM TRAPS

Other orifice steam trap conceptions utilize venturi nozzles and other flow limiting devices. There are problems in sizing these and installing them. They are high tech, labor intense installation procedures usually requiring an engineer. Some manufacturers still struggle with this problem.

Others, including some steam engineering consultants, have successfully forged a way to install nozzles in many types of heating systems without securing the machinery that the trap serves. One manufacturer designed provisions for testing a selection of steam nozzles, while providing a bypass built into the body of the trap. This permits flow from the machine while fitting one nozzle from the selection.

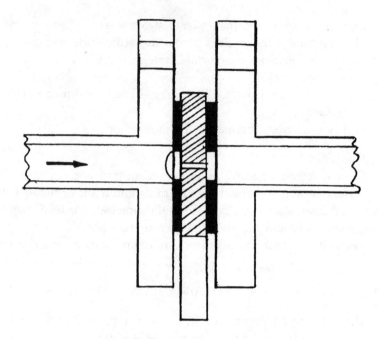

Figure E-1. Drain Flange Assembly.

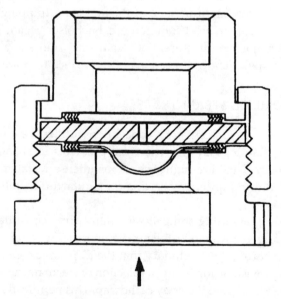

Figure E-2. Drain Union Assembly.

References

Mechanical Engineers Handbook, Baumeister and Marks
American Society of Heating Refrigeration and Air
    Conditioning Engineers Handbook
Industrial Pipework Engineers Handbook
American Society of Mechanical Engineers Steam Tables
Marine Engineers Handbook
Steam Trap Handbook, James F. McCauley, P.E.
Handbook of Air Conditioning Heating and Ventilation,
    Stempler and Kora

Index